VOLUME ONE HUNDRED AND THIRTY FIVE

# INTERNATIONAL REVIEW OF NEUROBIOLOGY

## Neurobiology of Chinese Herb Medicine

# INTERNATIONAL REVIEW OF NEUROBIOLOGY

VOLUME 135

## SERIES EDITOR


**PATRICIA JANAK**

*Janak Lab, Dunning Hall*
*Baltimore, MD, USA*

**PETER JENNER**

*Division of Pharmacology and Therapeutics*
*GKT School of Biomedical Sciences*
*King's College, London, UK*

VOLUME ONE HUNDRED AND THIRTY FIVE

# INTERNATIONAL REVIEW OF NEUROBIOLOGY

## Neurobiology of Chinese Herb Medicine

Edited by

**BAI-YUN ZENG**
*King's College London, London*

**KAICUN ZHAO**
*Middlesex University, London*

ELSEVIER

ACADEMIC PRESS
An imprint of Elsevier

Academic Press is an imprint of Elsevier
50 Hampshire Street, 5th Floor, Cambridge, MA 02139, United States
525 B Street, Suite 1800, San Diego, CA 92101-4495, United States
The Boulevard, Langford Lane, Kidlington, Oxford OX5 1GB, United Kingdom
125 London Wall, London, EC2Y 5AS, United Kingdom

First edition 2017



ISBN: 978-0-12-811779-8
ISSN: 0074-7742



*Publisher:* Zoe Kruze
*Acquisition Editor:* Kirsten Shankland
*Editorial Project Manager:* Andrea Gallego Ortiz
*Production Project Manager:* Surya Narayanan Jayachandran
*Cover Designer:* Alan Studholme

Typeset by SPi Global, India

# CONTENTS

# CONTRIBUTORS

**Sandra Appiah**
Faculty of Science and Technology, Middlesex University, The Burroughs, Hendon, London, United Kingdom

**Celia Bell**
Faculty of Science and Technology, Middlesex University, The Burroughs, Hendon, London, United Kingdom

**Gang Chen**
Center for Translational Systems Biology and Neuroscience, Key Laboratory of Integrative Biomedicine for Brain Diseases, Nanjing University of Chinese Medicine, Nanjing, Jiangsu, PR China

**Ting Chen**
The Second Affiliated Hospital of Zhejiang University School of Medicine; Zhejiang University Medical PET Center; Institute of Nuclear Medicine and Molecular Imaging of Zhejiang University; Key Laboratory of Medical Molecular Imaging of Zhejiang Province, Hangzhou, Zhejiang, China

**Kin Chiu**
The University of Hong Kong; State Key Laboratory of Brain and Cognitive Sciences, The University of Hong Kong, Hong Kong, SAR, China

**Alberto Coscione**
The Princess Alexandra Hospital NHS Trust, Harlow, United Kingdom

**Ghada Elayat**
Faculty of Science and Technology, Middlesex University, The Burroughs, Hendon, London, United Kingdom

**Rui Fang**
Royal Botanic Gardens, Kew, Richmond, United Kingdom

**Chong Gao**
School of Chinese Medicine, The University of Hong Kong, Hong Kong SAR, China

**Lucy Ghali**
Faculty of Science and Technology, Middlesex University, The Burroughs, Hendon, London, United Kingdom

**Xiaoyan Guo**
Center for Translational Systems Biology and Neuroscience, Key Laboratory of Integrative Biomedicine for Brain Diseases, Nanjing University of Chinese Medicine, Nanjing, Jiangsu, PR China

**Noor Hasan**
Faculty of Science and Technology, Middlesex University, The Burroughs, Hendon, London, United Kingdom

**Peter J. Houghton**
King's College London, London, United Kingdom

**Melanie-Jayne R. Howes**
Royal Botanic Gardens, Kew, Richmond; King's College London, London, United Kingdom

**Jing Huang**
The Second Affiliated Hospital of Zhejiang University School of Medicine; Zhejiang University Medical PET Center; Institute of Nuclear Medicine and Molecular Imaging of Zhejiang University; Key Laboratory of Medical Molecular Imaging of Zhejiang Province, Hangzhou, Zhejiang, China

**Yingjie Huang**
National Institute on Drug Dependence and Beijing Key Laboratory of Drug Dependence, Peking University, Beijing, China

**Dan Jiang**
Hallam Institution of TCM in Sheffield UK, Sheffield, United Kingdom

**Huw Jones**
Faculty of Science and Technology, Middlesex University, The Burroughs, Hendon, London, United Kingdom

**Lily Li**
St. Mary's Hospital Paddington, London, United Kingdom

**Lin Lu**
Institute of Mental Health, Peking University Sixth Hospital, and Key Laboratory of Mental Health, Beijing, China

**Abby L. Manthey**
The University of Hong Kong, Hong Kong, SAR, China

**Helen C. Owen**
Faculty of Science and Technology, Middlesex University, The Burroughs, Hendon, London, United Kingdom

**Mike Revitt**
Faculty of Science and Technology, Middlesex University, The Burroughs, Hendon, London, United Kingdom

**Jiangang Shen**
School of Chinese Medicine, The University of Hong Kong, Hong Kong SAR, China

**Amrinder Singh**
Traditional Chinese Herbal Medicine Programme, Middlesex University, The Borough, Hendon, London, United Kingdom

**Monique Simmonds**
Royal Botanic Gardens, Kew, Surrey, United Kingdom

**Kwok-Fai So**
The University of Hong Kong; State Key Laboratory of Brain and Cognitive Sciences, The University of Hong Kong, Hong Kong, SAR; Guangdong-Hong Kong-Macau Institute of CNS Regeneration and Guangdong Key Laboratory of Brain Function and Diseases; Ministry of Education Joint International Research Laboratory of CNS Regeneration, Jinan University, Guangzhou; Co-innovation Center of Neuroregeneration, Nantong University, Nantong, China

**Mei Tian**
The Second Affiliated Hospital of Zhejiang University School of Medicine; Zhejiang University Medical PET Center; Institute of Nuclear Medicine and Molecular Imaging of Zhejiang University; Key Laboratory of Medical Molecular Imaging of Zhejiang Province, Hangzhou, Zhejiang, China

**Milan Vu**
Faculty of Science and Technology, Middlesex University, The Burroughs, Hendon, London, United Kingdom

**Yao Yao**
The Second Affiliated Hospital of Zhejiang University School of Medicine; Zhejiang University Medical PET Center; Institute of Nuclear Medicine and Molecular Imaging of Zhejiang University; Key Laboratory of Medical Molecular Imaging of Zhejiang Province, Hangzhou, Zhejiang, China

**Bai-Yun Zeng**
Neurodegenerative Disease Research Group, Institute of Pharmaceutical Science, Faculty of Life Sciences & Medicine, King's College, London, United Kingdom

**Hong Zhang**
The Second Affiliated Hospital of Zhejiang University School of Medicine; Zhejiang University Medical PET Center; Institute of Nuclear Medicine and Molecular Imaging of Zhejiang University; Key Laboratory of Medical Molecular Imaging of Zhejiang Province, Hangzhou, Zhejiang, China

**Yinan Zhang**
National Institute on Drug Dependence and Beijing Key Laboratory of Drug Dependence, Peking University, Beijing, China

**Kaicun Zhao**
Traditional Chinese Herbal Medicine Programme, Middlesex University, The Borough, Hendon, London, United Kingdom

**Weili Zhu**
National Institute on Drug Dependence and Beijing Key Laboratory of Drug Dependence, Peking University, Beijing, China

# PREFACE

Traditional Chinese Medicine (TCM) is an integrative and systematic medical system. It has a long history in clinical practice in China. The ancient Chinese medicine text *The Yellow Emperor's Classic of Internal Medicine* (Huangdi Neijing, 黄帝内经) was published more than 2000 years ago. The fundamental TCM theories and therapeutic skills were established in the book. Until today, the principles and the treatment modalities stated in the ancient book are still followed by TCM practitioners. During the long history, TCM has been developing with continued improvement and perfection in both theories and practical treatment. From the *Treatise on Cold Pathogenic Disorders* (Shang Han Lun, 伤寒论) to *Treatise on Differentiation and Treatment of Seasonal Warm Diseases* (Wenbing Tiaobian, 温病条辩) and from *Key to Syndrome Identification and Treatment of Diseases in Infants* (Xiaoier Yaozheng Zhijue, 小儿药证直诀) to *Fu Qing-Zhu's Gynecology* (Fu Qingzhu's Nvke, 傅青主女科), TCM has been used widely in the treatment of various clinical conditions including internal, gynecological, and pediatric diseases.

TCM believes that human body exists as a unity of Yin and Yang which are reflected in all life processes and are mutually restricted and dependent. TCM recognizes that the Qi (energy) and blood are the essential materials for life, and all the Zang and Fu (organs) in the human body cooperated harmoniously through the meridian and collateral networks. Diseases will happen when the balance between Yin and Yang and the harmonizing among the Zang and Fu are broken by either internal or environmental factors. The clinical treatments of illnesses in TCM always take systematic and holistic approaches with the aim to balance the Yin and Yang, and harmonizing Qi–blood and Zang–Fu organs.

The major therapeutic modalities of TCM include acupuncture and Chinese herbal medicine. Chinese herbal medicine (CHM) is one of the most sophisticated traditional herbal medicine systems of the world with an unbroken tradition going back to the third century BC. Another ancient classical text published around the same time as that of *Huangdi Neijing*, *Shen Nong's Classics of Herbal Medicine* (Shennong Bencao Jing, 神农本草经), is the earliest Chinese herbal medicine book in which 365 commonly used medicinal herbs and their therapeutic functions are recorded. With the continuous development, more and more herbs with medical functions have been discovered. Nowadays, more than 5000 plant species have been recorded for use in CHM. Similar to acupuncture, CHM is still actively practiced in China and plays an important role in public healthcare. Furthermore, due to its holistic approach and clinical effectiveness, CHM has gained increasing acceptance and popularity in the West in the recent decades.

At the same time, there is a surge in scientific research of both acupuncture and CHM. For CHM, almost all of the commonly used Chinese medicinal herbs have been studied on their major chemical components and the biological activities. These studies have provided supportive evidence for the traditional use of the herbs and also discovered many new potential applications. In particular, the 2015 Nobel Prize in Physiology or Medicine was awarded to Professor Youyou Tu for her discovery of a novel antimalarial drug from Chinese herbal medicine *Artemisia annua* (Qinghao, 青蒿). This as example indicates that there is a great of potential innovations to be investigated and discovered from the traditional medicinal herbs. In 2013, a special volume of this series on the Neurobiology of Acupuncture has been published. This volume as a sister volume of the previous acupuncture book will focus on the neurobiological research of CHM with the aim to summarize the recent development in CHM and, in particular, the neurobiology of CHM and to promote the understanding and enhance CHM research in the future.

This volume contains a total of 13 chapters covering various aspects of Chinese herbal medicine from clinical applications for therapeutic treatment of neurological diseases to the underlying pharmacological action mechanisms. Chapter 1 provides a comprehensive discussion on the current understanding of the potential therapeutic effects of the polysaccharides from *Lycium barbarum* (Gouqizi, 枸杞子) for the treatment of various retinal diseases, including glaucoma, ischemia/reperfusion, age-related macular degeneration, retinitis pigmentosa, and diabetic retinopathy. The action mechanisms of the herb and the active component were evaluated. From Chapters 2–6, several commonly seen neurological disorders and the potential therapeutic effectiveness of CHM are discussed. Alzheimer's disease (AD) is a neurodegenerative disease and the treatment strategies are limited. Chapter 2 described some Chinese medicinal herbs and formulae that have been used traditionally for the treatment of AD or relieving AD symptoms. In this chapter, the authors also reviewed the available scientific evidence showing the active chemical components of the herbs and the pharmacological action mechanisms. Chapter 3 focused on the clinical effectiveness of Chinese herbal medicines in the treatment of Parkinson's disease. In this chapter, the mechanisms of action of a number of Chinese herbal compounds/extracts and Chinese herb formulas on the models of Parkinson's were reviewed. Clinical studies have shown that Chinese herb formulas as an adjunct improved both motor and nonmotor symptoms, and reduced the dose of dopaminergic drugs and the occurrence of dyskinesia. Another global health problem is depression which affects a large population. Due to the limitation of the currently used conventional antidepressants such as unsatisfied responses and remission rate, late onset of efficacy, and side effects, increasing attention has been drawn to traditional CHM. Chapter 4 introduced some Chinese herbal formulas and their clinical efficacy in the treatment of depression. The neurobiological mechanisms underlying antidepressant effects of Chinese herbal medicines, including enhancement of monoamine activities, inhibition of hyperactivation of hypothalamic–pituitary–adrenal axis, neuroimmune response, and neurotrophic factors, are discussed in the chapter. The rapid onset antidepressant effect of the herbal formula Yueju Wan (越鞠丸), as well as the associated neuroplastical mechanisms is very encouraging for the future development. Chapter 5 evaluated the research findings from both clinical trials and pharmacological studies on the Chinese herbal formulas and some key herbs used in the treatment of insomnia. The underlying pharmacological action mechanisms of the key herbs used in treatment of insomnia were evaluated. The major action mechanisms shared by most of the sedative herbs are to increase the level of γ-aminobutyric acid (GABA) or stimulate the activity of GABAA receptor. Some potentially useful herbs and formulas are summarized in the chapter. Adult neurogenesis plays important roles in cognitive and emotional behaviors. Accumulating evidence suggested that metabolic factors could be important modulators in the process of adult neurogenesis. Chapter 6 reviewed the current progress in research on the roles of metabolic factors in modulating adult neurogenesis and neural plasticity, and evaluated the potential effects of Chinese herbal medicines on modulation of adult neurogenesis through targeting certain metabolic factors. Many TCM formulas and their active ingredients originally used for metabolic disorders have promising effects on mediating neurogenesis and brain repair.

With increasingly wider use of Chinese herbal medicine in the world, some issues related to environment protection and ethics are raised. To maintain a sustainable development of Chinese herbal medicine, these issues have to be given an appropriate consideration. Bear bile is traditionally used in CHM. As several bear species now listed as endangered species and also there is an ethical issue in collecting the bile from living bear, it is important to seek and promote alternatives to the use of bear bile as medicine. Chapter 7 presented comprehensive investigations on the clinical uses and pharmacological activities of bear bile and proposed six plant species, either as single herb or in combination, and two herbal formulas and two individual phytochemicals, berberine and andrographolide, as potential alternatives to bear bile. Chapter 8 reported a new molecular imaging technology which has been applied recently to investigate the action mechanisms of CHMs in the treatment of neurological diseases. With this technique, it is possible to assess the cellular or molecular function in a living subject by using specific imaging probes. The development of the new technique provides perspectives on the future evaluations of therapeutic response of Chinese herbal medicine. Another important area covered in this book is herb–drug interactions. While some combination use of herbs and conventional drugs may produce beneficial effects, herb–drug interactions frequently bring safety issues, leading to either therapeutic failure or toxicities in clinical practice. Chapter 9 reviewed the available information and development in the area of herb–drug interactions. Herb–drug interactions can happen in both pharmacokinetics and pharmacodynamics. Abundant studies focused on pharmacokinetic interactions of herbs and drugs. Studies on pharmacodynamics interactions of herbs and drugs are still very limited. Information on the potential interactions among the commonly used Chinese medicinal herbs and conventional drugs, in particular the drugs used in the treatment of neurological diseases, is provided in this chapter.

Chapters 10 and 13 discussed the therapeutic effectiveness of Chinese herbal medicine in the treatment of female and male infertilities, respectively. Chapter 10 reviewed a number of studies published in past 15 years on the effects of herbs and multiple herb formulas either in various infertile models or in clinical studies for the treatment of different kinds of female infertility conditions. The research evidence showed that CHMs may work through either a specific single target or multitargets action mechanisms including regulating relevant hormone levels in female reproductive system, improving ovary function, and enhancing uterine receptivity. Chapter 13 summarized the recent development in CHM treatment of male infertility. Studies have shown that CHM is able to help improve sperm motility and quality, increase sperm count and rebalance the hormone levels, and adjust immune functions leading to successful fertility. In addition, evidence also showed that CHM in combination with conventional therapies exhibited significant synergism and enhanced the efficacy of conventional treatments with improved fertility rate.

Chapter 11 discussed the potentials of phytochemicals to help overcome chemoresistance of leukemia through the modulation of signaling pathways involved in the survival and death of leukemic clonal cells and/or leukemic stem cells. Several phytochemicals isolated from Chinese medicinal herbs have been shown to induce apoptosis or cause cell death via the induction of autophagy in leukemic cell lines. Some phytochemicals inhibited the Notch and Sonic hedgehog signaling pathways to induce apoptosis, indicating a potential to overcome chemoresistance by elimination of leukemic stem cells. In Chapter 12, a number of Chinese medicinal herbs were evaluated on their antiaddiction activities. Some herbs showed clinically useful in prevention of opioid abuse and dependence; some herbs have beneficial inhibitory effects on alcohol intake; some herbs antagonized the heroin-induced reinstatement of heroin seeking. Evidence has shown that CHMs may be used to complement current treatments for drug addiction.

Chinese herbal medicine is an old medical system but has been highly researched in the modern days. Literatures of scientific research on individual Chinese medicinal herbs and herbal formulas are increasing exponentially. It is frequently difficult to fully understand the action mechanisms of an herb due to the complexity of the chemical ingredients contained in the herb. In fact, Chinese herbal medicine is usually using a formula with multiple herbs in its clinical practice to treat various illnesses. This has made the research even more difficult. However, scientific evidence is accumulating, and research on Chinese medicinal herbs and formulas either in the pharmacological action mechanisms or in research methodologies is rapidly developing. Interactions of the herbs in an herbal formula and the integrative activities of an herbal formula are the areas to be further investigated in the future. This book focuses on the neurobiological activities of CHMs. We wish this book not only can provide information in research findings of CHM but also promote the progression of scientific research of Chinese herbal medicine.

Finally, we would like to thank the excellent group of research scientists and clinicians who contributed an exceptional set of chapters to make up this volume, and the dedicated editorial staff to help accomplish this book.

Bai-Yun Zeng
Kaicun Zhao

CHAPTER ONE

# Effects of *Lycium barbarum* on the Visual System

**Abby L. Manthey*, Kin Chiu*,†,1, Kwok-Fai So*,†,‡,§,¶,1**

*The University of Hong Kong, Hong Kong SAR, China
†State Key Laboratory of Brain and Cognitive Sciences, The University of Hong Kong, Hong Kong, SAR, China
‡Guangdong-Hong Kong-Macau Institute of CNS Regeneration and Guangdong Key Laboratory of Brain Function and Diseases, Jinan University, Guangzhou, China
§Ministry of Education Joint International Research Laboratory of CNS Regeneration, Jinan University, Guangzhou, China
¶Co-innovation Center of Neuroregeneration, Nantong University, Nantong, China
[1]Corresponding authors: e-mail address: datwai@hku.hk; hrmaskf@hku.hk

## Contents



## Abstract

*Lycium barbarum* (wolfberry, gogi berry, *gouqizi*, 枸杞) is one of the most widely used Chinese herbal medicines (CHMs) and is also one of the most scientifically studied. Indeed, the polysaccharide component of this berry (LBP) has been shown to have antioxidant, antiinflammatory, antiexcitotoxic, and antiapoptotic properties. These

*International Review of Neurobiology*, Volume 135
ISSN 0074-7742
http://dx.doi.org/10.1016/bs.irn.2017.02.002

properties make it a particularly useful treatment option for the ocular environment. Although there are a handful of studies investigating the use of LBP to treat diseases affecting the lens, the vast majority of the published literature investigating LBP in the visual system focus on the retina. In this chapter, we have described what is currently understood concerning the effects of LBP treatment on various retinal diseases, including glaucoma, ischemia/reperfusion, age-related macular degeneration, retinitis pigmentosa, and diabetic retinopathy. We then describe the functions attributed to LBP using other cellular contexts to elucidate the full mechanisms this CHM utilizes in the retina. By making connections between what is known about the function of LBP in a variety of tissues and its function as a therapy for retinal degenerative diseases, we hope to further emphasize the continued use of this CHM in clinical medicine in addition to providing a platform for additional study.

## 1. INTRODUCTION

Treatment with Chinese herbal medicines (CHMs) is a common aspect of the healthcare system in many Asian countries and traditionally focuses on the concept of balancing a person's yin and yang. When applied to the physiological functions and pathological changes in the human body, this theory utilizes a holistic approach, whereby various herbal supplements are administered to synergistically treat a disease. This model is in contrast to that of single synthetic drugs, which are often used in Western cultures. In fact, as the study and use of CHMs becomes increasingly accepted, more herbal treatments are being used, alone or in combination with classical drugs, to treat diseases worldwide. This has also lead to further scientific analysis, discovery, and quality control of various CHMs.

### 1.1 *Lycium barbarum*

*Lycium barbarum*, a reddish-orange berry (Fig. 1) also known as wolfberry, gogi berry, or *gouqizi* (枸杞) in Chinese, has been used as a preventative and curative agent in various forms for thousands of years (Wang et al., 2015). In fact, *L. barbarum* has been consumed as an herbal supplement in a number of forms, from tea to concentrated capsule, and has been linked to enhanced protection and nourishment of the kidney, liver, and eye (Junlin & Aicheng, 2002). It is also currently one of the most widely scientifically studied CHMs. Various active compounds have been isolated from *L. barbarum*, including various polyphenols/flavonoids and carotenoids (reviewed in Wang et al., 2015). Indeed, Zhou et al. (2016) recently identified 15 new compounds that appear to be dicaffeoylspermidine derivatives

**Fig. 1** Representative images showing *Lycium barbarum* fruit in its natural (*left panel*) and dried (*right panel*) forms.

that also make up a significant portion of the active constituents in *L. barbarum*. However, the most well-studied functional moiety involved in the observed *L. barbarum*-mediated health benefits appears to be the polysaccharide (LBP) component. This group of water-soluble glycoconjugates includes rhamnose, xylose, glucose, mannose, arabinose, and galactose (Benlloch et al., 2015; Liu et al., 2012).

The health benefits of LBPs have been classically linked to their antioxidative properties (Cheng & Kong, 2011; Cui et al., 2011; He, Yang, Jiao, Tian, & Zhao, 2012; Li, 2007; Shan, Zhou, Ma, & Chai, 2011; Xiao et al., 2012; Zhang, Chen, Zhao, & Xi, 2016; Zhou et al., 2016). LBPs have also been demonstrated to influence proliferation, cell cycle arrest, and apoptosis during carcinogenesis (Jin, Huang, Zhao, & Shang, 2013; Mao et al., 2011; Zhang et al., 2005). Other protective mechanisms related to *L. barbarum* include antiexcitotoxicity (Ho et al., 2009), antiinflammation (Chen, Soo, Srinivasan, Tan, & Chan, 2009; Xiao et al., 2012), and antiapoptosis (Ho et al., 2010; Li et al., 2011; Song, Roufogalis, & Huang, 2012). This CHM was also previously shown to play a neuroprotective role and help protect cells against amyloid-β neurotoxicity (Yu et al., 2007, 2005) and enhance cognitive function (Zhang, Du, et al., 2013; Zhou et al., 2016) in Alzheimer's disease-related neurodegeneration.

While the function of LBPs appears to be multifaceted, the underlying mechanisms involved may differ to a degree depending on physiological context and the specific cell type affected. In the current review, we have focused on the mechanism of LBPs in the visual system in addition to detailing their therapeutic use to treat various vision-related diseases.

## 1.2 The Visual System

Notably, the visual system is exceedingly sensitive to change, including increased oxidative stress (Kruk, Kubasik-Kladna, & Aboul-Enein, 2015), altered blood flow (Flammer et al., 2013), inflammation (Perez, Saeed, Tan, Urbieta, & Cruz-Guilloty, 2013), etc., implying that the mechanisms of LBPs in this cellular context may be similar to those demonstrated systemically or in other tissues. The eye is derived from three types of tissue during embryogenesis: the surface ectoderm, which generates the lens; the mesoderm, from which the cornea and sclera are produced; and the neural ectoderm, which gives rise to the retina and retinal pigmented epithelium (RPE) (for an excellent review of general ocular development, please see Heavner & Pevny, 2012). During development, signals between the surface ectoderm and the optic vesicle, an outgrowth from the developing central nervous system that will later become the optic nerve, initiate lens placode formation, invagination, and creation of the bilayered optic cup. These tissues will then differentiate into a number of specialized cell types that are essential for vision. Although some recent work has been conducted concerning the therapeutic effects of LBPs on both the formation of senile cataracts in the human lens (Kee et al., 2013 (clinicaltrials.gov identifier: NCT01142960)) and oxidatively damaged lens epithelial cells in culture (Qi et al., 2014), the majority of published literature concerning LBPs in the visual system focus on the retina.

The cells in the outer and inner layers of the optic cup formed during development will specifically differentiate into the RPE and neural retina, respectively (Fig. 2) (reviewed in Heavner & Pevny, 2012). The latter includes various cells types and is organized into three layers: the outer nuclear layer, composed of the rod and cone photoreceptors; the inner nuclear layer, containing the amacrine, horizontal, and bipolar cells as well as Müller glia; and the ganglion cell layer, which is composed of the retinal ganglion cells (RGCs). During vision, the dense layer of posteriorly located photoreceptors converts the incoming light into electrochemical signals involving a phototransduction cascade in the outer segment of the cell and downstream changes in cation channels (reviewed in Sung & Chuang, 2010). This hyperpolarization of the light-stimulated photoreceptor then causes altered neurotransmitter release that initiates an action potential-mediated chain reaction through the cells in the inner nuclear layer to the RGCs, which then transmit the signal to the brain via the optic nerve (reviewed in Baccus, 2007). Alternatively, the outer layer-derived RPE is a monolayer of cells that forms the blood–retinal barrier (BRB) and supports the photoreceptor cells via the phagocytosis of the outer

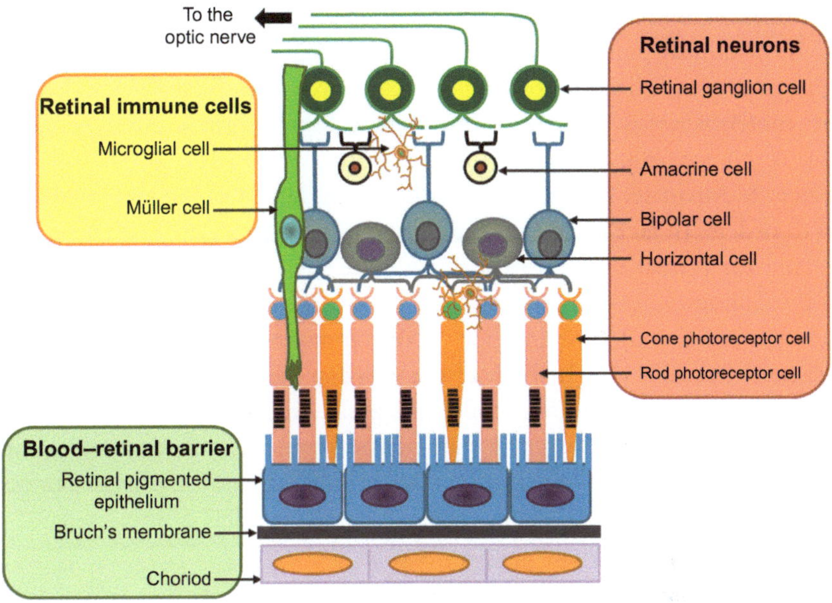


**Fig. 2** Diagram highlighting the organization and basic functions of various retinal cell types.

segments and secretion of various trophic factors (Heavner & Pevny, 2012). Finally, the Müller and retinal microglial cells are the resident immune cells in this tissue and eliminate foreign substances as well as dead/dying cells.

Under normal conditions, the retina internally balances the levels of reactive oxygen species (ROS) and inflammatory molecules, while limiting aberrant apoptosis and interference from the surrounding vasculature in order to process light signals properly for vision. Unfortunately, in a diseased state, these mechanisms may be inactive or dysfunctional leading to further problems. Thus, while some drugs or herbal supplements, including LBPs, may not have an effect on the visual system under normal conditions, they may have significant therapeutic benefits on retinal function under stress or after trauma-/diseased-induced retinal cell death or dysfunction.

## 2. EFFECT OF *L. BARBARUM* POLYSACCHARIDE TREATMENT IN ANIMAL MODELS OF RETINAL DISEASE

The loss of retinal neurons, including the RGCs and photoreceptors, either directly or as a secondary effect, can lead to significant deficits in visual acuity that have been associated with a number of retinal diseases, including glaucoma, ischemia/reperfusion (I/R), age-related macular degeneration (AMD), retinitis pigmentosa (RP), and diabetic retinopathy (DR). Although a number of clinical/preclinical studies have been performed in healthy humans (Amagase & Nance, 2008; Amagase, Sun, & Borek, 2009; Amagase, Sun, & Nance, 2009; Cheng, Chung, Szeto, & Benzie, 2005; Paul Hsu, Nance, & Amagase, 2012), few have been conducted on patients with retinal disorders. In fact, only three studies appear to focus on *L. barbarum* as a therapy for retinal diseases in humans, one involving RP patients (Chan et al., ongoing), one involving patients with normal tension glaucoma (Lai et al., ongoing), and another on development of AMD in the elderly (Bucheli et al., 2011). The bulk of the work in the current literature makes use of various animal models, which has greatly influenced our understanding of how LBP functions as a therapeutic supplement. Here, we outline these particular retinal diseases, focusing on their global prevalence and the general effects of LBP in the various animal models of the disease.

## 2.1 Glaucoma

Glaucoma is a retinal degenerative disease in which RGCs forming the optic nerve are selectively lost. This loss has been shown to occur in two phases: a primary phase that occurs due to direct damage to the retinal neurons and a secondary phase that is caused by the release of toxic factors during the primary wave of cell death (Levkovitch-Verbin et al., 2001, 2003; Yoles & Schwartz, 1998). Decreased thickness of the nerve fiber layer and increased intraocular pressure (IOP) are often associated with the progressive loss of vision experienced by glaucoma patients (Calkins, 2012). Notably, glaucoma is the most common cause of irreversible blindness, second only to cataract in terms of ocular diseases globally (Pascolini & Mariotti, 2012). Furthermore, glaucoma is expected to affect approximately 111.8 million people between 40 and 80 years of age by 2040 (Tham et al., 2014). Thus, continued research investigating this disease is essential.

Glaucoma etiology has two primary aspects: progressive disruption of RGC axonal transport and retinal ischemia. Several animal models have been used to study glaucoma-related disruption of the axons, including laser photocoagulation of the limbal and episcleral veins, complete and partial optic nerve transection, and optic nerve crush. Importantly, the detrimental effects observed following photocoagulation of the episcleral and limbal veins in Sprague-Dawley rats, a model of chronic ocular hypertension, were significantly prevented by feeding the rats LBPs for 1 week before laser treatment (Chan et al., 2007). Although the IOP was not affected in these animals, the level of RGC death was significantly decreased compared to control animals, an effect that may be modulated by various trophic factors (Chiu et al., 2010; Mi, Chiu, et al., 2012). These changes also appear to involve microglial cell function as treatment with factors that stimulate or inhibit activation altered the impact of LBP (Chiu et al., 2009). Notably, the use of this technique to model glaucoma and the effects of various drugs are continually being evaluated and advanced. In vivo manganese-enhanced magnetic resonance imaging, for example, was employed by Chan, Fu, Hui, So, and Wu (2008) to better understand the mechanisms underlying this disease. However, one aspect of glaucoma this model, as well as the complete optic nerve transection model, do not address fully is the differentiation between primary and secondary RGC loss. Fortunately, the recent development of partial optic nerve transection has allowed researchers to compare the effects of LBP treatment on these stages of progressive neuronal degradation. In fact, Li et al. (2013) administered LBPs for 1 week prior to either complete (highlighting primary degradation) or partial (highlighting secondary degradation) surgery and investigated the effects at various time points thereafter. In doing so, it appears that LBP treatment does not delay primary RGC degeneration in either model, but does decrease oxidative stress and increase the secretion of trophic factors (e.g., insulin-like growth factor 1 (IGF-1)), resulting in lower levels of secondary neuronal cell death. This decrease in secondary cell death after partial transection also appears to be related to an LBP-mediated decrease in microglial/macrophage activation in this model (Li et al., 2015). Importantly, these cellular changes modulated by LBP pretreatment were demonstrated to preserve retinal function, as highlighted by multifocal electroretinographic (ERG) analysis (Chu, Li, Chin, So, & Chan, 2013). While the aforementioned models focus on the disruption of axonal transport that occurs during glaucoma or acute optic nerve injury, various models have also been used to specifically study the ischemia-induced damaged that occurs during a number of retinal degenerative diseases.

## 2.2 Ischemia/Reperfusion Injury

Retinal ischemia occurs when the blood supply to the retina is inadequate, resulting in altered metabolic function due to the lack of oxygen. These changes ultimately lead to irreversible cell death. In fact, ischemia has been shown to play a role in neuronal cell death in a number of retinal disorders, including glaucoma, DR, ischemic optic neuropathies, etc. The most widely used method to study retinal ischemia is to induce acute ocular hypertension via the anterior chamber using an elevated hydrostatic reservoir. Doing so results in a jump in the IOP of the insulted eye to a level that is higher than that of the animal's systolic blood pressure, thus limiting the blood supply to the retina. Treating a mouse model of acute ocular hypertension with LBP for 7 days prior to increasing IOP (as well as postinsult for 4 or 7 days) resulted in a decreased loss of RGCs as well as less thinning of the nerve fiber layer, decreased immunoglobulin leakage, and increased blood vessel density compared to the control treated retinas (Mi, Feng, et al., 2012). A similar model using rats also showed an LBP-mediated delay in RGC death, with more than 50% of the cells remaining viable 7 days after the insult (He et al., 2014).

Animal models of middle cerebral artery occlusion, often used to study focal cerebral ischemia in the brain (i.e., ischemic stroke), have also been used to evaluate the effects of LBP in both the brain (Wang et al., 2013; Wang, Li, et al., 2014; Yang et al., 2012) and retina (Li et al., 2011). In this study, pretreatment with LBP for 1 week prior to ischemia was shown to protect the retina from neuronal cell death, glial cell activation, oxidative stress, retinal swelling, and disruption of the BRB at 48 h after reperfusion.

## 2.3 Age-Related Macular Degeneration

AMD, an aging-associated disease involving the progressive loss of vision, can be classified as wet or dry, with the latter being more common (approximately 80%–85% of cases). While the wet form involves the pathological growth of blood vessels into the subretinal space and physical disruption of retinal structure, the dry form is initiated by dysfunction in Bruch's membrane leading to the degeneration of the RPE and, subsequently, the loss of photoreceptors in the macular region of the retina (reviewed in Ambati, Ambati, Yoo, Ianchulev, & Adamis, 2003; Rattner & Nathans, 2006). Dry AMD has also been associated with the presence of amyloid-β deposits known as drusen. In regards to global prevalence, an estimated 5% of blindness is caused by AMD (WHO, 2012). However, with the proportion of elderly continually increasing, it is likely that this disease may become more common.

AMD appears to involve both genetic and environmental risk factors, making it a difficult disease to mirror in animal models. In order to study the end-point morphology of human AMD, light-induced retinal damage, which causes similar defects as those observed during advanced atrophic AMD, has been used in various rodent models (Marc et al., 2008; Organisciak & Vaughan, 2010; Zeiss, 2010). Several genetic models have also been used to mirror the symptoms associated with AMD (reviewed in Pennesi, Neuringer, & Courtney, 2012). However, while several human studies have evaluated the effects of LBP in the prevention of AMD symptoms in the elderly (Bucheli et al., 2011; Cheng et al., 2005), none of the aforementioned animal models have been used to directly test the effects of LBP treatment. Interestingly, the link between amyloid-β deposition in the brain and AMD retinas may prove to be indicative of the effects of LBPs during disease pathogenesis as cell cultures of primary cortical neurons pretreated with LBPs appear to be protected from amyloid-β-induced stress and cell death (Ho et al., 2007; Yu et al., 2007, 2005). Amyloid-β expression was also downregulated following LBP treatment in a mouse model of acute ocular hypertension (Mi, Feng, et al., 2012), a phenomena that may be mimicked during AMD.

## 2.4 Retinitis Pigmentosa

In RP, photoreceptors are lost due to inherited genetic mutations resulting in the loss of peripheral vision (initial loss of rod photoreceptors) followed by central vision (subsequent loss of cone photoreceptors) (Hartong, Berson, & Dryja, 2006). The disease is genetically and phenotypically heterogeneous and is estimated to affect 1 in 3000 to 1 in 7000 people worldwide (Ferrari et al., 2011). A number of animal models have been used to study RP, including P347L transgenic rabbits (which have rod-dominated, progressive photoreceptor degeneration with regional variations in the pattern of photoreceptor loss), RCS rats (which have a Merkd mutation related to RP pathogenesis), and $mer^{kd}$ mice (which share the same mutation as the RCS rats) as well as rd1 and rd10 mice (which have mutations in the phosphodiesterase 6b gene), etc.

Similar to the effects observed for LBP treatment in other retinal diseases, treating rd1 mice with LBP extract appears to decrease photoreceptor cell degeneration via reduced oxidative stress (Miranda et al., 2010). In addition to its capacity as an antioxidant, LBP treatment in rd10 mice also demonstrated antiinflammatory and neuroprotective functions (Wang, Xiao, et al., 2014). Recent work by Ni, Wei, Yin, Liu, and Liu (2013) using RCS rats also suggests that treatment with *L. barbarum* reduces apoptosis of photoreceptors in this model of RP; however, the authors hypothesize that this function is largely due to the carotenoid components of this CHM and additional work is necessary to evaluate these effects and mechanism.

### 2.5 Diabetic Retinopathy

The elevated glucose levels, or hyperglycemia, associated with systemic diabetes have been related to a number of complications/comorbidities. DR involves changes in the microvasculature of the retina leading to increased permeability of the BRB, ischemia, neovascularization, macular edema, inflammation, oxidative stress, and ultimately blindness. In 2010, DR accounted for 2.6% of blindness globally and appears to be progressively increasing compared to previous years (Leasher et al., 2016).

In db/db mice, a common animal model of DR, *L. barbarum* appears to attenuate hypoxia and mitochondrial stress (Tang et al., 2011; Yu et al., 2013), functions which have been attributed to the carotenoids in this medicinal herb, but need to be evaluated for the polysaccharide component. A streptozotocin-induced rat model of diabetes has also been used to study the effects of *L. barbarum* (Hu, Lee, Colitz, Chang, & Lin, 2012). In this study, retinal function was assessed using ERG and histopathological studies, and it appears that *L. barbarum* protected the function and viability of the retinal cells when administered to the rats after diabetes was confirmed.

## 3. UNDERLYING MECHANISMS OF *L. BARBARUM* POLYSACCHARIDES IN THE RETINA

The retinal diseases described in the previous sections, while different in genetics, phenotype, and prevalence, all share some common variables, namely, changes in the local retinal neurons, inflammation, and altered vascular function in addition to the overall increase in oxidative stress in the eye/retina. While the mechanistic effects of LBP treatment on these pathogenic processes could be dependent on the unique retinal microenvironment, it is also likely that some signaling pathway changes involved may mimic those employed in other cellular contexts. In fact, these functional pathways may overlap in a number of ways. In this section, we discuss the various mechanistic effects of LBP during retinal disease (Fig. 3), linking what has observed in this tissue to what is known in other parts of the body.

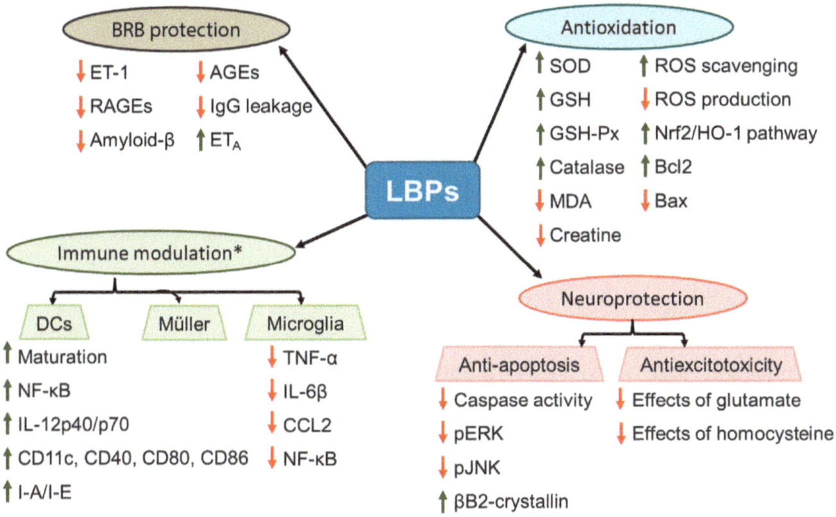


**Fig. 3** Mechanisms involved in *Lycium barbarum* polysaccharide (LBP) function in the diseased retina. During retinal disease, treatment with LBPs appears to affect four primary processes: antioxidation, neuroprotection, blood–retinal barrier (BRB) protection, and immune modulation. With the exception of immune system activation, the mechanisms underlying these LBP-mediated effects appear to be similar to those in nonocular tissues. The discrepancies in immune activation are likely related to the immune privileged nature of the retinal environment. As additional work is required to determine the retina-specific immune pathways affected, this process has been marked with an *asterisk* (*). *Abbreviations*: *AGEs*, advanced glycation endproducts; *Bax*, Bcl2-associated X protein; *Bcl2*, B-cell lymphoma 2; *CCL2*, C–C motif chemokine ligand 2; *CD*, cluster of differentiation; *ET-1*, endothelin 1; *ET$_A$*, endothelin receptor A; *GSH*, glutathione; *GSH-Px*, glutathione peroxidase; *I-A/I-E*, major histocompatibility complex class II I-A/I-E subregions; *IgG*, immunoglobulin; *IL*, interleukin; *MDA*, malondialdehyde; *NF-κB*, nuclear factor kappa B; *Nrf2/HO-1*, nuclear factor erythroid 1-related factor/heme oxygenase-1; *pERK*, phosphorylated extracellular signal-related kinase; *pJNK*, phosphorylated c-Jun N-terminal kinase; *RAGEs*, receptor for advanced glycation endproducts; *ROS*, reactive oxygen species; *SOD*, superoxide dismutase; *TNF-α*, tumor necrosis factor alpha.

## 3.1 Antioxidant Properties

Functionally, the beneficial mechanism of LBPs can be related to a number of potential functions that have been directly or indirectly linked to their antioxidant potential in various tissues (Cheng & Kong, 2011; Li, 2007; Li et al., 2011; Shan et al., 2011; Xiao et al., 2012; Zhang, Chen, et al., 2016), whereby the polysaccharides not only donate a hydrogen to various ROS (Jin, Lu, Huang, & Wang, 2011), but also activate enzymes involved in antioxidative activities. Indeed, in a 30 day randomized, double-blind, placebo-controlled clinical study, LBP treatment was shown to significantly increase the serum levels of superoxide dismutase (SOD) and glutathione peroxidase (GSH-Px), while decreasing serum malondialdehyde (MDA) in healthy Chinese adults, supporting its role as a therapeutic antioxidant (Amagase & Nance, 2008).

Notably, the formation of ROS during normal mitochondrial metabolism is expected in most tissues, including the retina; however, under normal conditions, elimination and repair mechanisms are active (Puertas et al., 1993). When the body is under stress or affected by disease, on the other hand, these mechanisms may be dysfunctional or inactive. High fat diets, exhaustive exercise, and subhealth conditions, for example, have all been shown to activate oxidative stress pathways. Notably, treating mice on a high fat diet with LBPs has been shown to increase the activity of SOD, GSH-Px, and catalase (Wu et al., 2010). Similarly, increased SOD and GSH-Px activity along with decreased MDA and creatine kinase activity was observed in the skeletal muscle of LBP-treated rats after intense exercise (Niu, Wu, Yu, & Wang, 2008; Shan et al., 2011). In addition to altering enzyme activity, LBPs also appear to inhibit ROS-producing reactions, including ultraviolet light-induced peroxidation and free radical generation by cytochrome *c* (Wang et al., 2002), that also play a significant role in oxidative stress in various tissues. Similar changes in peroxidation and decreased antioxidant activity, which are restored by LBP treatment, have been observed during normal aging processes (Li, Ma, & Liu, 2007) as well as chronic fatigue conditions (Zhao, Cai, Shao, & Ma, 2015).

In the retina, glutathione (GSH) and other GSH-related enzymes are the primary defenses against oxidative stress (Ganea & Harding, 2006). While the entire retina can be affected by disease/trauma-induced oxidative stress, the photoreceptors appear to be one of the primary targets along with other retinal neurons and the RPE (Beatty, Koh, Phil, Henson, & Boulton, 2000; Tanito et al., 2002). In rd1 mice, a model of RP photoreceptor cell death, treatment with a mixture of antioxidants that included LBPs was shown to increase GSH-Px and GSH concentration and decrease cell death (Miranda et al., 2010). However, these effects were not observed when the antioxidants were administered individually, indicating a synergistic effect. In RGCs, treatment with LBPs alone for 1 week prior to partial optic nerve transection significantly increased the expression of SOD 1 day after surgery (Li et al., 2013). In fact, the redox status appears to be maintained in RGCs through the activation of the nuclear factor erythroid 2-related factor (Nrf2)/heme oxygenase-1 (HO-1) antioxidant pathway (He et al., 2014). Furthermore, the effect of LBPs on stressed RPE cells in culture appears to involve the inhibition of ROS-induced downregulation of B-cell lymphoma 2 (Bcl2) and upregulation of Bcl2-associated X protein (Bax), indicating that the polysaccharides may function to not only decrease the levels of ROS via free radical scavenging and increased activity of antioxidative enzymes, but also by altering downstream gene function to prevent ROS-induced apoptosis (Liu et al., 2015).

## 3.2 Modulation of the Blood Vessels/Vascularization

The survival of neurons and other cells in a tissue microenvironment is, at least in part, dependent on the stability of the surrounding vasculature. The interaction between the retina and the vascular system is similar to that of the brain, whereby a cellular blood-barrier exists, known as the BRB and the blood–brain barrier (BBB), respectively. The BRB has two layers: an outer barrier formed by the RPE, Bruch's membrane, and the choriocapillaris; and an inner barrier formed by tight junctions between retinal endothelial cells, which functions similarly to the BBB. The presence of two barriers in the BRB limits the effect of vascularization on vision, allowing a clear path for light to travel to the photoreceptors. The primary function of these barriers is to regulate the transport of various molecules from the blood to the retina, functions that were recently outlined by Rizzolo (2014) and others (Steuer et al., 2005). Furthermore, disruption of the integrity of these barriers can lead to issues in the brain and retina, including leakage of serum inflammatory factors (e.g., immunoglobulin) (Cheung et al., 2005; Farrall & Wardlaw, 2009; Mi, Feng, et al., 2012), edema (Kimelberg, 1995), and neovascularization (Mi, Feng, et al., 2012).

In both the brain and retina, there is a strong relationship between the microvasculature, neurons, and associated glial cells (Yu et al., 2010). The cross talk between the astrocytes and vascular cells, for example, plays a role in maintaining barrier function and integrity (Zhang & Stone, 1997), whereby changes in blood flow/pressure to the tissue are modulated by altered expression of endothelin-1 (ET-1) and its associated receptors, $ET_A$ (which is expressed on vascular smooth muscle cells and is known to mediate blood vessel contraction) and $ET_B$ (which has been shown to be expressed in the endothelium and astrocytes and dilates the blood vessels) (Masaki, Vane, & Vanhoutte, 1994; Narayan, Prasanna, Krishnamoorthy, Zhang, & Yorio, 2003). During ischemic insult and disruption of the blood-barrier, the concentration of ET-1 increases along with that of $ET_B$ (Mi, Chiu, et al., 2012). Moreover, disruption of the BRB during ocular hypertension has been associated with overexpression of glial fibrillary acidic protein (GFAP) and aquaporin-4 (AQP4) (Kaur et al., 2007), as well as increased expression of receptor for advanced glycation endproducts (RAGEs), amyloid-β, and advanced glycation endproducts (AGEs) during ischemia (Mi, Feng, et al., 2012). Notably, pretreatment with LBP for 1 week prior to disrupting the BRB ameliorated these changes and prevented BRB disruption (Li et al., 2011). In fact, expression of ET-1, RAGEs, amyloid-β, and AGEs along with immunoglobulin leakage were all decreased during ocular hypertension after LBP treatment, while the expression of the $ET_A$ receptor was increased (Mi, Chiu, et al., 2012; Mi, Feng, et al., 2012). Thus, it appears that one of the primary mechanisms underlying the ocular health benefits of LBP is related to its ability to protect the structure and function of the retinal vasculature and BRB while also maintaining retinal homeostasis. This mechanism can also be related to LBP's neuroprotective qualities as retinal neuron viability and function is preserved.

## 3.3 Neuroprotection

In addition to balancing retinal homeostasis and limiting the effects of oxidative stress, apoptosis, and inflammation, LBPs have also been shown to have antiexcitotoxicity properties that directly protect neurons. Excitotoxicity occurs when receptors in the neurons are over activated by a surplus of extracellular neurotransmitters, such as glutamate, leading to the influx of ions into the cell and prolonged, excessive depolarization. These neurotoxic changes can in turn flood the tissue with more neurotransmitters, exacerbating the levels of cellular damage and death. Neuroprotection ultimately means saving the neuronal cells from apoptosis and preserving synapse function.

Glutamate excitotoxicity in particular has been identified as a possible pathogenic mechanism during various neurodegenerative diseases, including Alzheimer's disease, Parkinson's disease, and Huntington's disease (Arundine & Tymianski, 2004; Coyle & Puttfarcken, 1993; Ho et al., 2009; Won, Kim, & Gwag, 2002). Glutamate is also the primary neurotransmitter found in mammalian brain and retinal neurons. Thus, increased levels in these tissues often lead to over excitation. In the brain, LBP treatment appears to significantly reduce neuronal cell death caused by excess glutamate in a manner similar to that observed for conventional drugs such as *N*-methyl-D-aspartate receptor antagonists (Ho et al., 2009). Interestingly, ischemia, oxidative stress, inflammation, and accumulation of other neurotoxic amino acids and substances can also result in downstream glutamate excitotoxicity. For example, amyloid-β and homocysteine, both of which have been linked to neuronal damage in various diseases, have been shown to induce downstream glutamate excitotoxicity (Ho, Ortiz, Rogers, & Shea, 2002). Treatment with LBPs appears to reduce the levels of amyloid-β (Ho et al., 2007) and the effects of homocysteine in primary cortical neurons (Ho et al., 2010), while also reducing the downstream elevation in proapoptotic phosphorylated extracellular signal-related kinase (pERK) and c-Jun N-terminal kinase (pJNK) (Li et al., 2013) along with caspase-mediated apoptosis (Ho et al., 2007). In terms of neuronal function, Zhang et al. (2012) showed that LBP treatment prevented corticosterone-induced changes in the expression of postsynaptic density protein 95, an important protein in synapse maintenance, in the hippocampus.

In the retina, LBP treatment has been shown to decrease the phosphorylation of both JNK3 and c-Jun after partial optic nerve transection (Li et al., 2013). Furthermore, Chiu et al. (2010) previously identified multiple changes in the protein expression of various crystallins in LBP fed rats following induced ocular hypertension. While this family of proteins is well known for their structural role in the ocular lens, in other tissues they often act as molecular chaperones (please see Slingsby, Wistow, & Clark, 2013 and Thanos et al., 2014 for a review of crystallin function in the lens and other ocular tissues). It appears that LBP-mediated neuroprotection may, at least in part, function via the upregulation of these chaperones, specifically βB2-crystallin, during retinal stress (Chiu et al., 2010). Other studies investigating other retinal insults, including oxidative stress, ischemia, and inflammation, have also indicated decreases in neuronal cell death and restoration of function following LBP treatment.

## 3.4 Immunomodulation

LBPs have been show to induce a variety of changes in immune function, targeting dendritic cells, T- and B-lymphocytes, macrophage, and natural killer (NK) cells (for a review of these various target cell types, please see Zhang, Zhou, & Zhang, 2015). The eye is an immune privileged tissue, meaning that many of the cell types responsible for the cellular and humoral immune responses in the rest of the body are unable to participate in ocular immunity. Indeed, many of the classical cell types and signaling pathways involved in the inflammatory immune response to foreign antigens are suppressed or completely inhibited in this tissue. This is in part a consequence of the BRB, which acts as a physical barrier to a number of cell types and immune molecules. RPE cells have also been demonstrated to suppress T-cell activation and induce T regulatory cell production via surface-bound and secreted factors, including programmed death 1 (PD-1) (Sugita et al., 2009; Usui et al., 2008), FasL (Kaplan, Leibole, Tezel, & Ferguson, 1999), galectins, transforming growth factor β (TGFβ) (Sugita et al., 2008; Tanihara, Yoshida, Matsumoto, & Yoshimura, 1993), pigment epithelial-derived factor (PEDF) (Zamiri, Masli, Streilein, & Taylor, 2006), thrombospondin (Zamiri, Masli, Kitaichi, Taylor, & Streilein, 2005), and others (Ishida, Panjwani, Cao, & Streilein, 2003; Sugita, 2009). Furthermore, the RPE also appears to activate local macrophage in addition to suppressing inflammatory activity (Lau & Taylor, 2009).

The task of protecting the retina from foreign substances largely falls to the retinal Müller and microglial cells, although a small population of dendritic cells has been found and may respond to injured retinal neurons (Heuss, Lehmann, Norbury, McPherson, & Gregerson, 2012; Lehmann, Heuss, McPherson, Roehrich, & Gregerson, 2010; McPherson, Heuss, Pierson, & Gregerson, 2014; Xu, Dawson, Forrester, & Liversidge, 2007). These cells are found in all of the layers of the neural retina, with their morphology being largely associated with their function (Jonas et al., 2012; Saijo & Glass, 2011). In fact, a recent study using a mathematical modeling program further quantified retinal microglial cell morphology and orientation after optic nerve transection, indicating that changes in morphology are essential for function along with altered gene expression and cytokine secretion (Yuan, Liang, Peng, Lin, & So, 2015; Zhang, Peng, et al., 2016).

Recent work suggests that LBP treatment promotes maturation and antigen presentation in dendritic cells, specifically those derived from bone marrow (BMDCs), and increases the levels of interleukin (IL)-12p40, IL-12p70, cluster of differentiation (CD) 11c, CD40, CD80, CD86, and major histocompatibility complex class II I-A/I-E subregions (Chen, Lu, Srinivasan, Tan, & Chan, 2009; Zhu, Zhao, Zhao, & Chen, 2007). While the effects of LBPs on the small population of dendritic cells in the retina remain unstudied, it is possible that the mechanism of action is related to similar changes in cytokine secretion. Notably, in primary BMDC culture, LBPs appear to directly activate nuclear factor kappa B (NF-κB) via toll-like receptor (TLR)-2 and TLR4 (Zhu, Zhang, Shen, Zhou, & Yu, 2013).

Although the various roles of Müller cells in retinal disease have been well documented (reviewed in Bringmann & Wiedemann, 2012), to date, the effects of LBPs specifically on retinal Müller cell morphology and activity are largely unstudied. However, in an in vitro spinal cord injury model,

LBPs do not appear to directly affect astrocytes, which are phenotypically and functionally similar to Müller cells and astrocytes in the retina (Zhang, Wang, et al., 2013). In this same study, the authors sought to explain their findings by further investigating the role of microglial activation in their model. Retinal microglia, like those in the brain and spinal cord, are derived from myeloid cells and are the resident macrophage of this tissue. Their activation state, either M1 (proinflammatory) or M2 (tissue repair), has been associated with the pathogenesis of multiple retinal diseases (Gupta, Brown, & Milam, 2003; Zeiss & Johnson, 2004; Zeng et al., 2005). In fact, activated retinal microglia have also been demonstrated to interact with Müller cells (Fernandez-Bueno, Pastor, Gayoso, Alcalde, & Garcia, 2008; Wang, Ma, Zhao, Fariss, & Wong, 2011), leading to what is likely a synergistic signaling cascade involved in the immunomodulatory effects that occur following retinal insult.

It is important to note that the effect of LBPs on microglia/macrophage appears to vary depending on the cellular context, disease state/stimulus, and activation state of the cell at the time of treatment. For example, various studies indicate that LBPs enhance immunity by increasing the phagocytotic capabilities of the activated macrophages, as well as secretion of tumor necrosis factor (TNF)-α, nitric oxide, and other cytotoxic factors (Chen, Soo, et al., 2009). In the retina, on the other hand, LBPs have been shown to prevent microglial cell activation in rd10 mouse retinas, resulting in decreased production of inflammatory mediators such as TNF-α, IL6β, and C–C motif chemokine ligand 2 (CCL2) via the reduced expression of NF-κB (Wang, Xiao, et al., 2014). This neuroprotective state was also observed by Chiu et al. (2009) in a rat model of ocular hypertension after pretreatment with 1–100 mg/kg of LBPs. In this study, changes in the activation state of the microglia were highlighted by injecting either lipopolysaccharide or macrophage/microglia inhibitory factor immediately after the first laser treatment, leading to the stimulation or inhibition of microglial activation, respectively. The morphology of the LBP-activated retinal microglia also appears different from that induced by cytotoxic stimuli and ocular insult (Li et al., 2015). Thus, it is likely that LBPs may manipulate microglial activation in a way that increases activity in general, of both states, and it is the balance of these two pathways that leads to beneficial mechanisms and protection of the retinal neurons (Li, Lu, Tay, Moochhala, & He, 2007).

In addition to their focus on immune function, Wang, Xiao, et al. (2014) also demonstrated antioxidative mechanisms. Interestingly, these mechanisms may overlap as NF-κB and other transcription factors are redox sensitive and can be activated by oxidants. These activated signaling molecules can then, in turn, alter immune cell function. Immune cells themselves are also sensitive to oxidative stress as their membranes contain a high percentage of polyunsaturated fatty acids, and the presence of excessive ROS can alter the cell's redox state (Bennett & Griffiths, 2012). Taken together, these data indicate that the effect of LBP on immune cell function may also be linked to its antioxidative properties and vice versa.

### 3.5 Links Between the Visual System and Internal Organ Health in CHM

When discussing CHMs it is important to mention the connection between the liver and the eyes. Indeed, various systemic diseases affect both the liver and numerous ocular tissues (O'Neill, 1992), and improvements in hepatic function have been shown to heavily affect ocular health (Fatrai et al., 2015). Recent studies have demonstrated the effects of LBPs in acute liver injury, alcoholic liver disease, and nonalcoholic fatty liver disease, which are modulated through similar antioxidant, antiinflammatory, and neuroprotective mechanisms as those described for the retina (Cheng & Kong, 2011; Xiao, Fai So, Liong, & Tipoe, 2013; Xiao et al., 2012; Xiao, Wang, et al., 2014; Xiao, Xing, et al., 2014). Furthermore, diseases affecting the kidney and the eye also appear to have common pathogenesis in terms of genetic and phenotypic changes (Wong, Wong, Cheng, & Sabanayagam, 2014). Similar mechanistic functions are also apparent in kidney tissues following LBP treatment (Du, Hu, Kou, Zhang, & Zhang, 2016; Li, 2007). The connection between these organs likely plays a significant role in the similar mechanisms employed by LBPs during treatment. In traditional Chinese medicine, this interconnected pathway between the internal organs, particularly the liver and kidneys, and the eyes is described by Meridian Theory (O'Brien & Xue, 2016). This holistic philosophy explains that eye disorders are caused by liver blood deficiencies, liver heat, and internal liver wind as well as dysfunctional kidney function.

## 4. *L. BARBARUM* POLYSACCHARIDE AS A TREATMENT AND PREVENTATIVE MEDICINE: FUTURE PERSPECTIVES AND CONCLUSIONS

In Chinese medicine, a disease is traditionally treated as an issue of balance in the entire body. While the use of homeopathic compounds is becoming more mainstream worldwide, further scientific analysis is required in order to determine the mechanisms underlying the associated health benefits as well as to evaluate any detrimental side effects. As one of the most well-studied CHMs, *L. barbarum* has been used to treat a variety of diseases, including those of the retina, and a vast amount of literature is available outlining its mode of action as well as the affected downstream signaling pathways. These signaling pathways, which involve antioxidative, antiinflammatory, and neuroprotective facets, are in many ways linked together, whereby the effects of treatment are modulated by overlapping and synergistic changes in both the local tissue as well as in the global health of the patient/animal. With the exception of minor cases of allergic reactions, treatment with appropriate levels of *L. barbarum* does not appear to have any significant adverse effects. Therefore, as the quality and quantity of studies published concerning its use increases, in our laboratories and others, it is likely that *L. barbarum* will continue to be explored as a reliable, effective treatment option for retinal diseases.

## ACKNOWLEDGMENTS

Related research projects are supported by funding from the Funds of Leading Talents of Guangdong (2013), Programme of Introducing Talent of Discipline to Universities (B14063), the Fundamental Research Funds for the Central Universities Grant (21609101), and a key basic study and functional product research of wolfberry grant from Ningxia Province (all to K.-F.S.) as well as a Health and Medical Research Fund (HMRF) grant (number 14151281) to K.C.

## REFERENCES

Amagase, H., & Nance, D. M. (2008). A randomized, double-blind, placebo-controlled, clinical study of the general effects of a standardized *Lycium barbarum* (Goji) Juice, GoChi. *Journal of Alternative and Complementary Medicine*, *14*(4), 403–412.

Amagase, H., Sun, B., & Borek, C. (2009a). *Lycium barbarum* (goji) juice improves in vivo antioxidant biomarkers in serum of healthy adults. *Nutrition Research*, *29*(1), 19–25.

Amagase, H., Sun, B., & Nance, D. M. (2009b). Immunomodulatory effects of a standardized *Lycium barbarum* fruit juice in Chinese older healthy human subjects. *Journal of Medicinal Food*, *12*(5), 1159–1165.

Ambati, J., Ambati, B. K., Yoo, S. H., Ianchulev, S., & Adamis, A. P. (2003). Age-related macular degeneration: Etiology, pathogenesis, and therapeutic strategies. *Survey of Ophthalmology*, *48*(3), 257–293.

Arundine, M., & Tymianski, M. (2004). Molecular mechanisms of glutamate-dependent neurodegeneration in ischemia and traumatic brain injury. *Cellular and Molecular Life Sciences*, *61*(6), 657–668.

Baccus, S. A. (2007). Timing and computation in inner retinal circuitry. *Annual Review of Physiology*, *69*, 271–290.

Beatty, S., Koh, H., Phil, M., Henson, D., & Boulton, M. (2000). The role of oxidative stress in the pathogenesis of age-related macular degeneration. *Survey of Ophthalmology*, *45*(2), 115–134.

Benlloch, M., Muriach, M., Castellano, G., Sancho-Pelluz, F. J., Gonzalez-Garcia, E., Flores-Bellver, M., et al. (2015). Role of *Lycium barbarum* extracts in retinal diseases.

In R. C. C. Chang & K. F. So (Eds.), Lycium barbarum *and human health*. The Netherlands: Springer.

Bennett, S. J., & Griffiths, H. R. (2012). Regulation of T-cell functions by oxidative stress. In M. J. Alcaraz, O. Gualillo, & O. Sanchez-Pernaute (Eds.), *Studies on arthritis and joint disorders*. New York: Springer.

Bringmann, A., & Wiedemann, P. (2012). Muller glial cells in retinal disease. *Ophthalmologica*, *227*(1), 1–19.

Bucheli, P., Vidal, K., Shen, L., Gu, Z., Zhang, C., Miller, L. E., et al. (2011). Goji berry effects on macular characteristics and plasma antioxidant levels. *Optometry and Vision Science*, *88*(2), 257–262.

Calkins, D. J. (2012). Critical pathogenic events underlying progression of neurodegeneration in glaucoma. *Progress in Retinal and Eye Research*, *31*(6), 702–719.

Chan, H. H. L., So K.F., Chang R.C.C., Lai S.M., Do C.W., Benzie I.F.F., et al., (ongoing). Cone rescue in retinitis pigmentosa by the treatment of *Lycium barbarum*. Estimated completion date: March 2017. Clinicaltrials.gov identifier: NCT02244996.

Chan, H. C., Chang, R. C., Koon-Ching Ip, A., Chiu, K., Yuen, W. H., Zee, S. Y., et al. (2007). Neuroprotective effects of *Lycium barbarum* Lynn on protecting retinal ganglion cells in an ocular hypertension model of glaucoma. *Experimental Neurology*, *203*(1), 269–273.

Chan, K. C., Fu, Q. L., Hui, E. S., So, K. F., & Wu, E. X. (2008). Evaluation of the retina and optic nerve in a rat model of chronic glaucoma using in vivo manganese-enhanced magnetic resonance imaging. *NeuroImage*, *40*(3), 1166–1174.

Chen, Z., Lu, J., Srinivasan, N., Tan, B. K., & Chan, S. H. (2009). Polysaccharide-protein complex from *Lycium barbarum* L. is a novel stimulus of dendritic cell immunogenicity. *Journal of Immunology*, *182*(6), 3503–3509.

Chen, Z., Soo, M. Y., Srinivasan, N., Tan, B. K., & Chan, S. H. (2009). Activation of macrophages by polysaccharide-protein complex from *Lycium barbarum* L. *Phytotherapy Research*, *23*(8), 1116–1122.

Cheng, C. Y., Chung, W. Y., Szeto, Y. T., & Benzie, I. F. (2005). Fasting plasma zeaxanthin response to Fructus barbarum L. (wolfberry; Kei Tze) in a food-based human supplementation trial. *The British Journal of Nutrition*, *93*(1), 123–130.

Cheng, D., & Kong, H. (2011). The effect of *Lycium barbarum* polysaccharide on alcohol-induced oxidative stress in rats. *Molecules*, *16*(3), 2542–2550.

Cheung, A. K., Fung, M. K., Lo, A. C., Lam, T. T., So, K. F., Chung, S. S., et al. (2005). Aldose reductase deficiency prevents diabetes-induced blood-retinal barrier breakdown, apoptosis, and glial reactivation in the retina of db/db mice. *Diabetes*, *54*(11), 3119–3125.

Chiu, K., Chan, H. C., Yeung, S. C., Yuen, W. H., Zee, S. Y., Chang, R. C., et al. (2009). Erratum: Modulation of microglia by Wolfberry on the survival of retinal ganglion cells in a rat ocular hypertension model. *Journal of Ocular Biology, Diseases, and Informatics*, *2*(3), 127–136.

Chiu, K., Zhou, Y., Yeung, S. C., Lok, C. K., Chan, O. O., Chang, R. C., et al. (2010). Up-regulation of crystallins is involved in the neuroprotective effect of wolfberry on survival of retinal ganglion cells in rat ocular hypertension model. *Journal of Cellular Biochemistry*, *110*(2), 311–320.

Chu, P. H., Li, H. Y., Chin, M. P., So, K. F., & Chan, H. H. (2013). Effect of *Lycium barbarum* (wolfberry) polysaccharides on preserving retinal function after partial optic nerve transection. *PLoS One*, *8*(12), e81339.

Coyle, J. T., & Puttfarcken, P. (1993). Oxidative stress, glutamate, and neurodegenerative disorders. *Science*, *262*(5134), 689–695.

Cui, B., Liu, S., Lin, X., Wang, J., Li, S., & Wang, Q. (2011). Effects of *Lycium barbarum* aqueous and ethanol extracts on high-fat-diet induced oxidative stress in rat liver tissue. *Molecules*, *16*(11), 9116–9128.

Du, M., Hu, X., Kou, L., Zhang, B., & Zhang, C. (2016). *Lycium barbarum* polysaccharide mediated the antidiabetic and antinephritic effects in diet-streptozotocin-induced diabetic Sprague Dawley rats via regulation of NF-kappaB. *BioMed Research International, 2016*, 3140290.

Farrall, A. J., & Wardlaw, J. M. (2009). Blood-brain barrier: Ageing and microvascular disease—Systematic review and meta-analysis. *Neurobiology of Aging, 30*(3), 337–352.

Fatrai, A., Hempen, C.-H., Han, C., Leonhardy, H., Hummelsberger, J., Uhrig, S., et al. (2015). *Chinese ophthalmology: Acupuncture, herbal therapy, dietary therapy, tuina and qigong*. Wiesbaden, Germany: Tipani-Verlag.

Fernandez-Bueno, I., Pastor, J. C., Gayoso, M. J., Alcalde, I., & Garcia, M. T. (2008). Muller and macrophage-like cell interactions in an organotypic culture of porcine neuroretina. *Molecular Vision, 14*, 2148–2156.

Ferrari, S., Di Iorio, E., Barbaro, V., Ponzin, D., Sorrentino, F. S., & Parmeggiani, F. (2011). Retinitis pigmentosa: Genes and disease mechanisms. *Current Genomics, 12*(4), 238–249.

Flammer, J., Konieczka, K., Bruno, R. M., Virdis, A., Flammer, A. J., & Taddei, S. (2013). The eye and the heart. *European Heart Journal, 34*(17), 1270–1278.

Ganea, E., & Harding, J. J. (2006). Glutathione-related enzymes and the eye. *Current Eye Research, 31*(1), 1–11.

Gupta, N., Brown, K. E., & Milam, A. H. (2003). Activated microglia in human retinitis pigmentosa, late-onset retinal degeneration, and age-related macular degeneration. *Experimental Eye Research, 76*(4), 463–471.

Hartong, D. T., Berson, E. L., & Dryja, T. P. (2006). Retinitis pigmentosa. *Lancet, 368*(9549), 1795–1809.

He, M., Pan, H., Chang, R. C., So, K. F., Brecha, N. C., & Pu, M. (2014). Activation of the Nrf2/HO-1 antioxidant pathway contributes to the protective effects of *Lycium barbarum* polysaccharides in the rodent retina after ischemia-reperfusion-induced damage. *PLoS One, 9*(1), e84800.

He, N., Yang, X., Jiao, Y., Tian, L., & Zhao, Y. (2012). Characterisation of antioxidant and antiproliferative acidic polysaccharides from Chinese wolfberry fruits. *Food Chemistry, 133*(3), 978–989.

Heavner, W., & Pevny, L. (2012). Eye development and retinogenesis. *Cold Spring Harbor Perspectives in Biology, 4*(12), a008391.

Heuss, N. D., Lehmann, U., Norbury, C. C., McPherson, S. W., & Gregerson, D. S. (2012). Local activation of dendritic cells alters the pathogenesis of autoimmune disease in the retina. *Journal of Immunology, 188*(3), 1191–1200.

Ho, P. I., Ortiz, D., Rogers, E., & Shea, T. B. (2002). Multiple aspects of homocysteine neurotoxicity: Glutamate excitotoxicity, kinase hyperactivation and DNA damage. *Journal of Neuroscience Research, 70*(5), 694–702.

Ho, Y. S., Yu, M. S., Lai, C. S., So, K. F., Yuen, W. H., & Chang, R. C. (2007). Characterizing the neuroprotective effects of alkaline extract of *Lycium barbarum* on beta-amyloid peptide neurotoxicity. *Brain Research, 1158*, 123–134.

Ho, Y. S., Yu, M. S., Yang, X. F., So, K. F., Yuen, W. H., & Chang, R. C. (2010). Neuroprotective effects of polysaccharides from wolfberry, the fruits of *Lycium barbarum*, against homocysteine-induced toxicity in rat cortical neurons. *Journal of Alzheimer's Disease, 19*(3), 813–827.

Ho, Y. S., Yu, M. S., Yik, S. Y., So, K. F., Yuen, W. H., & Chang, R. C. (2009). Polysaccharides from wolfberry antagonizes glutamate excitotoxicity in rat cortical neurons. *Cellular and Molecular Neurobiology, 29*(8), 1233–1244.

Hu, C. K., Lee, Y. J., Colitz, C. M., Chang, C. J., & Lin, C. T. (2012). The protective effects of *Lycium barbarum* and Chrysanthemum morifolum on diabetic retinopathies in rats. *Veterinary Ophthalmology, 15*(Suppl. 2), 65–71.

Ishida, K., Panjwani, N., Cao, Z., & Streilein, J. W. (2003). Participation of pigment epithelium in ocular immune privilege. 3. Epithelia cultured from iris, ciliary body, and retina suppress T-cell activation by partially non-overlapping mechanisms. *Ocular Immunology and Inflammation*, *11*(2), 91–105.

Jin, M., Huang, Q., Zhao, K., & Shang, P. (2013). Biological activities and potential health benefit effects of polysaccharides isolated from *Lycium barbarum* L. *International Journal of Biological Macromolecules*, *54*, 16–23.

Jin, M., Lu, Z., Huang, M., & Wang, Y. (2011). Sulfated modification and antioxidant activity of exopolysaccahrides produced by Enterobacter cloacae Z0206. *International Journal of Biological Macromolecules*, *48*(4), 607–612.

Jonas, R. A., Yuan, T. F., Liang, Y. X., Jonas, J. B., Tay, D. K., & Ellis-Behnke, R. G. (2012). The spider effect: Morphological and orienting classification of microglia in response to stimuli in vivo. *PLoS One*, 7(2), e30763.

Junlin, L., & Aicheng, W. (2002). *Gou qi*. Beijing: Beijing Science and Technology Press.

Kaplan, H. J., Leibole, M. A., Tezel, T., & Ferguson, T. A. (1999). Fas ligand (CD95 ligand) controls angiogenesis beneath the retina. *Nature Medicine*, *5*(3), 292–297.

Kaur, C., Sivakumar, V., Yong, Z., Lu, J., Foulds, W. S., & Ling, E. A. (2007). Blood-retinal barrier disruption and ultrastructural changes in the hypoxic retina in adult rats: The beneficial effect of melatonin administration. *The Journal of Pathology*, *212*(4), 429–439.

Kee, C. S., Do, C. W., Cheong, M. Y., Wong, S. W., Law, K. M., Wong, C. Y., et al. (2013). Effects of *Lycium barbarum* supplement on the progression of senile cataract in a Hong Kong Chinese population (Abstract). In *Inaugural Symposium of HEALED Research Group "Health for Life"*.

Kimelberg, H. K. (1995). Current concepts of brain edema. Review of laboratory investigations. *Journal of Neurosurgery*, *83*(6), 1051–1059.

Kruk, J., Kubasik-Kladna, K., & Aboul-Enein, H. Y. (2015). The role oxidative stress in the pathogenesis of eye diseases: Current status and a dual role of physical activity. *Mini Reviews in Medicinal Chemistry*, *16*(3), 241–257.

Lai, J.S.M., So, K.F., Chang, R.C.C., Lee, J.W.Y., Chan, J.C.H., Lo, A., et al. (ongoing). Randomized control trial on the neuroprotective effects of oral Wolfberry supplement in the treatment of normal tension glaucoma. Estimated completion date: December 2018.

Lau, C. H., & Taylor, A. W. (2009). The immune privileged retina mediates an alternative activation of J774A.1 cells. *Ocular Immunology and Inflammation*, *17*(6), 380–389.

Leasher, J. L., Bourne, R. R., Flaxman, S. R., Jonas, J. B., Keeffe, J., Naidoo, K., et al. (2016). Global estimates on the number of people blind or visually impaired by diabetic retinopathy: A meta-analysis from 1990 to 2010. *Diabetes Care*, *39*(9), 1643–1649.

Lehmann, U., Heuss, N. D., McPherson, S. W., Roehrich, H., & Gregerson, D. S. (2010). Dendritic cells are early responders to retinal injury. *Neurobiology of Disease*, *40*(1), 177–184.

Levkovitch-Verbin, H., Quigley, H. A., Kerrigan-Baumrind, L. A., D'Anna, S. A., Kerrigan, D., & Pease, M. E. (2001). Optic nerve transection in monkeys may result in secondary degeneration of retinal ganglion cells. *Investigative Ophthalmology & Visual Science*, *42*(5), 975–982.

Levkovitch-Verbin, H., Quigley, H. A., Martin, K. R., Zack, D. J., Pease, M. E., & Valenta, D. F. (2003). A model to study differences between primary and secondary degeneration of retinal ganglion cells in rats by partial optic nerve transection. *Investigative Ophthalmology & Visual Science*, *44*(8), 3388–3393.

Li, X. M. (2007). Protective effect of *Lycium barbarum* polysaccharides on streptozotocin-induced oxidative stress in rats. *International Journal of Biological Macromolecules*, *40*(5), 461–465.

Li, H., Liang, Y., Chiu, K., Yuan, Q., Lin, B., Chang, R. C., et al. (2013). *Lycium barbarum* (wolfberry) reduces secondary degeneration and oxidative stress, and inhibits JNK pathway in retina after partial optic nerve transection. *PLoS One*, *8*(7), e68881.

Li, L., Lu, J., Tay, S. S., Moochhala, S. M., & He, B. P. (2007). The function of microglia, either neuroprotection or neurotoxicity, is determined by the equilibrium among factors released from activated microglia in vitro. *Brain Research*, *1159*, 8–17.

Li, X. M., Ma, Y. L., & Liu, X. J. (2007). Effect of the *Lycium barbarum* polysaccharides on age-related oxidative stress in aged mice. *Journal of Ethnopharmacology*, *111*(3), 504–511.

Li, H. Y., Ruan, Y. W., Kau, P. W., Chiu, K., Chang, R. C., Chan, H. H., et al. (2015). Effect of *Lycium barbarum* (Wolfberry) on alleviating axonal degeneration after partial optic nerve transection. *Cell Transplantation*, *24*(3), 403–417.

Li, S. Y., Yang, D., Yeung, C. M., Yu, W. Y., Chang, R. C., So, K. F., et al. (2011). *Lycium barbarum* polysaccharides reduce neuronal damage, blood-retinal barrier disruption and oxidative stress in retinal ischemia/reperfusion injury. *PLoS One*, *6*(1), e16380.

Liu, H., Fan, Y., Wang, W., Liu, N., Zhang, H., Zhu, Z., et al. (2012). Polysaccharides from *Lycium barbarum* leaves: Isolation, characterization and splenocyte proliferation activity. *International Journal of Biological Macromolecules*, *51*(4), 417–422.

Liu, L., Lao, W., Ji, Q. S., Yang, Z. H., Yu, G. C., & Zhong, J. X. (2015). *Lycium barbarum* polysaccharides protected human retinal pigment epithelial cells against oxidative stress-induced apoptosis. *International Journal of Ophthalmology*, *8*(1), 11–16.

Mao, F., Xiao, B., Jiang, Z., Zhao, J., Huang, X., & Guo, J. (2011). Anticancer effect of *Lycium barbarum* polysaccharides on colon cancer cells involves G0/G1 phase arrest. *Medical Oncology*, *28*(1), 121–126.

Marc, R. E., Jones, B. W., Watt, C. B., Vazquez-Chona, F., Vaughan, D. K., & Organisciak, D. T. (2008). Extreme retinal remodeling triggered by light damage: Implications for age related macular degeneration. *Molecular Vision*, *14*, 782–806.

Masaki, T., Vane, J. R., & Vanhoutte, P. M. (1994). International Union of Pharmacology nomenclature of endothelin receptors. *Pharmacological Reviews*, *46*(2), 137–142.

McPherson, S. W., Heuss, N. D., Pierson, M. J., & Gregerson, D. S. (2014). Retinal antigen-specific regulatory T cells protect against spontaneous and induced autoimmunity and require local dendritic cells. *Journal of Neuroinflammation*, *11*, 205.

Mi, X. S., Chiu, K., Van, G., Leung, J. W., Lo, A. C., Chung, S. K., et al. (2012). Effect of *Lycium barbarum* polysaccharides on the expression of endothelin-1 and its receptors in an ocular hypertension model of rat glaucoma. *Neural Regeneration Research*, 7(9), 645–651.

Mi, X. S., Feng, Q., Lo, A. C., Chang, R. C., Lin, B., Chung, S. K., et al. (2012). Protection of retinal ganglion cells and retinal vasculature by *Lycium barbarum* polysaccharides in a mouse model of acute ocular hypertension. *PLoS One*, 7(10), e45469.

Miranda, M., Arnal, E., Ahuja, S., Alvarez-Nolting, R., Lopez-Pedrajas, R., Ekstrom, P., et al. (2010). Antioxidants rescue photoreceptors in rd1 mice: Relationship with thiol metabolism. *Free Radical Biology & Medicine*, *48*(2), 216–222.

Narayan, S., Prasanna, G., Krishnamoorthy, R. R., Zhang, X., & Yorio, T. (2003). Endothelin-1 synthesis and secretion in human retinal pigment epithelial cells (ARPE-19): Differential regulation by cholinergics and TNF-alpha. *Investigative Ophthalmology & Visual Science*, *44*(11), 4885–4894.

Ni, T., Wei, G., Yin, X., Liu, X., & Liu, D. (2013). Neuroprotective effect of *Lycium barbarum* on retina of Royal College of Surgeons (RCS) rats: A preliminary study. *Folia Neuropathologica*, *51*(2), 158–163.

Niu, A. J., Wu, J. M., Yu, D. H., & Wang, R. (2008). Protective effect of *Lycium barbarum* polysaccharides on oxidative damage in skeletal muscle of exhaustive exercise rats. *International Journal of Biological Macromolecules*, *42*(5), 447–449.

O'Brien, K. A., & Xue, C. C. (2016). The theoretical framework of Chinese medicine. In P.-C. Leung (Ed.), *A comprehensive guide to Chinese medicine*. Singapore: World Scientific Publishing Co. Pte. Ltd.

O'Neill, D. P. (1992). The eye and liver disorders. *Eye (London, England)*, *6*(Pt. 4), 366–370.

Organisciak, D. T., & Vaughan, D. K. (2010). Retinal light damage: Mechanisms and protection. *Progress in Retinal and Eye Research*, *29*(2), 113–134.

Pascolini, D., & Mariotti, S. P. (2012). Global estimates of visual impairment: 2010. *The British Journal of Ophthalmology*, *96*(5), 614–618.

Paul Hsu, C. H., Nance, D. M., & Amagase, H. (2012). A meta-analysis of clinical improvements of general well-being by a standardized *Lycium barbarum*. *Journal of Medicinal Food*, *15*(11), 1006–1014.

Pennesi, M. E., Neuringer, M., & Courtney, R. J. (2012). Animal models of age related macular degeneration. *Molecular Aspects of Medicine*, *33*(4), 487–509.

Perez, V. L., Saeed, A. M., Tan, Y., Urbieta, M., & Cruz-Guilloty, F. (2013). The eye: A window to the soul of the immune system. *Journal of Autoimmunity*, *45*, 7–14.

Puertas, F. J., Diaz-Llopis, M., Chipont, E., Roma, J., Raya, A., & Romero, F. J. (1993). Glutathione system of human retina: Enzymatic conjugation of lipid peroxidation products. *Free Radical Biology & Medicine*, *14*(5), 549–551.

Qi, B., Ji, Q., Wen, Y., Liu, L., Guo, X., Hou, G., et al. (2014). *Lycium barbarum* polysaccharides protect human lens epithelial cells against oxidative stress-induced apoptosis and senescence. *PLoS One*, *9*(10), e110275.

Rattner, A., & Nathans, J. (2006). Macular degeneration: Recent advances and therapeutic opportunities. *Nature Reviews. Neuroscience*, 7(11), 860–872.

Rizzolo, L. J. (2014). Barrier properties of cultured retinal pigment epithelium. *Experimental Eye Research*, *126*, 16–26.

Saijo, K., & Glass, C. K. (2011). Microglial cell origin and phenotypes in health and disease. *Nature Reviews. Immunology*, *11*(11), 775–787.

Shan, X., Zhou, J., Ma, T., & Chai, Q. (2011). *Lycium barbarum* polysaccharides reduce exercise-induced oxidative stress. *International Journal of Molecular Sciences*, *12*(2), 1081–1088.

Slingsby, C., Wistow, G. J., & Clark, A. R. (2013). Evolution of crystallins for a role in the vertebrate eye lens. *Protein Science*, *22*(4), 367–380.

Song, M. K., Roufogalis, B. D., & Huang, T. H. (2012). Reversal of the caspase-dependent apoptotic cytotoxicity pathway by taurine from *Lycium barbarum* (goji berry) in human retinal pigment epithelial cells: Potential benefit in diabetic retinopathy. *Evidence-based Complementary and Alternative Medicine*, *2012*, 323784.

Steuer, H., Jaworski, A., Elger, B., Kaussmann, M., Keldenich, J., Schneider, H., et al. (2005). Functional characterization and comparison of the outer blood-retina barrier and the blood-brain barrier. *Investigative Ophthalmology & Visual Science*, *46*(3), 1047–1053.

Sugita, S. (2009). Role of ocular pigment epithelial cells in immune privilege. *Archivum Immunologiae et Therapiae Experimentalis*, *57*(4), 263–268.

Sugita, S., Horie, S., Nakamura, O., Futagami, Y., Takase, H., Keino, H., et al. (2008). Retinal pigment epithelium-derived CTLA-2alpha induces TGFbeta-producing T regulatory cells. *Journal of Immunology*, *181*(11), 7525–7536.

Sugita, S., Usui, Y., Horie, S., Futagami, Y., Aburatani, H., Okazaki, T., et al. (2009). T-cell suppression by programmed cell death 1 ligand 1 on retinal pigment epithelium during inflammatory conditions. *Investigative Ophthalmology & Visual Science*, *50*(6), 2862–2870.

Sung, C. H., & Chuang, J. Z. (2010). The cell biology of vision. *The Journal of Cell Biology*, *190*(6), 953–963.

Tang, L., Zhang, Y., Jiang, Y., Willard, L., Ortiz, E., Wark, L., et al. (2011). Dietary wolfberry ameliorates retinal structure abnormalities in db/db mice at the early stage of diabetes. *Experimental Biology and Medicine (Maywood, N.J.)*, *236*(9), 1051–1063.

Tanihara, H., Yoshida, M., Matsumoto, M., & Yoshimura, N. (1993). Identification of transforming growth factor-beta expressed in cultured human retinal pigment epithelial cells. *Investigative Ophthalmology & Visual Science*, *34*(2), 413–419.

Tanito, M., Nishiyama, A., Tanaka, T., Masutani, H., Nakamura, H., Yodoi, J., et al. (2002). Change of redox status and modulation by thiol replenishment in retinal photooxidative damage. *Investigative Ophthalmology & Visual Science, 43*(7), 2392–2400.

Tham, Y. C., Li, X., Wong, T. Y., Quigley, H. A., Aung, T., & Cheng, C. Y. (2014). Global prevalence of glaucoma and projections of glaucoma burden through 2040: A systematic review and meta-analysis. *Ophthalmology, 121*(11), 2081–2090.

Thanos, S., Bohm, M. R., Meyer zu Horste, M., Prokosch-Willing, V., Hennig, M., Bauer, D., et al. (2014). Role of crystallins in ocular neuroprotection and axonal regeneration. *Progress in Retinal and Eye Research, 42*, 145–161.

Usui, Y., Okunuki, Y., Hattori, T., Kezuka, T., Keino, H., Ebihara, N., et al. (2008). Functional expression of B7H1 on retinal pigment epithelial cells. *Experimental Eye Research, 86*(1), 52–59.

Wang, Y., Chen, H., Wu, M., Zeng, S., Liu, Y., & Dong, J. (2015). Chemical and genetic diversity of wolfberry. In R. C. C. Chang & K. F. So (Eds.), Lycium barbarum *and human health*. The Netherlands: Springer.

Wang, H. B., Li, Y. X., Hao, Y. J., Wang, T. F., Lei, Z., Wu, Y., et al. (2013). Neuroprotective effects of LBP on brain ischemic reperfusion neurodegeneration. *European Review for Medical and Pharmacological Sciences, 17*(20), 2760–2765.

Wang, T., Li, Y., Wang, Y., Zhou, R., Ma, L., Hao, Y., et al. (2014). *Lycium barbarum* polysaccharide prevents focal cerebral ischemic injury by inhibiting neuronal apoptosis in mice. *PLoS One, 9*(3), e90780.

Wang, M., Ma, W., Zhao, L., Fariss, R. N., & Wong, W. T. (2011). Adaptive Muller cell responses to microglial activation mediate neuroprotection and coordinate inflammation in the retina. *Journal of Neuroinflammation, 8*, 173.

Wang, K., Xiao, J., Peng, B., Xing, F., So, K. F., Tipoe, G. L., et al. (2014). Retinal structure and function preservation by polysaccharides of wolfberry in a mouse model of retinal degeneration. *Scientific Reports, 4*, 7601.

Wang, Y., Zhao, H., Sheng, X., Gambino, P. E., Costello, B., & Bojanowski, K. (2002). Protective effect of Fructus Lycii polysaccharides against time and hyperthermia-induced damage in cultured seminiferous epithelium. *Journal of Ethnopharmacology, 82*(2–3), 169–175.

WHO. (2012). *Global data on visual impairments 2010*. World Health Organization. Available from http://www.iapb.org/vision-2020/global-facts.

Won, S. J., Kim, D. Y., & Gwag, B. J. (2002). Cellular and molecular pathways of ischemic neuronal death. *Journal of Biochemistry and Molecular Biology, 35*(1), 67–86.

Wong, C. W., Wong, T. Y., Cheng, C. Y., & Sabanayagam, C. (2014). Kidney and eye diseases: Common risk factors, etiological mechanisms, and pathways. *Kidney International, 85*(6), 1290–1302.

Wu, H. T., He, X. J., Hong, Y. K., Ma, T., Xu, Y. P., & Li, H. H. (2010). Chemical characterization of *Lycium barbarum* polysaccharides and its inhibition against liver oxidative injury of high-fat mice. *International Journal of Biological Macromolecules, 46*(5), 540–543.

Xiao, J., Fai So, K., Liong, E. C., & Tipoe, G. L. (2013). Recent advances in the herbal treatment of non-alcoholic fatty liver disease. *Journal of Traditional and Complementary Medicine, 3*(2), 88–94.

Xiao, J., Liong, E. C., Ching, Y. P., Chang, R. C., So, K. F., Fung, M. L., et al. (2012). *Lycium barbarum* polysaccharides protect mice liver from carbon tetrachloride-induced oxidative stress and necroinflammation. *Journal of Ethnopharmacology, 139*(2), 462–470.

Xiao, J., Wang, J., Xing, F., Han, T., Jiao, R., Liong, E. C., et al. (2014). Zeaxanthin dipalmitate therapeutically improves hepatic functions in an alcoholic fatty liver disease model through modulating MAPK pathway. *PLoS One, 9*(4), e95214.

Xiao, J., Xing, F., Huo, J., Fung, M. L., Liong, E. C., Ching, Y. P., et al. (2014). *Lycium barbarum* polysaccharides therapeutically improve hepatic functions in non-alcoholic steatohepatitis rats and cellular steatosis model. *Scientific Reports, 4*, 5587.

Xu, H., Dawson, R., Forrester, J. V., & Liversidge, J. (2007). Identification of novel dendritic cell populations in normal mouse retina. *Investigative Ophthalmology & Visual Science, 48*(4), 1701–1710.

Yang, D., Li, S. Y., Yeung, C. M., Chang, R. C., So, K. F., Wong, D., et al. (2012). *Lycium barbarum* extracts protect the brain from blood-brain barrier disruption and cerebral edema in experimental stroke. *PLoS One*, 7(3), e33596.

Yoles, E., & Schwartz, M. (1998). Degeneration of spared axons following partial white matter lesion: Implications for optic nerve neuropathies. *Experimental Neurology, 153*(1), 1–7.

Yu, P. K., Balaratnasingam, C., Morgan, W. H., Cringle, S. J., McAllister, I. L., & Yu, D. Y. (2010). The structural relationship between the microvasculature, neurons, and glia in the human retina. *Investigative Ophthalmology & Visual Science, 51*(1), 447–458.

Yu, M. S., Lai, C. S., Ho, Y. S., Zee, S. Y., So, K. F., Yuen, W. H., et al. (2007). Characterization of the effects of anti-aging medicine Fructus lycii on beta-amyloid peptide neurotoxicity. *International Journal of Molecular Medicine, 20*(2), 261–268.

Yu, M. S., Leung, S. K., Lai, S. W., Che, C. M., Zee, S. Y., So, K. F., et al. (2005). Neuroprotective effects of anti-aging oriental medicine *Lycium barbarum* against beta-amyloid peptide neurotoxicity. *Experimental Gerontology, 40*(8–9), 716–727.

Yu, H., Wark, L., Ji, H., Willard, L., Jaing, Y., Han, J., et al. (2013). Dietary wolfberry upregulates carotenoid metabolic genes and enhances mitochondrial biogenesis in the retina of db/db diabetic mice. *Molecular Nutrition & Food Research, 57*(7), 1158–1169.

Yuan, T. F., Liang, Y. X., Peng, B., Lin, B., & So, K. F. (2015). Local proliferation is the main source of rod microglia after optic nerve transection. *Scientific Reports, 5*, 10788.

Zamiri, P., Masli, S., Kitaichi, N., Taylor, A. W., & Streilein, J. W. (2005). Thrombospondin plays a vital role in the immune privilege of the eye. *Investigative Ophthalmology & Visual Science, 46*(3), 908–919.

Zamiri, P., Masli, S., Streilein, J. W., & Taylor, A. W. (2006). Pigment epithelial growth factor suppresses inflammation by modulating macrophage activation. *Investigative Ophthalmology & Visual Science, 47*(9), 3912–3918.

Zeiss, C. J. (2010). Animals as models of age-related macular degeneration: An imperfect measure of the truth. *Veterinary Pathology, 47*(3), 396–413.

Zeiss, C. J., & Johnson, E. A. (2004). Proliferation of microglia, but not photoreceptors, in the outer nuclear layer of the rd-1 mouse. *Investigative Ophthalmology & Visual Science, 45*(3), 971–976.

Zeng, H. Y., Zhu, X. A., Zhang, C., Yang, L. P., Wu, L. M., & Tso, M. O. (2005). Identification of sequential events and factors associated with microglial activation, migration, and cytotoxicity in retinal degeneration in rd mice. *Investigative Ophthalmology & Visual Science, 46*(8), 2992–2999.

Zhang, M., Chen, H., Huang, J., Li, Z., Zhu, C., & Zhang, S. (2005). Effect of *Lycium barbarum* polysaccharide on human hepatoma QGY7703 cells: Inhibition of proliferation and induction of apoptosis. *Life Sciences, 76*(18), 2115–2124.

Zhang, Q., Chen, W., Zhao, J., & Xi, W. (2016). Functional constituents and antioxidant activities of eight Chinese native goji genotypes. *Food Chemistry, 200*, 230–236.

Zhang, Q., Du, X., Xu, Y., Dang, L., Xiang, L., & Zhang, J. (2013). The effects of Gouqi extracts on Morris maze learning in the APP/PS1 double transgenic mouse model of Alzheimer's disease. *Experimental and Therapeutic Medicine, 5*(5), 1528–1530.

Zhang, Y., Peng, B., Wang, S., Liang, Y. X., Yang, J., So, K. F., et al. (2016). Image processing methods to elucidate spatial characteristics of retinal microglia after optic nerve transection. *Scientific Reports, 6*, 21816.

Zhang, Y., & Stone, J. (1997). Role of astrocytes in the control of developing retinal vessels. *Investigative Ophthalmology & Visual Science, 38*(9), 1653–1666.

Zhang, Y. K., Wang, J., Liu, L., Chang, R. C., So, K. F., & Ju, G. (2013). The effect of *Lycium barbarum* on spinal cord injury, particularly its relationship with M1 and M2 macrophage in rats. *BMC Complementary and Alternative Medicine, 13*, 67.

Zhang, E., Yau, S. Y., Lau, B. W., Ma, H., Lee, T. M., Chang, R. C., et al. (2012). Synaptic plasticity, but not hippocampal neurogenesis, mediated the counteractive effect of wolfberry on depression in rats(1). *Cell Transplantation, 21*(12), 2635–2649.

Zhang, X., Zhou, W., & Zhang, Y. (2015). Immunoregulation and *Lycium barbarum*. In R. C. C. Chang & K. F. So (Eds.), Lycium barbarum *and human health*. The Netherlands: Springer.

Zhao, R., Cai, Y., Shao, X., & Ma, B. (2015). Improving the activity of *Lycium barbarum* polysaccharide on sub-health mice. *Food & Function, 6*(6), 2033–2040.

Zhou, Z. Q., Fan, H. X., He, R. R., Xiao, J., Tsoi, B., Lan, K. H., et al. (2016). Lycibarbarspermidines A-O, new dicaffeoylspermidine derivatives from wolfberry, with activities against Alzheimer's disease and oxidation. *Journal of Agricultural and Food Chemistry, 64*(11), 2223–2237.

Zhu, J., Zhang, Y., Shen, Y., Zhou, H., & Yu, X. (2013). *Lycium barbarum* polysaccharides induce toll-like receptor 2- and 4-mediated phenotypic and functional maturation of murine dendritic cells via activation of NF-kappaB. *Molecular Medicine Reports, 8*(4), 1216–1220.

Zhu, J., Zhao, L. H., Zhao, X. P., & Chen, Z. (2007). *Lycium barbarum* polysaccharides regulate phenotypic and functional maturation of murine dendritic cells. *Cell Biology International, 31*(6), 615–619.

CHAPTER TWO

# Effect of Chinese Herbal Medicine on Alzheimer's Disease

**Melanie-Jayne R. Howes**[*,†,1], **Rui Fang**[*], **Peter J. Houghton**[†]
[*]Royal Botanic Gardens, Kew, Richmond, United Kingdom
[†]King's College London, London, United Kingdom
[1]Corresponding author: e-mail address: m.howes@kew.org

## Contents



## Abstract

Alzheimer's disease (AD) is reaching epidemic proportions yet treatment strategies are limited and are restricted to providing symptomatic relief for the cognitive and behavioral and psychological symptoms of dementia (BPSD). Chinese herbal medicine (CHM) has been a valuable source of medicines for centuries and research has burgeoned in recent years to understand the scientific basis for their use. Some plants have been used in CHM for AD symptoms (e.g., *Polygala tenuifolia*), while others are CHMs for different conditions, but they show mechanistic effects relevant to AD (e.g., *Salvia miltiorrhiza*).

*International Review of Neurobiology*, Volume 135
ISSN 0074-7742
http://dx.doi.org/10.1016/bs.irn.2017.02.003

Some CHMs (e.g., *Ginkgo biloba* extract, and huperzine A from *Huperzia serrata*) show pharmacological activities relevant to AD, and promising effects on cognitive functions in clinical trials. Other CHMs show effects relevant to BPSD (e.g., *Crocus sativus*). This chapter discusses available scientific evidence for CHM plants and formulae that have been used both traditionally for AD, and those that have been used traditionally but not specifically for AD symptoms, and encompasses chemical, pharmacological and clinical studies. The ethnopharmacological approach to understanding the use of CHMs for AD is also discussed.

# 1. INTRODUCTION

## 1.1 Alzheimer's Disease

Alzheimer's disease (AD) is the most common form of dementia and is reaching epidemic proportions throughout the world. It is a major cause of disability and dependency among older people, having major physical, psychological, social and economic impact on families, caregivers, and society. Symptoms include cognitive impairments, particularly memory, attention, and executive functions, in addition to behavioral and psychological symptoms of dementia (BPSD). These include psychosis, agitation, anxiety, sleep disorders, and depression. AD pathology involves β-amyloidosis, resulting in amyloid plaques in the central nervous system (CNS) and abnormal tau, which forms neurofibrillary tangles, leading to neurodegeneration. Neurotransmitter abnormalities are characterized by cholinergic deficits, but other neurotransmitter systems, including glutamatergic and serotonergic, are also impaired in the CNS. Inflammatory mechanisms, oxidative damage, apoptosis, and attenuated neuroplasticity have also been implicated in AD pathology (Perry & Howes, 2011). Current treatment strategies are limited and only provide symptomatic relief. In Western medicine, licensed drugs include those that improve cholinergic function in the CNS to improve cognitive functions by inhibiting acetylcholinesterase (AChE) to reduce the hydrolysis of the neurotransmitter acetylcholine (ACh) by this enzyme. AChE inhibitor drugs are donepezil, galantamine, and rivastigmine, with the latter two drugs derived from natural products (Howes, 2013). Other available drugs used in AD include memantine [a glutamate *N*-methyl-D-aspartate (NMDA) receptor antagonist] that targets cognitive functions, while antipsychotics, antidepressants, tranquilizers, and hypnotics are prescribed for BPSD (Perry & Howes, 2011).

## 1.2 Ethnopharmacology

The scientific study of traditional medicines is named "ethnopharmacology" and much research in this field has been published in recent years (Chadwick & Marsh, 1994; Heinrich, Barnes, Gibbons, & Williamson, 2012; Heinrich & Jäger, 2015; Houghton, 2009). Every human society has a traditional medicine system, with flowering plants making the biggest contribution to the materials used in every case. Traditional Chinese medicine (TCM), which includes Chinese herbal medicine (CHM), is no exception and in recent years has received much scientific attention.

Pharmaceutical interest in ethnopharmacological research has two major foci, the establishment of a scientific rationale for the traditional use of the plant material and the discovery of new leads for conventional "single-chemical entity" (SCE). SCE pharmaceuticals might consist of the "active compound" determined by the research or its derivative, or a synthetic compound based on the structure of the naturally occurring compound. CHM has yielded some important SCE drugs used worldwide, including ephedrine from Ma Huang (*Ephedra* species), but probably the most notable example is the antimalarial artemisinin from Qing Hao (*Artemisia annua* L.) herb. The Chinese scientist Youyou Tu was awarded the Nobel Prize in Physiology or Medicine in 2015 for the part she played in its discovery and development.

Although the regulatory schemes for medicines in most countries worldwide are very much biased towards the SCE model, the clinical use of plant material itself or, more commonly, extracts, is nowadays receiving much attention. It is important to note that high quality clinical trials have been carried out on only a small number of extracts, with even less being subjected to systematic clinical reviews (e.g., Cochrane studies). Such products contain a complex mixture of compounds and in many cases, thorough investigations with the aim of identifying the "active constituent" have been fruitless, and the extract appears to give a "better" result, either in terms of increased efficacy or increased safety, than any one of the compounds isolated from it. Two phenomena are given as explanations of this situation. These are synergy, where an effect given by a combination of substances is greater than would have been expected from a consideration of individual contributions; and polyvalence, where different types of constituents within the plant have different pharmacological effects, all contributing to the overall therapeutic effect observed (Houghton, Howes, Lee, & Steventon, 2007; Williamson, 2009).

With such polyvalent medicines, the treatment of AD might prove to be more useful than if a SCE drug is used, since more than one target is affected (Howes & Houghton, 2012; Li, Zhang, & Yang, 2016). However, if the plant material used has not been authenticated, or if the profile for constituents in the extract has not been defined, any investigation for activity is relatively worthless. Legislation in some parts of the world, including Europe, has sought to address issues of authenticity, quality, and chemical characterization of plant extracts used for medicinal applications, although a more coordinated and robust approach to these issues is still needed (Howes & Simmonds, 2015).

## 1.3 TCM and AD

One of the many difficulties in ethnopharmacological studies is identifying the disease in question from old records, either because the local culture describes the condition differently from Western medicine, or because the disease may not be common in that context. This is particularly true with AD, which was identified as a disease syndrome only about 100 years ago. Thus relevant symptoms have to be identified (e.g., age-related memory loss) and used when searching the literature or in talking with local practitioners.

CHM ethnopharmacological research has the advantage that there is a wide range of literature sources, some dating back hundreds of years, and also that CHM is still a widely practiced, with a large number of practitioners who can provide information. However, comprehension of the written and spoken language presents a problem for many non-Chinese researchers and the alignment of diagnosis and symptoms according to the traditional Chinese philosophy of medicine with that of AD as understood by modern science also poses problems.

The etiology and pathogenesis of senile dementia has long been recognized by ancient Chinese physicians, although there is no distinction between the different causes of neurodegenerative dementia (such as AD, Parkinson's disease, and vascular dementia). Treatment strategies for dementia in TCM were systematically introduced by a physician *Zhang Jing Yue* (张景岳) in his medical book *Jing Yue Quan Shu—Za Zheng Mo* (《景岳全书·杂证谟》A.D.1624) during the Ming Dynasty (Xie & Qu, 2008). The ancient TCM approach for "Tonifying the kidney essence" is the basic principle of dementia treatment. According to TCM theory, senile dementia is caused by shortage of kidney essence (精), stagnation of blood, marrow deficiency, and it also involves weakness of the heart, liver, spleen, and lungs. The main symptoms are considered as blood stasis blocking the brain, forgetfulness, phlegm-damp (痰湿), obstruction, and rheumatism (Wang, Guo, Dou, & Wang, 2011). The associated symptoms of cognitive impairment, muscle rigidity, and movement disorders are considered to overlap with some dementia, or neurodegenerative disease, symptoms (Su et al., 2014). Thus, TCMs used for dementia are not specifically targeted at the nervous system and there is a more multifunctional approach to dementia treatment (Ho, So, & Chang, 2011).

Vast progress in TCM, particularly CHM, research has been made, particularly since the 1980s, and is discussed later. CHM has been the source of SCE drugs but also crude extracts, where synergistic and polyvalent factors can occur. Most of the plants and compounds discussed have been subjected to some in vivo as well as relevant in vitro studies, while clinical data are also discussed for those plant species and SCEs that have been evaluated in studies with humans.

## 2. PLANTS USED TRADITIONALLY FOR AD-LIKE SYMPTOMS

### 2.1 *Acorus calamus*

"Acori Calami Rhizoma" is the dried rhizome of *Acorus calamus* L. (Acoraceae), which although not indicated for AD-like symptoms in the Pharmacopeia of the People's Republic of China (CPC, 2010), is one of the most frequently reported ingredients of classical CHM formulas for memory impairment associated with aging (May et al., 2016). The rhizomes of *A. calamus* [Shi Chang Pu (石菖蒲)] are used in Far Eastern traditional medicine for memory improvement (Oh, Houghton, Whang, & Cho, 2004). Investigations arising from a similar use in Ayurvedic medicine led to the identification of β-asarone (Fig. 1) as an AChE inhibitor (Mukherjee, Kumar, Mal, & Houghton, 2007). The related α-asarone, also present in *A. calamus*, was shown to improve memory in rats injected with β-amyloid (Aβ) (Limón et al., 2009) and displays antioxidant properties in the CNS (Manikandan & Devi, 2005; Pages et al., 2010). α-Asarone enhances cognition in amnesic mice due to both cholinergic and antioxidant effects (Kumar et al., 2012), while both α- and β-asarone are sedative in vivo (Howes & Houghton, 2003), so may be useful for BPSD. Oral administration to rats of *A. calamus* suspension showed significant improvement in hyoscine-induced memory loss (Barua et al., 2015). However, β-asarone

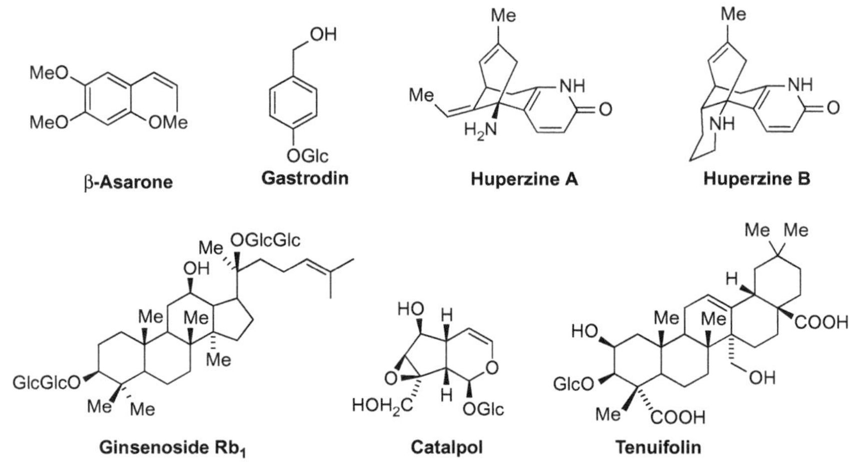


**Fig. 1** Chemical structures of phytochemicals from some plants used traditionally for Alzheimer's disease-like symptoms.

has not been tested in animals, nor has the plant extract or isolated compounds been tested clinically.

It should be considered that the genotoxic and hepatocarcinogenic effects of α- and β-asarone may restrict the therapeutic use of *A. calamus* and these alkenylbenzenes for AD, and it has been recommended that their concentration in herbal medicinal products should be reduced to a minimum and diploid varieties should be preferred (HMPC, 2005) to reduce potential risks.

## 2.2 *Gastrodia elata*

The CHM "Rhizoma Gastrodiae" [Tian Ma (天麻)] is from the tuber of *Gastrodia elata* Blume (Orchidaceae). It is traditionally used to "stop endogenous wind" (antispasmodic), to "expel wind" (to relieve rheumatic pain), and to "dredge the meridians." It is indicated for convulsions, tetanus, headaches, epilepsy, dizziness, numbness, rheumatism, and paralysis (CPC, 2015). It is usually combined with other CHMs as a compound prescription (Tang & Xun, 2003) and is a component of traditional formulas for dementia (May et al., 2016).

Studies in vivo demonstrated that *G. elata* could significantly improve spatial memory in animal models of AD, associated with raised choline acetyltransferase (ChAT) expression and AChE inhibition in the brain (Huang et al., 2013). Further evidence that *G. elata* improves cholinergic function to mediate cognitive effects was revealed in a study which concluded that prolonged intake (6 weeks) of *G. elata* extract improved learning and memory capabilities in scopolamine-treated rats (Park et al., 2015). *G. elata* tuber extract also increased superoxide dismutase activity and reduced tau protein expression in the brains of experimental dementia rats (Huang, Luo, & Yu, 2007).

Studies in vitro concluded that tuber extracts were neuroprotective against fragments of amyloid precursor protein (APP) (Kim, Kim, Lyu, Lee, & Kang, 2006) and against Aβ, possibly via regulation of protein-folding mechanisms, with the tuber constituents gastrodin (Fig. 1) and 4-hydroxybenzylalcohol associated with these effects (Lee, Kim, Han, Bhandary, et al., 2012). Thus, modulation of amyloid appears to be a promising target for *G. elata* and its constituents. Other studies have revealed that a tuber extract enhances α-secretase-mediated processing of APP and this action may contribute to the neuroprotective and cognitive-enhancing effects of *G. elata* (Mishra et al., 2011).

In addition to gastrodin and 4-hydroxybenzylalcohol, other tuber constituents, including 4-hydroxybenzaldehyde, vanillin, and vanillyl alcohol, are also considered as active components with antioxidant, antiinflammatory, and neuroprotective effects (Jang et al., 2015). Gastrodin in particular appears to be an important constituent for mediating the cognitive-enhancing effects of *G. elata*. This phenolic glucoside improved learning and memory in animal models of brain ischemia (Li & Zhang, 2015) and AD, with the latter observation associated with neuroprotection via antiinflammatory and anti-amyloidogenic effects (Hu, Li, & Shen, 2014). Furthermore, gastrodin attenuated apoptosis in neural progenitor cells in vitro and improved hippocampal neurogenesis in mice injected with Aβ (Li & Qian, 2016), while it also suppressed β-site APP-cleaving enzyme 1 (BACE1) expression to alleviate memory deficits in vivo (Zhang, Zhou, et al., 2016). *G. elata* and its constituents have not been evaluated in randomized-controlled trials (RCTs) to evaluate potential efficacy in AD.

It is important to consider that overharvesting of plant material in the wild has resulted in the decline of populations of *G. elata*, such that this species is described in the International Union for Conservation of Nature and Natural Resources (IUCN) Red List as vulnerable (IUCN, 2016), while trade is regulated under the Convention on International Trade in Endangered Species of Wild Flora and Fauna (CITES, 2016). Therefore further evaluation of the medicinal properties of *G. elata* must also respect conservation strategies for this species to protect it from extinction.

## 2.3 *Huperzia serrata*

Qian Ceng Ta (千层塔) is prepared from *Huperzia serrata* (Thunb.) Trevis. (Lycopodiaceae), which has been used in CHM to alleviate memory loss, to promote circulation and to counteract fever and inflammation (Howes & Houghton, 2009a). Huperzine A (Fig. 1), a major quinolizidine-related alkaloid present, reversibly inhibits AChE in vitro and in vivo (Williams, Sorribas, & Howes, 2011) and it may also favorably affect other neurotransmitter systems to improve memory, so has potential in treating AD symptoms (Howes, 2013). Huperzine B (Fig. 1) also inhibits AChE, although is less potent than huperzine A (Houghton, Ren, & Howes, 2006).

Huperzine A is neuroprotective against Aβ (Xiao, Zhang, & Tang, 2002), oxygen-free radicals (Xiao, Yang, & Tang, 1999), and glutamate (Ved, Koenig, Dave, & Doctor, 1997), and recent reviews highlight this varied protective effect and its mechanisms relevant to AD (Howes, 2013; Howes & Houghton, 2009a, 2009b; Williams et al., 2011). Additional activities include antagonism at NMDA receptors in the cerebral cortex (Wang, Zhang, Yang, & Hu, 1999) and a reduction in brain iron levels (Huang et al., 2014). Huperzine A improved memory retention processes in vivo (Howes & Houghton, 2009b); while RCTs have been conducted, e.g., a multicenter, double-blind trial showed that huperzine A significantly improved memory and behavior in AD patients, being more selective for AChE than butylcholinesterase and less toxic than donepezil and tacrine (Shu, 1998; Small et al., 1997).

Huperzine A is used clinically in China for AD (Ban et al., 2016) and several RCTs have been carried out with AD patients. However, different meta-analyses have concluded that the published trial data were of poor quality (Rafii et al., 2011; Yang, Wang, Tian, & Liu, 2013; Yue et al., 2012) and toxicity data are considered as inadequate (Ha, Wong, & Zhang, 2011). However, a more recent meta-analysis concluded that eight trials studied were of sufficient quality to be valid and that huperzine A was significantly beneficial in several aspects of AD and was well tolerated (Xing, Zhu, Zhang, & An, 2014). In another recent clinical study with AD patients, when cholinesterase inhibitors (ChEIs) were added to memantine, it was shown that huperzine A gave the best results (Shao, 2015).

Recently there has been considerable interest in huprines, synthetic compounds combining structural features of both huperzine A and tacrine, the first ChEI to be used clinically. Huprine X showed AChE inhibitory effects in vitro (Camps et al., 2000) and displayed beneficial effects in reducing amyloidogenic processes and increasing synaptophysin levels (Hedberg et al., 2010), but no clinical studies using this compound have yet been reported.

## 2.4 *Panax ginseng*

"Ginseng Radix et Rhizoma" [Ren Shen (人参)] is the dried root of *Panax ginseng* C.A.Mey (Araliaceae) and it is indicated for conditions such as frailty caused by long-term illness, and is considered to tranquilize the mind and replenish wisdom (CPC, 2010). It has also been an ingredient of multiple CHM formulas indicated for memory impairment in the context of aging (May et al., 2016). The chemistry of *P. ginseng* has been extensively studied and the major constituents are the triterpene saponins, designated as ginsenosides (e.g., Re, $Rg_2$, Rf) (Court, 2000; Kite, Howes, Leon, & Simmonds, 2003).

In RCTs with normal young volunteers, ginseng improved memory (Kennedy, Haskell, Wesnes, & Scholey, 2004), however, frequent use for extended periods ($\leq$2 years) by healthy subjects did not improve memory (Radad, Gille, Liu, & Rausch, 2006). The majority of other RCTs with AD subjects and those with mild cognitive impairment (MCI) concluded ginseng could improve cognitive functions, although trials were mainly open-label or single-blind and could have been more methodologically robust (Howes & Perry, 2011; Perry & Howes, 2011).

Mechanistically, ginsensosides show a range of biological activities relevant to AD pathology, including antioxidant, inhibition of glutamate- and Aβ-induced cytotoxicity ($Rb_1$, $Rh_2$, $Rg_2$, $Rg_3$), and Aβ-induced tau phosphorylation ($Rb_1$; Fig. 1), they are neuroprotective ($Rg_3$), antagonize NMDA receptors ($Rg_3$, $Rh_2$), modulate ACh release, and increase ChAT ($Rb_1$, $Rg_1$); they also decrease Aβ in transgenic AD mice (Re, $Rg_1$, $Rg_3$) and improve cognition in vivo (Howes & Perry, 2011).

## 2.5 *Polygala* Species

"Polygalae Radix" [Yuan Zhi (远志)] is the dried root of *Polygala tenuifolia* Willd. or *P. sibirica* L. (Polygalaceae) and its indications in CHM include insomnia, dream-disturbed sleep and forgetfulness (CPC, 2010). *P. tenuiflora* root has also been an ingredient of multiple CHM formulas indicated for memory impairment in the context of aging (May et al., 2016).

*P. tenuifolia* root constituents have been associated with a range of mechanistic effects relevant to AD in vitro and in vivo, suggesting extracts from this species may act via polyvalency. For example, tenuigenin is anti-apoptotic and inhibits Aβ secretion, while tenuifolin (Fig. 1) displays the latter effect and it modulates neurotransmitter systems to improve learning and memory in vivo (Howes & Houghton, 2012). Onjisaponins increase nerve growth factor (NGF) synthesis and are neuroprotective against Aβ, while onjisaponin B also reduces Aβ production and improves memory; and polygalasaponins are anxiolytic and sedative in vivo, with one such compound (XXXII) also improving learning abilities (Howes & Houghton, 2012; Li, Cui, et al., 2016). However, more studies on the bioavailability and pharmacokinetics of *Polygala* constituents are needed, with consideration of the potential for saponin components to be hydrolyzed in the gastrointestinal (GI) tract, prior to absorption into the systemic circulation. In this context, one of the main triterpene aglycones produced following hydrolysis of *Polygala* saponins is polygalacic acid, which modulates cholinergic function and neuroinflammation in vivo to mediate a neuroprotective effect against cognitive impairment (Guo, Shen, Meng, Yang, & Li, 2016).

Extracts from this species also improve memory and behavioral disorders are neuroprotective and promote stem cell proliferation in vivo (Chen, Hsieh, Wu, & Lin, 2004; Park et al., 2002, 2008). Standardized extracts have not been investigated for efficacy in AD patients, although two RCTs in normal and elderly subjects concluded that *P. tenuifolia* extract (BT-11) could improve cognitive functions (Lee et al., 2009; Shin et al., 2009).

## 2.6 *Rehmannia glutinosa*

"Rehmanniae Radix" [Di Huang (地黄)] is the root tuber of *Rehmannia glutinosa* (Gaertn.) DC. (Plantaginaceae) and although it is not indicated in the current Pharmacopeia for symptoms associated with AD (CPC, 2010), it is one the most frequently reported ingredients of multiple CHM formulas indicated for memory impairment associated with aging (May et al., 2016). Root extracts improve memory in vivo and also increase expression of NGF in hippocampal neurons (Lin et al., 2012), while in a fly (*Drosophila*) model of AD, *R. glutinosa* extract was protective against Aβ neurotoxicity (Liu, Lee, et al., 2015).

Iridoids, catalpol (Fig. 1) in particular, isolated from the root of *R. glutinosa* have shown mechanistic effects that appear to contribute to the actions of root extracts in studies relevant to AD. Catalpol attenuates Aβ-induced neurotoxicity in vitro (Jiang, Du, Liu, Bao, & An, 2008) and improves memory in vivo, possibly via favorable effects on the cholinergic system (Xia, Zhang, Wu, Xia, & Hu, 2012). Other studies revealed catalpol has antioxidant, antiinflammatory, neurotrophic, and antiapoptotic effects (Lin et al., 2012), so may mediate a range of actions. Catalpol is also neuroprotective in vivo (Zhang et al., 2008), while a different root constituent, 5-hydroxymethyl furfural, was neuroprotective in vitro (Zhang, Jin, Zhang, Gong, & Gu, 2014). Although root extracts and constituents have shown some promising mechanistic effects that are potentially useful in AD, there is a lack of RCTs investigating their efficacy in humans, particularly in AD. However, *R. glutinosa* has been explored for its effects on cognitive functions as a component of CHM formulae (see later), although the specific contribution of this species to clinical outcomes, and any polyvalent or synergistic effects, remain to be evaluated.

## 3. PLANTS USED TRADITIONALLY BUT NOT FOR AD-LIKE SYMPTOMS: EVIDENCE FOR RELEVANT ACTIVITY

### 3.1 *Crocus sativus*

"Croci Stigma" [Xi Hong Hua (西红花)] is the dried stigma of *Crocus sativus* L. (Iridaceae), often described by the vernacular name "saffron" and in CHM it is used to treat depression, fear, confusion, menstrual difficulties, abdominal pain, and to activate blood circulation. Long-term use is considered to relieve anxiety and create feelings of joy (Zhao, Dai, & Chen, 2006); in CHM, it is also indicated for conditions that include psychosis and is considered to tranquilize the mind and to relieve depression (CPC, 2010); thus, some traditional uses appear to suggest some relevance to BPSD (although not used traditionally for dementia).

Clinical studies concluded that saffron extracts (30 mg daily for 6 weeks) improve mild-to-moderate depression in patients compared to placebo, and were as effective as the antidepressant drugs imipramine and fluoxetine (Akhondzadeh, Fallah-Pour, Afkham, Jamshidi, & Khalighi-Cigaroudi, 2004, Akhondzadeh et al., 2005; Noorbala, Akhondzadeh, Tahmacebi-Pour, & Jamshidi, 2005; Nourbala, Tahmasebipour, Akhoundzadeh, Khani, & Jamshidi, 2004). Further evidence for potential usefulness in BPSD was suggested as the component safranal was anxiolytic and hypnotic in vivo (Hosseinzadeh & Noraei, 2009). RCTs also indicate saffron extracts (30 mg daily for 16 or 22 weeks) can improve cognitive functions in AD patients when compared to the AChE inhibitor drug donepezil, and caused fewer adverse effects (Akhondzadeh et al., 2010a, 2010b). Another RCT concluded the same dose of saffron taken for 1 year was comparable to memantine in reducing cognitive decline in AD patients (Farokhnia et al., 2014).

Although the extracts investigated in these clinical studies were not chemically characterized, mechanistic studies suggest the crocin constituents (a series of different sugar esters of the dicarboxylic acid crocetin) may contribute to the observed effects. Crocin is antiapoptotic, neuroprotective, and antioxidant in vitro, and improves memory in vivo; crocetin (Fig. 2) and crocin are antiinflammatory in microglia, while

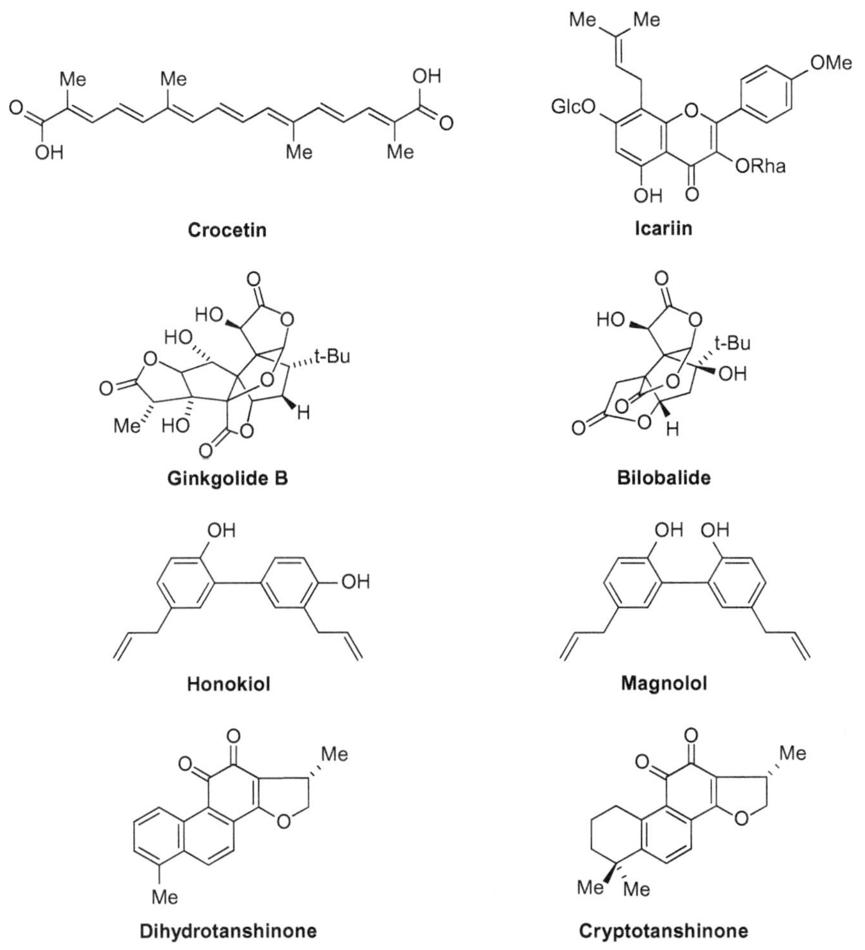


**Fig. 2** Chemical structures of phytochemicals from some plants used traditionally but not for Alzheimer's disease-like symptoms.

*trans*-crocin-4 inhibits Aβ fibrillogenesis (Howes & Houghton, 2012; Howes & Perry, 2011). Further studies are needed to characterize the "active" constituents of saffron and their modes of action relevant to AD, particularly as the crocins show poor intestinal absorption in vivo and are hydrolyzed to crocetin in the GI tract (Lechtenberg et al., 2008).

## 3.2 *Epimedium* Species

Leaves from *Epimedium brevicornum* Maxim. (accepted name: *E. brevicornu* Maxim.), *E. sagittatum* (Siebold & Zucc.) Maxim., *E. pubescens* Maxim., and *E. koreanum* Nakai (Berberidaceae) are indicated in CHM as "Epimedii Folium" [Yin Yang Huo (淫羊藿)] for debilitation of kidney yang, impotence, sinew and bone disorders, "painful impediment caused by wind-dampness," numbness, and spasms (CPC, 2010). Various species of *Epimedium*, a genus of small herbaceous plants, have been used in Far Eastern medicine, and it is of interest that *E. koreanum* is a traditional remedy for forgetfulness in Korea (Ma et al., 2011).

Much interest has been shown in the flavonoid glycosides present in *Epimedium* species, particularly icariin (Fig. 2). Icariin protects PC-12 cells from apoptosis induced by Aβ (Wang, Zhang, Wang, Qi, & Lou, 2007; Zhang, Wang, et al., 2015) and shows a dose-dependent protective effect at low concentrations on neurons damaged by ischemia/reperfusion, while in vivo, it improved learning and memory, possibly via some antioxidant effect (Li, Zhou, & Shi, 2005; Xu et al., 2009). Another in vivo study using the 5xFAD mouse, considered to be a better model for AD, showed markedly less damage in axons and dendrites caused by Aβ in the groups treated with icariin (Urano & Tohda, 2010). The related compound icariside II has recently been shown to improve cognition in rats treated with streptozocin, which has a deleterious effect in the hippocampus (Yin et al., 2016). A possible extra contributing factor to the usefulness of *Epimedium* species in treating AD symptoms is the fact that the aporphine alkaloid epimediphine, present in *E. koreanum*, has AChE inhibitory activities (Zhang et al., 2013).

A clinical trial of a CHM formula comprising six herbs, including *Epimedium*, showed that the herbal mixture was better than donepezil at improving scores for mild AD (Zhang, Lin, et al., 2015), but studies with *Epimedium* alone or with icariin are needed to establish clinical efficacy of this herb and its constituents.

## 3.3 *Ginkgo biloba*

"Ginkgo Folium" [Yin Xing Ye (银杏叶)] is the dried leaf of *Ginkgo biloba* L. (Ginkgoaceae), which is indicated in CHM for a range of conditions, including those associated with respiratory disorders and for "heart pain" (CPC, 2010). In recent years, leaf extracts have become very popular worldwide to alleviate conditions associated with reduced microvascular flow, especially in the CNS (Heinrich et al., 2012). *G. biloba* leaves contain a mixture of phytochemical classes, each of which displays several activities which could be relevant to alleviating AD symptoms, so *G. biloba* extract is an excellent example of an herb displaying polyvalence. Antiinflammatory flavonoids are present, e.g., ginkgetin, but these are also antioxidant and may also be vasodilatory (Kubota et al., 2001); and the flavonol quercetin inhibits Aβ-induced cell death (Shi et al., 2009) and stimulates neurite outgrowth (Tchantchou et al., 2009). The diterpene ginkgolides, e.g., ginkgolide B (Fig. 2), are unique to this species and are potent platelet aggregation factor (PAF) antagonists (Braquet, Hosford, & Koltai, 1994) as well as being antioxidant scavengers of reactive oxygen species. More recently they have been shown to inhibit Aβ-induced cell death (Bate, Tayebi, & Williams, 2008). The related sesquiterpene bilobalide (Fig. 2) also has strong antioxidant activity, promotes expression of growth factors, leading to neural outgrowth (Tchantchou et al., 2009) and reduces levels of proinflammatory nuclear factor (NF)-κB (Li et al., 2008).

Studies on *G. biloba* extracts in animals and healthy elderly humans gave encouraging results which demonstrated significant improvement in markers of cognitive function; e.g., a meta-analysis in 2002 covered eight RCTs, each showing significant improvement in memory tests, anxiety, dizziness, headaches, and tinnitus, with no serious adverse effects (Canter & Ernst, 2001). These findings stimulated interest in *G. biloba* for possible prevention, or alleviation of symptoms, of AD. The first double-blind RCT study reported encouraging results with patients given *G. biloba* extract showing stabilization and, in some cases improvement in cognitive performance and social functioning with no adverse effects observed (Le Bars et al., 1997). However, other RCTs were more equivocal about any improvements and the conclusions of meta-analyses about the value of *G. biloba* for treating AD have become increasingly negative (Birks & Grimley Evans, 2002, 2009), with the latter review stating that "There is no convincing evidence that *G. biloba* is efficacious for dementia and cognitive impairment." However some reported RCTs still show a positive outcome (e.g., Hashiguchi, Ohta, Shimizu, Maruyama, & Mochizuki, 2015; Wang et al., 2010; Yang, Wang, Sun, Zhang, & Liu, 2016), although another RCT showing some positive effects in AD concluded that the ChEI rivastigmine gave a better outcome than *G. biloba* (Nasab, Bahrammi, Nikpour, Rahim, & Naghibi, 2012). Thus, *G. biloba* does appear to have more potential than some other plant species in having clinical usefulness for AD. Most published trials note that it displays fewer adverse effects than other agents used clinically, although it has been associated with anticoagulant effects (Howes & Houghton, 2009a), so might interact with medication such as warfarin, causing excessive bleeding.

## 3.4 *Magnolia officinalis*

"Magnoliae Officinalis Cortex" [Hou Po (厚朴)] is the dried stem bark, root bark, or branch bark of *Magnolia officinalis* Rehder & E.H.Wilson or *M. officinalis* var. *biloba* Rehder & E.H.Wilson (Magnoliaceae) and it is indicated in CHM for conditions that include vomiting and diarrhea, abdominal disorders, and cough (CPC, 2010). Extracts of *M. officinalis*, and its isolated compounds, particularly the biphenolic lignans honokiol and magnolol (Fig. 2), have been evaluated for activities relevant to AD, although RCTs are lacking.

Studies have shown that honokiol and magnolol increased ChAT activity, inhibited AChE activity in vitro, and increased hippocampal ACh release in vivo (Hou, Chao, & Chen, 2000). Other activities relevant to AD pathology have also been identified. Antioxidant effects have been shown for *M. officinalis* extract (Zhou & Xu, 1992), magnolol (Chen et al., 2001), and honokiol (Lo, Teng, Chen, Chen, & Hong, 1994). Magnolol showed neuroprotective properties in cortical neuron-astrocyte cultures (Lee, Hseih, Kuo, Yeh, & Huang, 1998), while both magnolol and honokiol protect against Aβ damage (Hoi, Ho, Baum, & Chow, 2010; Xian, Ip, Mao, & Lin, 2016). Honokiol also promoted neurite outgrowth in cultured PC-12 cells (Fukuyama et al., 2002), while magnolol showed antiinflammatory activity perhaps via inhibition of cyclooxygenase and 5-lipoxygenase (Wang, Ho, Chang, & Chen, 1995). Thus, the extract of *M. officinalis*, and honokiol and magnolol, apparently display polyvalence, having multiple actions relevant to AD therapy.

The promise for amelioration of symptoms associated with AD held out by such in vitro studies has been fulfilled by several in vivo experiments, using the extract of the bark (Lee, Choi, Han, Kim, et al., 2012), magnolol

(Li et al., 2013) and honokiol (Hu et al., 2013; Xian et al., 2015) when improved memory, reduced accumulation and toxicity of Aβ, and reduced oxidative damage were noted. Another constituent, 4-*O*-methylhonokiol, has also shown similar activities in vivo (Jung, Lee, Choi, & Hong, 2014; Lee, Choi, Choi, et al., 2012; Lee, Choi, Lee, et al., 2012). With respect to BPSD, honokiol and magnolol have also shown anxiolytic effects, associated with modulation of GABAergic neurotransmission; and when combined in a prescription with other CHM herbs (Banxia Houpu), *M. officinalis* was antidepressant in vivo (Howes, Perry, & Houghton, 2003).

### 3.5 *Salvia miltiorrhiza*

"Salviae Miltiorrhizae Radix et Rhizoma" [Dan Shen (丹参)] is the dried root and rhizome of *Salvia miltiorrhiza* Bunge (Lamiaceae) and its main uses in CHM include for cardiovascular and bleeding disorders (CPC, 2010). Root extracts have shown a range of mechanistic effects in the CNS, with many studies showing some promising effects relevant to vascular and ischemic disorders in particular (Houghton & Howes, 2005; Perry, Howes, Houghton, & Perry, 2000). More recently, root extracts have shown neuroprotection against Aβ in vitro (Yu et al., 2014) and could improve learning and memory in an animal model of AD (Liu, Guo, et al., 2015), although RCTs in AD patients are lacking. It is also important to consider that *S. miltiorrhiza* preparations may interact with other medicines, such as anticoagulant drugs (Zhou, Chan, & Yeung, 2012).

There has been much research on the chemical constituents of *S. miltiorrhiza* root and their modes of action. Salvianolic acid constituents protect against ischemia-induced memory impairment in vivo, with salvianolic acids A and B also protecting against Aβ-induced neurotoxicity in vitro via antioxidant effects (Cao et al., 2013; Howes & Houghton, 2012), and they inhibit aggregation of Aβ in vitro (Zhang, Qian, Zhang, & Wang, 2016). Salvianolic acid B also improves memory in animal models of cholinergic impairment and Aβ-induced cognitive deficits (Bonaccini, Karioti, Bergonzi, & Bilia, 2015). Recent studies suggest root extracts and salvianolic acids A and B may enhance neural stem cell proliferation and differentiation (Zhang, Qian, et al., 2016), although more research on potential applications in vivo is needed.

Other antioxidant components of the root include the diterpene constituents (tanshinones), which are also considered to explain some of the antiinflammatory mechanisms of root extracts (Howes & Houghton, 2009a), including inhibition of eicosanoid generation (Howes, Houghton, & Hoult, 2000), inhibition of other inflammatory mediators, including interleukin (IL)-8, IL-10, and tumor necrosis factor (TNF)-α (Liang et al., 2013), and modulation of the NF-κB pathway (Akaberi, Iranshahi, & Mehri, 2016). Both antioxidant and antiinflammatory mechanisms are suggested as important to mediate the observed neuroprotective effects of the tanshinones (Bonaccini et al., 2015).

Tanshinone diterpene constituents display a range of other biological activities that appear to provide further scientific evidence for the potential usefulness of root extracts in AD. Tanshinones inhibit AChE, with the dihydrofurans dihydrotanshinone and cryptotanshinone (Fig. 2) being more potent ChEIs than the furans, tanshinones I and IIa (Ren, Houghton, Hider, & Howes, 2004). Cryptotanshinone also ameliorates cognitive impairments in vivo, induced by scopolamine (Wong et al., 2010) and by Aβ, while also reducing plaque deposition of the latter (Zhang, Qian, et al., 2016). These findings suggest this diterpene can cross the blood–brain barrier to mediate cognitive effects via the cholinergic system and via antiamyloidosis. Other studies revealed tanshinone IIa is neuroprotective against cerebral ischemia/reperfusion injury in vivo via an antiapoptotic action (Chen et al., 2012), and against Aβ-induced neurotoxicity in vitro (Qian, Xiao, & Xu, 2012), while tanshinone I enhances neurogenesis in the mouse hippocampus (Chen et al., 2016).

## 4. CHM FORMULAS RELEVANT TO AD

CHM formulas are often composed of complex mixtures of different herbal and other substances, which introduces challenges for their quality control and for understanding their mechanistic and clinical effects, in the context of the role of polyvalency and synergy. The plant species discussed earlier are also included in CHM formulas, with some of these (and others) evaluated in pharmacological and clinical studies. Examples include "DX-9386," which is composed of *Polygala tenuifolia*, *Panax ginseng*, *Acorus gramineus*, and *Poria cocos* (1:1:25:50), and it improves learning and memory in different animal models (Howes & Houghton, 2003; Howes et al., 2003); and "ZiBuPiYin" [滋补脾阴], which is a modification of the "Zicheng decoction" [资成汤], traditionally used for memory loss and composed of 12 components (including roots from red ginseng, white peony, and Dan Shen) that has been suggested to prevent diabetes-related cognitive decline in vivo (Chen et al., 2014). *Panax ginseng* is also a component of "Yizhi capsule" [益智胶囊], in addition to

*Polygala tenuifolia* and other components, which has been investigated for efficacy in vascular dementia, although RCT outcomes to date are inconclusive with regard to efficacy (Howes & Perry, 2011).

The genus *Acorus* appears to be particularly important in CHM formulas for dementia. In a recent data-mining study, 104 CHM formulas were analyzed, including 147 CHMs for senile dementia, and the results enabled the medication rules on the CHM inheritance auxiliary system to be summarized. It was concluded that CHM prescriptions for senile dementia are mainly composed of a combination of three types of herbs: tonifying, activating blood circulation, and resuscitating, to reflect the TCM theory for senile dementia. The most frequently used herbs, accounting for 28% of all the relevant formulas, were *Ligusticum striatum* DC. rhizome ["Chuanxiong Rhizoma," Chuan Xong (川芎)] and *Acorus calamus* var. *angustatus* Besser ["Acori Tatarinowii," Shi Chang Pu (石菖蒲)] (Zong, Ji, Wei, & Shi, 2014).

The formula "Sailuotong" [塞络通] has been systematically studied in the laboratory and in RCTs. This formula is composed of *Panax ginseng*, *Ginkgo biloba*, and *Crocus sativus* and produced small improvements in working memory when taken by healthy adults for one week in a small RCT (Steiner et al., 2016), and it improved the cognitive and memory impairments associated with vascular dementia (Steiner et al., 2015). Other formulas investigated in small RCTs include "Liuweidihuang-tang" [六味地黄汤], consisting of six components including *Rehmannia* Radix, which enhanced cognitive ability in normal young adults (Park et al., 2005) and "Ba wei di huang wan" [八味地黄丸] which has muscarinic receptor effects, improves memory in vivo, and cognitive functions in patients with mild-to-severe dementia (Howes & Perry, 2011). Also of potential relevance for use in AD is the formula "Jia wei wu zi yan zong ke li" [加味五子衍宗颗粒], composed of six CHMs, including *Epimedium grandiflorum* C.Morren herb, since when a granule preparation was taken for 3 months by volunteers with MCI, it was concluded as superior to the placebo in improving cognitive functions (May et al., 2009).

## 5. CONCLUSION

There has been much research to investigate the scientific basis to understand the traditional uses of CHMs for dementia. Other research has associated plant species not used traditionally for dementia with mechanistic effects suggesting potential relevance for use in AD. For plant species investigated in both circumstances, there is considerable variation in the extent of the research performed. Available data range from extensive pharmacological and clinical studies on plant extracts, including those standardized to contain specific levels of constituents (e.g., *Ginkgo biloba*), and on SCEs (e.g., huperzine A), to studies that are limited to pharmacological tests with such species not adequately evaluated for safety and efficacy in robust RCTs (e.g., *Acorus calamus*). In addition to the need for more extensive pharmacological studies and RCTs to investigate efficacy of promising CHMs for dementia, evaluation of their safety is also needed.

Other challenges for the investigation of CHMs for their potential relevance in AD include their standardization to known active constituents. Relatively few plants that modify cognitive functions, or those with relevance to BPSD, have been extensively studied to determine the compounds responsible to mediate such effects. More research is needed on their modes of action, and on whether they are therapeutically active and bioavailable at appropriate clinical doses. Other complexities include understanding polyvalent and synergistic effects, which is particularly difficult to elucidate for CHM formulae composed of a mixture of different species. Other factors to consider are the effects of traditional processing methods on the chemical and pharmacological profiles of CHMs and how this may impact on their potential for therapeutic use in AD. It is under these circumstances that the appropriate qualitative and quantitative chemical profiles of specific CHMs must be investigated to determine their safety and efficacy.

In conclusion, CHMs provide a vast resource for biologically active constituents that may be therapeutically useful in AD, either in the form of standardized extracts or as SCEs. Further research on the role of CHMs in AD therapy is important to explore further, and should encompass respect for TCM theory and traditions, while also involving robust chemical and pharmacological tests, reliable RCTs, and consideration of the conservation of biodiversity and the sustainable uses of plants.

## ACKNOWLEDGMENTS

The authors thank the late Dr. Nigel C. Veitch, Royal Botanic Gardens, Kew, for expertly drawing some of the chemical structures.

## REFERENCES

Akaberi, M., Iranshahi, M., & Mehri, S. (2016). Molecular signaling pathways behind the biological effects of *Salvia* species diterpenes in neuropharmacology and cardiology. *Phytotherapy Research*, *30*, 878–893.

Akhondzadeh, S., Fallah-Pour, H., Afkham, K., Jamshidi, A.-H., & Khalighi-Cigaroudi, F. (2004). Comparison of *Crocus sativus* L. and imipramine in the treatment of mild to moderate depression: A pilot double-blind randomized trial [ISRCTN45683816]. *BMC Complementary and Alternative Medicine*, *4*, 12.

Akhondzadeh, S., Sabet, M. S., Harirchian, M. H., Togha, M., Cheraghmakani, H., Razeghi, S., et al. (2010a). A 22-week, multicenter, randomized, double-blind controlled trial of *Crocus sativus* in the treatment of mild-to-moderate Alzheimer's disease. *Psychopharmacology*, *207*, 637–643.

Akhondzadeh, S., Sabet, M. S., Harirchian, M., Togha, M., Cheraghmakani, H., Razeghi, S., et al. (2010b). Saffron in the treatment of patients with mild to moderate Alzheimer's disease: A 16-week, randomized and placebo-controlled trial. *Journal of Clinical Pharmacy and Therapeutics*, *35*, 581–588.

Akhondzadeh, S., Tahmacebi-Pour, N., Noorbala, A. A., Amini, H., Fallah-Pour, H., Jamshidi, A. H., et al. (2005). *Crocus sativus* L. in the treatment of mild to moderate depression: A double-blind, randomized and placebo-controlled trial. *Phytotherapy Research*, *19*, 148–151.

Ban, C.-X., Xiao, S.-F., Lin, X., Wang, T., Qiu, Q., Zhu, M.-J., et al. (2016). Clinicians' prescription preferences for treating patients with Alzheimer's disease in Shanghai. *Translational Neurodegeneration*, *5*, 8.

Barua, C. C., Haloi, P., Patowary, P., Bora, M., Barua, A. G., Bordoloi, M. J., et al. (2015). Evaluation of anti-amnestic activity of few medicinal plants against scopolamine induced amnesia. *Indian Journal of Traditional Knowledge*, *14*, 581–589.

Bate, C., Tayebi, M., & Williams, A. (2008). Ginkgolides protect against amyloid-β 1–42-mediated synapse damage in vitro. *Molecular Neurodegeneration*, *3*, 1.

Birks, J., & Grimley Evans, J. (2002). *Ginkgo biloba* for cognitive impairment and dementia. *The Cochrane Database of Systematic Reviews*, *4*, CD003120.

Birks, J., & Grimley Evans, J. (2009). *Ginkgo biloba* for cognitive impairment and dementia. *The Cochrane Database of Systematic Reviews*, *1*, CD003120.

Bonaccini, L., Karioti, A., Bergonzi, M. C., & Bilia, A. R. (2015). Effects of *Salvia miltiorrhiza* on CNS neuronal injury and degeneration: A plausible complementary role of tanshinones and depsides. *Planta Medica*, *81*, 1003–1016.

Braquet, P., Hosford, D., & Koltai, M. (1994). Ginkgolide B (BN 52021): A natural PAF receptor antagonist. *Neuropsychopharmacology*, *10*, 693S.

Camps, P., Cusack, B., Mallender, W. D., El Achab, R., Morral, J., Muñoz-Torrero, D., et al. (2000). Huprine X is a novel high-affinity inhibitor of acetylcholinesterase that is of interest for treatment of Alzheimer's disease. *Molecular Pharmacology*, *57*, 409–417.

Canter, P., & Ernst, E. (2001). *Ginkgo biloba*: A smart drug? A systematic review of controlled trials of the cognitive effects of ginkgo biloba extracts in healthy people. *Psychopharmacology Bulletin*, *36*, 108–123.

Cao, Y. Y., Wang, L., Ge, H., Lu, X. L., Pei, Z., Gu, Q., et al. (2013). Salvianolic acid A, a polyphenolic derivative from *Salvia miltiorrhiza* Bunge, as a multifunctional agent for the treatment of Alzheimer's disease. *Molecular Diversity*, *17*, 515–524.

Chadwick, D. J., & Marsh, J. (Eds.), (1994). *Ethnobotany and the search for new drugs*. Chichester, UK: Wiley.

Chen, Y.-L., Hsieh, C.-L., Wu, P.-H. B., & Lin, J.-G. (2004). Effect of *Polygala tenuifolia* root on behavioral disorders by lesioning nucleus basalis magnocellularis in rat. *Journal of Ethnopharmacology*, *95*, 47–55.

Chen, J., Liang, L., Zhan, L., Zhou, Y., Zheng, L., Sun, X., et al. (2014). ZiBuPiY in recipe protects db/db mice from diabetes-associated cognitive decline through improving multiple pathological changes. *PloS One*, *9*, e91680.

Chen, Y.-L., Lin, K.-F., Shiao, M.-S., Chen, Y.-T., Hong, C.-Y., & Lin, S.-J. (2001). Magnolol, a potent antioxidant from *Magnolia officinalis*, attenuates intimal thickening and MCP-1 expression after balloon injury of the aorta in cholesterol-fed rabbits. *Basic Research in Cardiology, 96*, 353–363.

Chen, B. H., Park, J. H., Cho, J. H., Kim, I. H., Lee, J. C., Lee, T.-K., et al. (2016). Tanshinone I enhances neurogenesis in the mouse hippocampal dentate gyrus via increasing Wnt-3, phosphorylated glycogen synthase kinase-3β and β-catenin immunoreactivities. *Neurochemical Research, 41*, 1958–1968.

Chen, Y., Wu, X., Yu, S., Fauzee, N. J. S., Wu, J., Li, L., et al. (2012). Neuroprotective capabilities of tanshinone IIA against cerebral ischemia/reperfusion injury via anti-apoptotic pathway in rats. *Biological and Pharmaceutical Bulletin, 35*, 164–170.

CITES: Convention on International Trade in Endangered Species of Wild Fauna and Flora. (2016). *Appendices I, II and III*. http://www.cites.org. Retrieved on 06/09/16.

Court, W. E. (Ed.), (2000). *Ginseng: The genus Panax*. Amsterdam: Harwood Academic.

CPC: Chinese Pharmacopoeia Commission. (2010). *Pharmacopoeia of the People's Republic of China: Vol. 1*. Beijing: China Medical Science Press.

CPC: Chinese Pharmacopoeia Commission. (2015). *Pharmacopoeia of the People's Republic of China*. Beijing: China Medical Science Press.

Farokhnia, M., Shafiee Sabet, M., Iranpour, N., Gougol, A., Yekehtaz, H., Alimardani, R., et al. (2014). Comparing the efficacy and safety of *Crocus sativus* L. with memantine in patients with moderate to severe Alzheimer's disease: A double-blind randomized clinical trial. *Human Psychopharmacology: Clinical and Experimental, 29*, 351–359.

Fukuyama, Y., Nakade, K., Minoshima, Y., Yokoyama, R., Zhai, H., & Mitsumoto, Y. (2002). Neurotrophic activity of honokiol on the cultures of fetal rat cortical neurons. *Bioorganic & Medicinal Chemistry Letters, 12*, 1163–1166.

Guo, C., Shen, J., Meng, Z., Yang, X., & Li, F. (2016). Neuroprotective effects of polygalacic acid on scopolamine-induced memory deficits in mice. *Phytomedicine, 23*, 149–155.

Ha, G. T., Wong, R. K., & Zhang, Y. (2011). Huperzine A as potential treatment of Alzheimer's disease: An assessment on chemistry, pharmacology, and clinical studies. *Chemistry & Biodiversity, 8*, 1189–1204.

Hashiguchi, M., Ohta, Y., Shimizu, M., Maruyama, J., & Mochizuki, M. (2015). Meta-analysis of the efficacy and safety of *Ginkgo biloba* extract for the treatment of dementia. *Journal of Pharmaceutical Health Care and Sciences, 1*, 14.

Hedberg, M. M., Clos, M. V., Ratia, M., Gonzalez, D., Unger Lithner, C., Camps, P., et al. (2010). Effect of huprine X on β-amyloid, synaptophysin and α7 neuronal nicotinic acetylcholine receptors in the brain of 3xTg-AD and APPswe transgenic mice. *Neurodegenerative Diseases*, 7, 379–388.

Heinrich, M., Barnes, J., Gibbons, S., & Williamson, E. M. (2012). Ethnobotany and ethnopharmacology. In M. Heinrich, J. Barnes, S. Gibbons, & E. M. Williamson (Eds.), *Fundamentals of pharmacognosy and phytotherapy* (2nd ed., pp. 49–59). London: Churchill Livingstone. 249–250.

Heinrich, M., & Jäger, A. K. (Eds.), (2015). *Ethnopharmacology*. Chichester, UK: Wiley.

HMPC: Committee on Herbal Medicinal Products. (2005). *Public statement on the use of herbal medicinal products containing asarone*. London: European Medicines Agency. EMEA/HMPC/139215/2005.

Ho, Y.-S., So, K.-F., & Chang, R. C. (2011). Drug discovery from Chinese medicine against neurodegeneration in Alzheimer's and vascular dementia. *Chinese Medicine, 6*, 15.

Hoi, C. P., Ho, Y. P., Baum, L., & Chow, A. H. (2010). Neuroprotective effect of honokiol and magnolol, compounds from *Magnolia officinalis*, on β-amyloid-induced toxicity in PC12 cells. *Phytotherapy Research, 24*, 1538–1542.

Hosseinzadeh, H., & Noraei, N. B. (2009). Anxiolytic and hypnotic effect of *Crocus sativus* aqueous extract and its constituents, crocin and safranal, in mice. *Phytotherapy Research, 23*, 768–774.

Hou, Y.-C., Chao, P.-D. L., & Chen, S.-Y. (2000). Honokiol and magnolol increased hippocampal acetylcholine release in freely-moving rats. *The American Journal of Chinese Medicine*, *28*, 379–384.

Houghton, P. J. (2009). Traditional plant medicines as a source of new drugs. In W. C. Evans (Ed.), *Trease and Evans pharmacognosy* (16th ed., pp. 62–74). London: Elsevier.

Houghton, P., & Howes, M.-J. (2005). Natural products and derivatives affecting neurotransmission relevant to Alzheimer's and Parkinson's disease. *Neurosignals*, *14*, 6–22.

Houghton, P., Howes, M.-J., Lee, C., & Steventon, G. (2007). Uses and abuses of in vitro tests in ethnopharmacology: Visualizing an elephant. *Journal of Ethnopharmacology*, *110*, 391–400.

Houghton, P. J., Ren, Y., & Howes, M.-J. (2006). Acetylcholinesterase inhibitors from plants and fungi. *Natural Product Reports*, *23*, 181–199.

Howes, M.-J. R. (2013). Alkaloids and drug discovery for neurodegenerative diseases. In K. G. Ramawat & J.-M. Mérillon (Eds.), *Natural products. Phytochemistry, botany and metabolism of alkaloids, phenolics and terpenes* (pp. 1331–1365). Berlin, Heidelberg: Springer-Verlag.

Howes, M.-J. R., & Houghton, P. J. (2003). Plants used in Chinese and Indian traditional medicine for improvement of memory and cognitive function. *Pharmacology, Biochemistry, and Behavior*, *75*, 513–527.

Howes, M.-J. R., & Houghton, P. J. (2009a). Traditional medicine for memory enhancement. In K. G. Ramawat (Ed.), *Herbal drugs: Ethnomedicine to modern medicine* (pp. 239–291). Berlin, Heidelberg: Springer-Verlag.

Howes, M.-J. R., & Houghton, P. J. (2009b). Acetylcholinesterase inhibitors of natural origin. *International Journal of Biomedical and Pharmaceutical Sciences*, *3*, 67–86.

Howes, M.-J. R., & Houghton, P. J. (2012). Ethnobotanical treatment strategies against Alzheimer's disease. *Current Alzheimer Research*, *9*, 67–85.

Howes, M.-J., Houghton, P., & Hoult, J. (2000). Activity of *Salvia miltiorrhiza* root used traditionally for memory improvement, against leukocyte eicosanoid formation. *Journal of Pharmacy and Pharmacology*, *52*(Suppl.), 272.

Howes, M.-J. R., & Perry, E. (2011). The role of phytochemicals in the treatment and prevention of dementia. *Drugs & Aging*, *28*, 439–468.

Howes, M. J. R., Perry, N. S., & Houghton, P. J. (2003). Plants with traditional uses and activities, relevant to the management of Alzheimer's disease and other cognitive disorders. *Phytotherapy Research*, *17*, 1–18.

Howes, M.-J. R., & Simmonds, M. S. (2015). Pharmacognosy in the United Kingdom: Past, present, and future. *Science*, *350*, S57–S58.

Hu, Z., Bian, X., Liu, X., Zhu, Y., Zhang, X., Chen, S., et al. (2013). Honokiol protects brain against ischemia–reperfusion injury in rats through disrupting PSD95–nNOS interaction. *Brain Research*, *1491*, 204–212.

Hu, Y., Li, C., & Shen, W. (2014). Gastrodin alleviates memory deficits and reduces neuropathology in a mouse model of Alzheimer's disease. *Neuropathology*, *34*, 370–377.

Huang, L.-Y., Luo, H.-B., & Yu, J. (2007). Effect of *Gastrodia elata* Blume on tau protein, superoxide lipid in brain tissue of dementia mice. *Chinese Journal of Public Health*, *10*, 19.

Huang, X.-T., Qian, Z.-M., He, X., Gong, Q., Wu, K.-C., Jiang, L.-R., et al. (2014). Reducing iron in the brain: A novel pharmacologic mechanism of huperzine A in the treatment of Alzheimer's disease. *Neurobiology of Aging*, *35*, 1045–1054.

Huang, G. B., Zhao, T., Muna, S. S., Jin, H. M., Park, J. I., Jo, K. S., et al. (2013). Therapeutic potential of *Gastrodia elata* Blume for the treatment of Alzheimer's disease. *Neural Regeneration Research*, *8*, 1061–1070.

IUCN: International Union for Conservation of Nature. (2016). *The IUCN red list of threatened species (version 2016-2)*. http://www.iucnredlist.org. Retrieved on 06/09/16.

Jang, J. H., Son, Y., Kang, S. S., Bae, C. S., Kim, J. C., Kim, S. H., et al. (2015). Neuropharmacological potential of *Gastrodia elata* Blume and its components. *Evidence-Based Complementary and Alternative Medicine, 2015*, 309261.

Jiang, B., Du, J., Liu, J.-H., Bao, Y.-M., & An, L.-J. (2008). Catalpol attenuates the neurotoxicity induced by β-amyloid 1-42 in cortical neuron-glia cultures. *Brain Research, 1188*, 139–147.

Jung, Y.-Y., Lee, Y.-J., Choi, D.-Y., & Hong, J. T. (2014). Amelioration of cognitive dysfunction in APP/ps1 double transgenic mice by long-term treatment of 4-O-methylhonokiol. *Biomolecules & Therapeutics, 22*, 232–238.

Kennedy, D., Haskell, C., Wesnes, K., & Scholey, A. (2004). Improved cognitive performance in human volunteers following administration of guarana (*Paullinia cupana*) extract: Comparison and interaction with *Panax ginseng*. *Pharmacology, Biochemistry, and Behavior, 79*, 401–411.

Kim, S. T., Kim, J. D., Lyu, Y.-S., Lee, M.-Y., & Kang, H.-W. (2006). Neuroprotective effect of some plant extracts in cultured CT105-induced PC12 cells. *Biological and Pharmaceutical Bulletin, 29*, 2021–2024.

Kite, G. C., Howes, M.-J. R., Leon, C. J., & Simmonds, M. S. (2003). Liquid chromatography/mass spectrometry of malonyl-ginsenosides in the authentication of ginseng. *Rapid Communications in Mass Spectrometry, 17*, 238–244.

Kubota, Y., Tanaka, N., Umegaki, K., Takenaka, H., Mizuno, H., Nakamura, K., et al. (2001). *Ginkgo biloba* extract-induced relaxation of rat aorta is associated with increase in endothelial intracellular calcium level. *Life Sciences, 69*, 2327–2336.

Kumar, H., Kim, B.-W., Song, S.-Y., Kim, J.-S., Kim, I.-S., Kwon, Y.-S., et al. (2012). Cognitive enhancing effects of alpha asarone in amnesic mice by influencing cholinergic and antioxidant defense mechanisms. *Bioscience, Biotechnology, and Biochemistry, 76*, 1518–1522.

Le Bars, P. L., Katz, M. M., Berman, N., Itil, T. M., Freedman, A. M., & Schatzberg, A. F. (1997). A placebo-controlled, double-blind, randomized trial of an extract of *Ginkgo biloba* for dementia. *Journal of the American Medical Association, 278*, 1327–1332.

Lechtenberg, M., Schepmann, D., Niehues, M., Hellenbrand, N., Wünsch, B., & Hensel, A. (2008). Quality and functionality of saffron: Quality control, species assortment and affinity of extract and isolated saffron compounds to NMDA and σ1 receptors. *Planta Medica, 74*, 764–772.

Lee, Y.-J., Choi, D.-Y., Choi, I. S., Kim, K. H., Kim, Y. H., Kim, H. M., et al. (2012a). Inhibitory effect of 4-O-methylhonokiol on lipopolysaccharide-induced neuroinflammation, amyloidogenesis and memory impairment via inhibition of nuclear factor-kappa β in vitro and in vivo models. *Journal of Neuroinflammation, 9*, 35.

Lee, Y. J., Choi, D. Y., Han, S. B., Kim, Y. H., Kim, K. H., Hwang, B. Y., et al. (2012b). Inhibitory effect of ethanol extract of *Magnolia officinalis* on memory impairment and amyloidogenesis in a transgenic mouse model of Alzheimer's disease via regulating β-secretase activity. *Phytotherapy Research, 26*, 1884–1892.

Lee, Y.-J., Choi, D.-Y., Lee, Y. K., Lee, Y. M., Han, S. B., Kim, Y. H., et al. (2012c). 4-O-Methylhonokiol prevents memory impairment in the Tg2576 transgenic mice model of Alzheimer's disease via regulation of β-secretase activity. *Journal of Alzheimer's Disease, 29*, 677–690.

Lee, M.-M., Hseih, M.-T., Kuo, J.-S., Yeh, F.-T., & Huang, H.-M. (1998). Magnolol protects cortical neuronal cells from chemical hypoxia in rats. *Neuroreport, 9*, 3451–3456.

Lee, G. H., Kim, H. R., Han, S. Y., Bhandary, B., Kim, D. S., Kim, M. G., et al. (2012). *Gastrodia elata* Blume and its pure compounds protect BV-2 microglial-derived cell lines against β-amyloid: The involvement of GRP78 and CHOP. *Biological Research, 45*, 403–410.

Lee, J.-Y., Kim, K. Y., Shin, K. Y., Won, B. Y., Jung, H. Y., & Suh, Y.-H. (2009). Effects of BT-11 on memory in healthy humans. *Neuroscience Letters*, *454*, 111–114.
Li, X., Cui, J., Yu, Y., Li, W., Hou, Y., Wang, X., et al. (2016). Traditional Chinese nootropic medicine Radix Polygalae and its active constituent onjisaponin B reduce β-amyloid production and improve cognitive impairments. *PloS One*, *11*, e0151147.
Li, Y. S., Hong, Y. F., He, J., Lin, J. X., Shan, Y. L., Fu, D. Y., et al. (2013). Effects of magnolol on impairment of learning and memory abilities induced by scopolamine in mice. *Biological and Pharmaceutical Bulletin*, *36*, 764–771.
Li, M., & Qian, S. (2016). Gastrodin protects neural progenitor cells against amyloid β (1–42)-induced neurotoxicity and improves hippocampal neurogenesis in amyloid β (1–42)-injected Mice. *Journal of Molecular Neuroscience*, *60*, 21–32.
Li, Y., & Zhang, Z. (2015). Gastrodin improves cognitive dysfunction and decreases oxidative stress in vascular dementia rats induced by chronic ischemia. *International Journal of Clinical and Experimental Pathology*, *8*, 14099–14109.
Li, L., Zhang, L., & Yang, C. C. (2016). Multi-target strategy and experimental studies of traditional Chinese medicine for Alzheimer's disease therapy. *Current Topics in Medicinal Chemistry*, *16*, 537–548.
Li, L. Y., Zhao, X. L., Fei, X. F., Gu, Z. L., Qin, Z. H., & Liang, Z. Q. (2008). Bilobalide inhibits 6-OHDA-induced activation of NF-κB and loss of dopaminergic neurons in rat substantia nigra. *Acta Pharmacologica Sinica*, *29*, 539–547.
Li, L., Zhou, Q.-X., & Shi, J.-S. (2005). Protective effects of icariin on neurons injured by cerebral ischemia/reperfusion. *Chinese Medical Journal*, *118*, 1637–1643.
Liang, X.-Y., Li, H.-N., Yang, X.-Y., Zhou, W.-Y., Niu, J.-G., & Chen, B.-D. (2013). Effect of Danshen aqueous extract on serum hs-CRP, IL-8, IL-10, TNF-α levels, and IL-10 mRNA, TNF-α mRNA expression levels, cerebral TGF-β1 positive expression level and its neuroprotective mechanisms in CIR rats. *Molecular Biology Reports*, *40*, 3419–3427.
Limón, I. D., Mendieta, L., Díaz, A., Chamorro, G., Espinosa, B., Zenteno, E., et al. (2009). Neuroprotective effect of alpha-asarone on spatial memory and nitric oxide levels in rats injected with amyloid-β (25-35). *Neuroscience Letters*, *453*, 98–103.
Lin, Z., Gu, J., Xiu, J., Mi, T., Dong, J., & Tiwari, J. K. (2012). Traditional Chinese medicine for senile dementia. *Evidence-Based Complementary and Alternative Medicine*, *2012*, 692621.
Liu, M., Guo, H., Li, C., Wang, D., Wu, J., Wang, C., et al. (2015). Cognitive improvement of compound danshen in an Aβ25-35 peptide-induced rat model of Alzheimer's disease. *BMC Complementary and Alternative Medicine*, *15*, 382.
Liu, Q. F., Lee, J. H., Kim, Y.-M., Lee, S., Hong, Y. K., Hwang, S., et al. (2015). In vivo screening of traditional medicinal plants for neuroprotective activity against Aβ42 cytotoxicity by using *Drosophila* models of Alzheimer's disease. *Biological and Pharmaceutical Bulletin*, *38*, 1891–1901.
Lo, Y. C., Teng, C. M., Chen, C. F., Chen, C. C., & Hong, C. Y. (1994). Magnolol and honokiol isolated from *Magnolia officinalis* protect rat heart mitochondria against lipid peroxidation. *Biochemical Pharmacology*, *47*, 549–553.
Ma, H., He, X., Yang, Y., Li, M., Hao, D., & Jia, Z. (2011). The genus *Epimedium*: An ethnopharmacological and phytochemical review. *Journal of Ethnopharmacology*, *134*, 519–541.
Manikandan, S., & Devi, R. S. (2005). Antioxidant property of α-asarone against noise-stress-induced changes in different regions of rat brain. *Pharmacological Research*, *52*, 467–474.
May, B. H., Feng, M., Zhou, I. W., Chang, S.-Y., Lu, S.-C., Zhang, A. L., et al. (2016). Memory impairment, dementia, and Alzheimer's disease in classical and contemporary traditional Chinese medicine. *Journal of Alternative and Complementary Medicine (New York, N.Y.)*, *22*, 695–705.

May, B. H., Yang, A. W., Zhang, A. L., Owens, M. D., Bennett, L., Head, R., et al. (2009). Chinese herbal medicine for mild cognitive impairment and age associated memory impairment: A review of randomised controlled trials. *Biogerontology*, *10*, 109–123.

Mishra, M., Huang, J., Lee, Y. Y., Chua, D. S. K., Lin, X., Hu, J.-M., et al. (2011). *Gastrodia elata* modulates amyloid precursor protein cleavage and cognitive functions in mice. *Bioscience Trends*, *5*, 129–138.

Mukherjee, P. K., Kumar, V., Mal, M., & Houghton, P. J. (2007). *Acorus calamus*: Scientific validation of Ayurvedic tradition from natural resources. *Pharmaceutical Biology*, *45*, 651–666.

Nasab, N. M., Bahrammi, M. A., Nikpour, M. R. A., Rahim, F., & Naghibi, S. N. (2012). Efficacy of rivastigmine in comparison to ginkgo for treating Alzheimer's dementia. *Journal of the Pakistan Medical Association*, *62*, 677–680.

Noorbala, A., Akhondzadeh, S., Tahmacebi-Pour, N., & Jamshidi, A. (2005). Hydro-alcoholic extract of *Crocus sativus* L. versus fluoxetine in the treatment of mild to moderate depression: A double-blind, randomized pilot trial. *Journal of Ethnopharmacology*, *97*, 281–284.

Nourbala, A., Tahmasebipour, N., Akhoundzadeh, S., Khani, M., & Jamshidi, A. (2004). *Crocus sativus* L. in the treatment of mild to moderate depression: A double-blind, randomised and placebo controlled trial. *Journal of Medicinal Plants*, *3*, 31–38.

Oh, M., Houghton, P., Whang, W., & Cho, J. (2004). Screening of Korean herbal medicines used to improve cognitive function for anti-cholinesterase activity. *Phytomedicine*, *11*, 544–548.

Pages, N., Maurois, P., Delplanque, B., Bac, P., Stables, J. P., Tamariz, J., et al. (2010). Activities of α-asarone in various animal seizure models and in biochemical assays might be essentially accounted for by antioxidant properties. *Neuroscience Research*, *68*, 337–344.

Park, C. H., Choi, S. H., Koo, J. W., Seo, J. H., Kim, H. S., Jeong, S. J., et al. (2002). Novel cognitive improving and neuroprotective activities of *Polygala tenuifolia* Willdenow extract, BT-11. *Journal of Neuroscience Research*, *70*, 484–492.

Park, E., Kang, M., Oh, J.-W., Jung, M., Park, C., Cho, C., et al. (2005). Yukmijihwang-tang derivatives enhance cognitive processing in normal young adults: A double-blinded, placebo-controlled trial. *The American Journal of Chinese Medicine*, *33*, 107–115.

Park, H. J., Lee, K., Heo, H., Lee, M., Kim, J. W., Whang, W. W., et al. (2008). Effects of *Polygala tenuifolia* root extract on proliferation of neural stem cells in the hippocampal CA1 region. *Phytotherapy Research*, *22*, 1324–1329.

Park, Y.-M., Lee, B.-G., Park, S.-H., Oh, H.-G., Kang, Y.-G., Kim, O.-J., et al. (2015). Prolonged oral administration of *Gastrodia elata* extract improves spatial learning and memory of scopolamine-treated rats. *Laboratory Animal Research*, *31*, 69–77.

Perry, E., & Howes, M.-J. R. (2011). Medicinal plants and dementia therapy: Herbal hopes for brain aging? *CNS Neuroscience & Therapeutics*, *17*, 683–698.

Perry, N. S. L., Howes, M.-J. R., Houghton, P. J., & Perry, E. (2000). Why sage may be a wise remedy: Effects of *Salvia* on the nervous system. In S. E. Kintzios (Ed.), *Sage: The genus Salvia* (pp. 207–223). Amsterdam: Harwood Academic Publishers.

Qian, Y.-H., Xiao, Q., & Xu, J. (2012). The protective effects of tanshinone IIA on β-amyloid protein (1–42)-induced cytotoxicity via activation of the Bcl-xL pathway in neuron. *Brain Research Bulletin*, *88*, 354–358.

Radad, K., Gille, G., Liu, L., & Rausch, W.-D. (2006). Use of ginseng in medicine with emphasis on neurodegenerative disorders. *Journal of Pharmacological Sciences*, *100*, 175–186.

Rafii, M., Walsh, S., Little, J., Behan, K., Reynolds, B., Ward, C., et al. (2011). A phase II trial of huperzine A in mild to moderate Alzheimer disease. *Neurology*, *76*, 1389–1394.

Ren, Y., Houghton, P. J., Hider, R. C., & Howes, M.-J. R. (2004). Novel diterpenoid acetylcholinesterase inhibitors from *Salvia miltiorhiza*. *Planta Medica*, *70*, 201–204.

Shao, Z.-Q. (2015). Comparison of the efficacy of four cholinesterase inhibitors in combination with memantine for the treatment of Alzheimer's disease. *International Journal of Clinical and Experimental Medicine*, *8*, 2944–2948.
Shi, C., Zhao, L., Zhu, B., Li, Q., Yew, D. T., Yao, Z., et al. (2009). Protective effects of *Ginkgo biloba* extract (EGb761) and its constituents quercetin and ginkgolide B against β-amyloid peptide-induced toxicity in SH-SY5Y cells. *Chemico-Biological Interactions*, *181*, 115–123.
Shin, K. Y., Lee, J.-Y., Won, B. Y., Jung, H. Y., Chang, K.-A., Koppula, S., et al. (2009). BT-11 is effective for enhancing cognitive functions in the elderly humans. *Neuroscience Letters*, *465*, 157–159.
Shu, Y.-Z. (1998). Recent natural products based drug development: A pharmaceutical industry perspective. *Journal of Natural Products*, *61*, 1053–1071.
Small, G. W., Rabins, P. V., Barry, P. P., Buckholtz, N. S., DeKosky, S. T., Ferris, S. H., et al. (1997). Diagnosis and treatment of Alzheimer disease and related disorders: Consensus statement of the American Association for Geriatric Psychiatry, the Alzheimer's Association, and the American Geriatrics Society. *Journal of the American Medical Association*, *278*, 1363–1371.
Steiner, G. Z., Yeung, A., Camfield, D. A., Pipingas, A., Scholey, A. B., Stough, C., et al. (2015). A pilot study on Sailuotong, a Chinese herbal medicine formula, for cognition and brain activity. *Advances in Integrative Medicine*, *2*, 115.
Steiner, G. Z., Yeung, A., Liu, J.-X., Camfield, D. A., De Blasio, F. M., Pipingas, A., et al. (2016). The effect of Sailuotong (SLT) on neurocognitive and cardiovascular function in healthy adults: A randomised, double-blind, placebo controlled crossover pilot trial. *BMC Complementary and Alternative Medicine*, *16*, 15.
Su, Y., Wang, Q., Wang, C., Chan, K., Sun, Y., & Kuang, H. (2014). The treatment of Alzheimer's disease using Chinese medicinal plants: From disease models to potential clinical applications. *Journal of Ethnopharmacology*, *152*, 403–423.
Tang, D., & Xun, J. (2003). *Science of Chinese materia medica*. Shanghai: Publishing House of Shanghai University of Traditional Chinese Medicine.
Tchantchou, F., Lacor, P. N., Cao, Z., Lao, L., Hou, Y., Cui, C., et al. (2009). Stimulation of neurogenesis and synaptogenesis by bilobalide and quercetin via common final pathway in hippocampal neurons. *Journal of Alzheimer's Disease: JAD*, *18*, 787–798.
Urano, T., & Tohda, C. (2010). Icariin improves memory impairment in Alzheimer's disease model mice (5xFAD) and attenuates amyloid β-induced neurite atrophy. *Phytotherapy Research*, *24*, 1658–1663.
Ved, H. S., Koenig, M. L., Dave, J. R., & Doctor, B. P. (1997). Huperzine A, a potential therapeutic agent for dementia, reduces neuronal cell death caused by glutamate. *Neuroreport*, *8*, 963–968.
Wang, Y.-B., Guo, L., Dou, Z.-F., & Wang, Y.-X. (2011). Discussion on the etiology and pathogenesis of the Alzheimer's disease in Chinese medicine. *Chinese Archives of Traditional Chinese Medicine*, *29*, 743–745.
Wang, J. P., Ho, T. F., Chang, L. C., & Chen, C. C. (1995). Anti-inflammatory effect of magnolol, isolated from *Magnolia officinalis*, on A23187-induced pleurisy in mice. *Journal of Pharmacy and Pharmacology*, *47*, 857–860.
Wang, B.-S., Wang, H., Song, Y.-Y., Qi, H., Rong, Z.-X., Zhang, L., et al. (2010). Effectiveness of standardized *Ginkgo biloba* extract on cognitive symptoms of dementia with a six-month treatment: A bivariate random effect meta-analysis. *Pharmacopsychiatry*, *43*, 86–91.
Wang, Z., Zhang, X., Wang, H., Qi, L., & Lou, Y. (2007). Neuroprotective effects of icaritin against beta amyloid-induced neurotoxicity in primary cultured rat neuronal cells via estrogen-dependent pathway. *Neuroscience*, *145*, 911–922.

Wang, X., Zhang, J., Yang, H., & Hu, G. (1999). Modulation of NMDA receptor by huperzine A in rat cerebral cortex. *Acta Pharmacologica Sinica, 20*, 31–35.

Williams, P., Sorribas, A., & Howes, M.-J. R. (2011). Natural products as a source of Alzheimer's drug leads. *Natural Product Reports, 28*, 48–77.

Williamson, E. M. (2009). Synergy in relation to the pharmacological action of phytomedicinals. In W. C. Evans (Ed.), *Trease and Evans pharmacognosy* (16th ed., pp. 53–61). London: Elsevier.

Wong, K. K.-K., Ho, M. T.-W., Lin, H. Q., Lau, K.-F., Rudd, J. A., Chung, R. C.-K., et al. (2010). Cryptotanshinone, an acetylcholinesterase inhibitor from *Salvia miltiorrhiza*, ameliorates scopolamine-induced amnesia in Morris water maze task. *Planta Medica, 76*, 228–234.

Xia, Z., Zhang, R., Wu, P., Xia, Z., & Hu, Y. (2012). Memory defect induced by β-amyloid plus glutamate receptor agonist is alleviated by catalpol and donepezil through different mechanisms. *Brain Research, 1441*, 27–37.

Xian, Y.-F., Ip, S.-P., Mao, Q.-Q., & Lin, Z.-X. (2016). Neuroprotective effects of honokiol against β-amyloid-induced neurotoxicity via GSK-3β and β-catenin signaling pathway in PC12 cells. *Neurochemistry International, 97*, 8–14.

Xian, Y.-F., Ip, S.-P., Mao, Q.-Q., Su, Z. R., Chen, J. N., Lai, X. P., et al. (2015). Honokiol improves learning and memory impairments induced by scopolamine in mice. *European Journal of Pharmacology, 760*, 88–95.

Xiao, X. Q., Yang, J. W., & Tang, X. C. (1999). Huperzine A protects rat pheochromocytoma cells against hydrogen peroxide-induced injury. *Neuroscience Letters, 275*, 73–76.

Xiao, X. Q., Zhang, H. Y., & Tang, X. C. (2002). Huperzine A attenuates amyloid β-peptide fragment 25-35-induced apoptosis in rat cortical neurons via inhibiting reactive oxygen species formation and caspase-3 activation. *Journal of Neuroscience Research, 67*, 30–36.

Xie, J., & Qu, R. (2008). Symptoms and pathogenesis of senile dementia in ancient Chinese medical literature Li Shi Zhen. *Medicine and Materia Medica Research, 19*, 3073–3074.

Xing, S. H., Zhu, C. X., Zhang, R., & An, L. (2014). Huperzine A in the treatment of Alzheimer's disease and vascular dementia: A meta-analysis. *Evidence-Based Complementary and Alternative Medicine, 2014*, 363985.

Xu, R. X., Wu, Q., Luo, Y., Gong, Q. H., Yu, L. M., Huang, X. N., et al. (2009). Protective effects of icariin on cognitive deficits induced by chronic cerebral hypoperfusion in rats. *Clinical and Experimental Pharmacology and Physiology, 36*, 810–815.

Yang, G., Wang, Y., Sun, J., Zhang, K., & Liu, J. (2016). *Ginkgo biloba* for mild cognitive impairment and Alzheimer's disease: A systematic review and meta-analysis of randomized controlled trials. *Current Topics in Medicinal Chemistry, 16*, 520–528.

Yang, G., Wang, Y., Tian, J., & Liu, J.-P. (2013). Huperzine A for Alzheimer's disease: A systematic review and meta-analysis of randomized clinical trials. *PloS One, 8*, e74916.

Yin, C., Deng, Y., Gao, J., Li, X., Liu, Y., & Gong, Q. (2016). Icariside II, a novel phosphodiesterase-5 inhibitor, attenuates streptozotocin-induced cognitive deficits in rats. *Neuroscience, 328*, 69–79.

Yu, H., Yao, L., Zhou, H., Qu, S., Zeng, X., Zhou, D., et al. (2014). Neuroprotection against Aβ 25-35-induced apoptosis by *Salvia miltiorrhiza* extract in SH-SY5Y cells. *Neurochemistry International, 75*, 89–95.

Yue, J., Dong, B. R., Lin, X., Yang, M., Wu, H. M., & Wu, T. (2012). Huperzine A for mild cognitive impairment. *Cochrane Database of Systematic Reviews, 12*, CD008827.

Zhang, L.-N., Jin, G.-Q., Zhang, X.-L., Gong, Z.-B., & Gu, C.-Y. (2014). Effects of 5-hydroxymethyl furfural extracted from *Rehmannia glutinosa* Libosch on the expression of signaling molecules relevant to learning and memory among hippocampal neurons exposed to high concentration of corticosterone. *Chinese Journal of Integrative Medicine, 20*, 844–849.

Zhang, Y., Lin, C., Zhang, L., Cui, Y., Gu, Y., Guo, J., et al. (2015). Cognitive improvement during treatment for mild Alzheimer's disease with a Chinese herbal formula: A randomized controlled trial. *PloS One*, *10*, e0130353.

Zhang, X., Oh, M., Kim, S., Kim, J., Kim, H., Kim, S., et al. (2013). Epimediphine, a novel alkaloid from *Epimedium koreanum* inhibits acetylcholinesterase. *Natural Product Research*, *27*, 1067–1074.

Zhang, X.-Z., Qian, S.-S., Zhang, Y.-J., & Wang, R.-Q. (2016). *Salvia miltiorrhiza*: A source for anti-Alzheimer's disease drugs. *Pharmaceutical Biology*, *54*, 18–24.

Zhang, D., Wang, Z., Sheng, C., Peng, W., Hui, S., Gong, W., et al. (2015). Icariin prevents amyloid β-induced apoptosis via the PI3K/Akt pathway in PC-12 cells. *Evidence-Based Complementary and Alternative Medicine*, *2015*, 235265.

Zhang, X., Zhang, A., Jiang, B., Bao, Y., Wang, J., & An, L. (2008). Further pharmacological evidence of the neuroprotective effect of catalpol from *Rehmannia glutinosa*. *Phytomedicine*, *15*, 484–490.

Zhang, J.-S., Zhou, S.-F., Wang, Q., Guo, J.-N., Liang, H.-M., Deng, J.-B., et al. (2016). Gastrodin suppresses BACE1 expression under oxidative stress condition via inhibition of the PKR/eIF2α pathway in Alzheimer's disease. *Neuroscience*, *325*, 1–9.

Zhao, G. P., Dai, S., & Chen, R. S. (2006). In *Zhong Yao Da Ci Dian: Vol. 2*. Shanghai: Shanghai Scientific & Technical Publishers.

Zhou, X., Chan, K., & Yeung, J. H. (2012). Herb-drug interactions with Danshen (*Salvia miltiorrhiza*): A review on the role of cytochrome P450 enzymes. *Drug Metabolism and Drug Interactions*, *27*, 9–18.

Zhou, Y., & Xu, R. (1992). Antioxidative effect of Chinese drugs. *China Journal of Chinese Materia Medica*, *17*, 368–369.

Zong, X., Ji, X. M., Wei, F. Q., & Shi, Z. R. (2014). Analysis on prescription rules of treating senile dementia based on traditional Chinese medicine inheritance auxiliary systems. *China Journal of Chinese Materia Medica*, *39*, 640–643.

CHAPTER THREE

# Effect and Mechanism of Chinese Herbal Medicine on Parkinson's Disease

**Bai-Yun Zeng**[1]
Neurodegenerative Disease Research Group, Institute of Pharmaceutical Science, Faculty of Life Science & Medicine, King's College, London, United Kingdom
[1]Corresponding author: e-mail address: b.zeng@kcl.ac.uk

## Contents



## Abstract

Parkinson's disease is a progressive neurodegenerative disorder. Although both genetic and environmental factors are implicated in the development of Parkinson's disease, the cause of the disease is still unclear. So far conventional treatments to Parkinson's are symptomatic relief and focused mainly on motor symptoms. Chinese herbal medicine has been used to treat many conditions in China, Korea, Japan, and many Southeast Asian countries for 1000 years. During past a few decades, Chinese herbal medicine has gained wider and increasing acceptance within both public and medical profession due to its effectiveness on many conditions in western countries. In this chapter, mechanisms of action of many Chinese herbal compounds/extracts and Chinese herb formulas on the models of Parkinson's were reviewed. Further, reports of effectiveness of Chinese herb formulas on patients with Parkinson's were summarized. It was shown that both Chinese herbal compounds/extracts and herb formulas have either specific target mechanisms of action or multitargets mechanisms of action, as antioxidant,

*International Review of Neurobiology*, Volume 135
ISSN 0074-7742
http://dx.doi.org/10.1016/bs.irn.2017.02.004

antiinflammatory, and antiapoptosis agents. Clinical studies showed that Chinese herb formulas as an adjunct improved both motor and nonmotor symptoms, and reduced dose of dopaminergic drugs and occurrence of dyskinesia. The evidence from the studies suggests that Chinese herb medicine has potential, acting as neuroprotective to slow down the progression of Parkinson's, and it is able to simultaneously treat both motor and nonmotor symptoms of Parkinson's. More studies are needed to explore the new compounds/extracts derived from Chinese herbs, in particular, their mechanisms of action. It is hopeful that new drugs developed from Chinese herb compounds/extracts and Chinese herb formulas will lead to better and complimentary therapy to PD.

## 1. INTRODUCTION

Parkinson's disease (PD) is a progressive neurodegenerative disorder. Although both environmental and genetic factors are linked with the development of PD, the cause of disease is still unknown. The motor symptoms of PD are characterized by slowness of movement, resting tremor, rigidity of muscles, and joints. The severe loss of dopaminergic neurons in the substance nigra of middle brain and subsequent depletion of dopamine in the striatum are often associated with development of motor symptoms (Hornykiewicz, 2001; Obeso et al., 2008). The nonmotor symptoms of PD are found in the vast majority of patients with PD, consisting of neuropsychiatric disturbance, autonomic dysfunction, sleep disorders, gastrointestinal symptoms, and many others (Chaudhuri, Healy, & Schapira, 2006; Chaudhuri, Odin, Antonini, & Martinez-Martin, 2011). The nonmotor symptoms of PD and related disorders usually respond poorly to dopaminergic treatment. This implies, in addition to the loss of dopamine, the changes in other neurotransmitters, such as norepinephrine (NE), serotonin (5-HT), glutamate, and gamma-aminobutyric acid, in the different brain regions and peripheral nervous systems, are implicated in the pathogenesis of many nonmotor symptoms (Ferrer, 2011). The pathogenesis of PD is thought to be involved in oxidative stress, impaired mitochondria function, inflammation, apoptosis, and dysfunction of proteolysis and loss of neurotrophic factors (Hirsch, Jenner, & Przedborski, 2013). There is currently no cure for PD. The prevalence medication for motor symptoms, such as dopamine replacement therapy, in particular levodopa treatment, is only symptomatic relief with limited effect, but has many adverse effects, such as hallucination and involuntary movement (Connolly & Lang, 2014; Olanow & Schapira, 2013).

Chinese herbal medicine (CHM) is a fundamental part of Chinese medicine, which is one of the most ancient and still largely used traditional medicines in the world. Herbal products have been used for PD treatment worldwide in traditional medicine (Manyam & Sanchez-Ramos, 1999). In China, records of herbal medicine for treating PD dated back to about 2200 years ago (Zheng, 2009), and recent literature reports reflected a continuous and active use of CHM for the treatment of PD (Li, Zhao, & Bezard, 2006; Zhang & Liu, 2006). During past a few decades, biological activities and potential molecular targets of many Chinese herbal extracts have been discovered (Li, Zhang, Liu, & Lu, 2013). Recently, the biological effects of active ingredients/extracts of many Chinese herbs are investigated on the models of PD (Li et al., 2013; Song et al., 2012). In this chapter, studies about mechanism of antiparkinsonian action of Chinese herbs and their extracts on PD models and clinical application of CHM in PD were summarized.

## 2. MECHANISM OF ACTION OF EXTRACTS/COMPOUNDS FROM CHINESE HERBS ON PARKINSON'S MODEL

Many extracts or components of Chinese herbs possess anti-Parkinson's potentials. However, they are reported to exert their therapeutic effect through different mechanisms of action (Li et al., 2013; Song et al., 2012).

### 2.1 Antioxidant

*Ginkgo biloba* [Yin xing (银杏)] extract was widely used to treat many conditions for a long time (Nash & Shah, 2015). Pretreatment of *G. biloba* extract to 6-hydroxydopamine (6-OHDA)-induced semiparkinsonian rats (50, 100, or 150 mg/kg) daily for 3 weeks, dose dependently attenuated 6-OHDA-induced motor deficits, improved muscular coordination, neutralized the increased generation of thiobarbituric acid reactive substances, byproduct of lipid peroxidation, and depletion of reduced glutathione (GSH), and increased tyrosine hydroxylase-immunoreactive (TH-IR) density in the lesioned substantia nigra (SN) (Ahmad, Saleem, Ahmad, Yousuf, et al., 2005). Further, *G. biloba* extract pretreatment dose dependently restored the activities of GSH-dependent enzymes, catalase, and superoxide dismutase (SOD), and levels of dopamine D2 receptor expression in the striatum, which had been significantly altered by toxin (Ahmad, Saleem, Ahmad, Yousuf, et al., 2005). Similarly, pretreatment of *Withania somnifera* [Shui jia (睡茄)] extract showed markedly antioxidant property in

6-OHDA-treated rats (Ahmad, Saleem, Ahmad, Ansari, et al., 2005). In 1-methyl-4-phenyl-1,2,3,6-tetrahydropyridine (MPTP)-treated mice, treatment with EGb761, a standardized formulation of *G. biloba*, for 16 days significantly improved MPTP-induced motor impairment, markedly attenuated MPTP-induced loss of striatal dopamine levels and TH-IR density in the striatum and SN. The neuroprotective effect of EGb761 against MPTP neurotoxicity was associated with reduction of superoxide radical production and blockade of lipid peroxidation (Rojas et al., 2001, 2004, 2008). In another study, coadministration of *Centella asiatica* [Ma ti cao (马蹄草)] extract with MPTP to rats significantly restored the changes caused by MPTP toxicity in oxidative biomarkers such as lipid peroxidase, total antioxidants enzyme levels in the striatum and hippocampus compared with MPTP-treated group (Haleagrahara & Ponnusamy, 2010). Similarly, cotreatment of *Rhizoma curcumae longae* [Jiang huang (姜黄)] with MPTP to mice markedly blocked MPTP-induced GSH depletion and lipid peroxidation, and increased in SOD and catalase expression in both striatum and SN (Rajeswari, 2006). Recently, studies showed that treatment with Quercetin extracted from *Quercus dentata* [Hu shu (槲树)] to 6-OHDA-lesioned rats significantly attenuated toxin-induced alterations in malondialdehyde (MDA) level, the activity of SOD, catalase, and glutathione peroxidase (GPx) compared with control groups (Sriraksa et al., 2012). Pretreatment of quercetin to MPTP-induced parkinsonian mice (50, 100, or 200 mg/kg) daily for 14 days dose dependently significantly improved motor balance and coordination in MPTP-treated mice and markedly reduced the level of 4-hydroxy-2-nonenal (4-HNE) in the striatum compared with MPTP alone treated group (Lv et al., 2012). Further, quercetin administration significantly increased activities of antioxidant markers such as GPx, SOD, and the content of dopamine in the brain (Lv et al., 2012). Administration of *Radix scutellariae* [Huang qin (黄芩)] to MPTP-treated mice (140, 280, or 560 mg/kg) daily for 7 days dose dependently significantly protected dopamine neurones against toxin and improved motor function, and this was accompanied by the restoring antioxidant biomarker expression in the striatum (Mu, He, Yuan, Li, & Du, 2011).

## 2.2 Antiapoptosis

Accumulated studies showed the antiapoptotic property of many Chinese herb extracts. Pretreatment of ginsenoside Rg1 from *Radix ginseng* [Ren shen (人参)] either 3 or 15 days to MPTP-treated mice increased TH-positive cells in the SN and decreased the TUNEL-positive ratio, enhanced the expression of antiapoptotic Bcl-2 protein and Bcl-2 mRNA, but reduced the expression of proapoptotic Bax, Bax mRNA, and inducible nitric oxide synthase (iNOS) protein compared with the MPTP alone group (Chen et al., 2005; Xu et al., 2005). In another study, pretreatment of echinacoside extracted from *Herba cistanches* [Rou cong rong (肉苁蓉)] for 15 days to MPTP mice protected TH-positive cells in the SN and inhibited dopamine depletion in the striatum, and improved motor function (Geng, Tian, Tu, & Pu, 2007). This was associated with the attenuation of MPTP-induced activation of proapoptotic caspase-3 and caspase-8 signaling pathways (Geng et al., 2007). Studies of pretreatment of *Fructus mori* [Sang shen (桑椹)] extract to MPTP-treated mouse showed that pretreatment significantly improved MPTP-induced motor deficit, reduced the loss of TH-positive cells in the SN, and attenuated the decrease in TH-positive density in the striatum compared with MPTP alone group (Kim et al., 2010). In vitro study, pretreatment of mulberry fruit extract significantly protected SHSY5Y cells by neutralizing toxin-induced apoptotic signaling pathway (Kim et al., 2010). Pretreatment with *Radix notoginseng* [San qi (三七)] extract markedly attenuated the decrease in TH expression, modulated expression of proapoptotic signaling proteins procaspase-9, procaspase-12, and caspase-3 compared with control group; in addition, it enhanced thioredoxin-1 expression, an antioxidative stress molecule and suppressed proinflammatory enzyme, cyclooxygenase-2 (COX-2) expression (Luo et al., 2011). Tetramethylpyrazine (TMP) is an alkaloid extracted from *Radix chuanxiong* [Chuan xiong (川芎)], a Chinese medicinal herb, has been extensively used to treat heart, kidney, and brain diseases in Asian countries for hundreds of years (Tan, 2009). Administration of tetramethylpyrazine bis-nitrone (TN-2), a TMP bis-nitrone derivative to MPTP-treated mice (20, 40, or 80 mg/kg) daily for 14 days dose dependently prevented the loss of TH-positive nigral cells and attenuated striatal dopamine depletion compared with control group (Xu, Duan, et al., 2014; Xu, Zhang, et al., 2014). Further, TN-2 modulated dysregulation of Bax and Bcl-2 expression and release of cytochrome *c* from the mitochondria to the cytoplasm, and activation of caspase proteins (Xu, Duan, et al., 2014; Xu, Zhang, et al., 2014).

## 2.3 Antiinflammation

Administration of esveratrol extracted from *Rhizoma polygoni cuspidati* [Hu zhang (虎杖)] to 6-OHDA-induced hemiparkinsonian rats (10, 20, or 40 mg/kg) daily for 10 weeks dose dependently reduced

6-OHDA-induced rotations, neutralized elevated expression of proinflammatory cytokines, tumor necrosis factor alpha (TNF-α), and proinflammatory enzyme COX-2; further, ultrastructural analysis showed that esveratrol improved 6-OHDA-induced mitochondrial tumefaction, chromatin condensation, and vacuolization of dopaminergic neurons in rat SN (Jin, Wu, Lu, Gong, & Shi, 2008). In another study, treatment of baicalein, from *R. scutellariae* [Huang qin (黄芩)] with 6-OHDA-induced hemiparkinsonian rats markedly reduced 6-OHDA-induced muscle tremor, decreased the number of glial fibrillary acidic protein-positive astrocytes, and reduced the loss of TH-positive neurones in the SN, although less effective on rotational behavioral test compared with control group (Mu et al., 2009). Cotreatment of tanshinone I, one of compounds from *Radix salviae miltiorrhizae* [Dan shen (丹参)] with MPTP to mice showed that tanshinone I modulated MPTP-induced microglial activation and the expressions of cytokines and attenuated the increase of TNF-α, in the SN and striatum of MPTP-intoxicated mice compared with control group (Xu, Duan, et al., 2014; Xu, Zhang, et al., 2014).

## 2.4 Monoamine Oxidase-B Inhibition

In C57 black mice, $MPP^+$ treatment increased monoamine oxidase (MAO) activity in the striatum by 30% 6 h after treatment; pretreatment with *G. biloba* [Yin xing (银杏)] extract EGb761 to mice daily for 17 days attenuated $MPP^+$-induced increased MAO activity and partially prevented dopamine depletion (Rojas et al., 2004). Similarly, *Fructus forsythiae* [Lian qiao (连翘)] extract decreased striatal MAO-B activity compared with control group (Mohanasundari & Sabesan, 2007). In another study, administration of methanolic extract of *Semen hyoscyami* [Tian xian zi (天仙子)] to MPTP-induced parkinsonian mice for 2 days markedly improved motor deficits such as akinesia, catalepsy, and reduction in swim score, and decreased striatal dopamine loss in MPTP-treated mice compared with control group (Sengupta et al., 2011). Further, *S. hyoscyami* [Tian xian zi (天仙子)] extract significantly reduced striatal MAO-B activity and hydroxyl radical ($^{\bullet}OH$) generation in isolated mitochondria caused by MPTP toxicity (Sengupta et al., 2011).

## 2.5 Antidyskinesia

Chronic administration of L-3,4-dihydroxyphenylalanine (L-DOPA) to 6-OHDA-treated hemiparkinsonian rats led to the development of abnormal involuntary movements, AIMs (Bido, Marti, & Morari, 2011).

L-DOPA-induced rodent AIMs is equivalent to L-DOPA-induced dyskinesia (LID) in PD patients, which is a severe side effect of L-DOPA therapy to majority of patients with PD after 3–5-year medication (Stocchi, Jenner, & Obeso, 2010). L-Stepholidine (L-SPD) is an active compound from *Stephania intermedia* [Qian jin teng (千金藤)] (Jin & Sun, 1995). It has been reported that treatment of L-SPD to parkinsonian rats induced stable contralateral rotational behavior without eliciting significant AIMs (Mo et al., 2010). Repeated coadministration of L-SPD with L-DOPA significantly ameliorated AIMs without compromising the therapeutic potency of L-DOPA, indicating that L-SPD attenuated LID development; this was accompanied by restoring expression of the mRNA level of 5-HT1A receptor mRNA and adenosine A2a receptor mRNA on the lesioned striatum (Mo et al., 2010). This suggested that antidyskinetic effect of L-SPD might be mediated by regulation of 5-HT1A receptor and A2a receptor mRNA expression. In LID mouse model, coadministration of *Rhizoma gastrodiae* [Tian ma (天麻)] extract with L-DOPA to LID mouse (200, 400, or 800 mg/kg) daily for 10 days dose dependently ameliorated AIMs compared with control group, and this was associated with the attenuation of L-DOPA-induced increase of pERK1/2 and FosB, the immediate early genes in the striatum, that were thought to be involved in the development of LID (Doo et al., 2014).

## 2.6 Multiple Targets

Because of the complex nature of PD, it may be advantageous or even essential that therapeutics for PD treatment acts through multiple mechanisms of action, i.e., by targeting multiple pathological pathways at the same time.

It has been shown that neuroprotective effects exerted by TN-2 were not only via inhibition of mitochondrial-dependent apoptotic pathway, elevation of SOD activity, and GSH levels but also via acting on some key steps of mitochondrial-dependent apoptotic pathway, including overproduction of intracellular reactive oxygen species (ROS) in MPTP mice (Xu, Duan, et al., 2014; Xu, Zhang, et al., 2014). It was shown, in vitro, that TN-2 decreased excessive intracellular ROS, prevented the loss of mitochondrial membrane potential, blocked the release of mitochondrial cytochrome *c*, and inhibited the activation of caspase-3 and caspase-9 (Xu, Duan, et al., 2014; Xu, Zhang, et al., 2014). Neuroprotective effects of *G. biloba* [Yin xing (银杏)] extract, e.g., EGb761 on dopamine neurones in Parkinson's models appeared to act via antioxidant, free radical scavenging, MAO-B-inhibiting, and modulated expression of dopamine-related genes (Rojas, Montes, Rojas, Serrano-García, & Rojas-Castañeda, 2012; Rojas,

Ruiz-Sánchez, Rojas, & Ogren, 2012; Tanaka, Galduróz, Gobbi, & Galduróz, 2013). It was reported that Resveratrol of *Rhizoma polygoni cuspidati* [Hu zhang (虎杖)] protected dopamine neurons against toxic substance in Parkinson's models through antiinflammation, antioxidant, and antiapoptosis (Blanchet et al., 2008; Jin et al., 2008; Lu et al., 2008). Similarly, Baicalein extracted from *R. scutellariae* [Huang qin (黄芩)] counteracted toxic effect on nigral–striatal neuronal pathway via similar mechanisms of action (Mu et al., 2009, 2011).

In the rat model of early phase of parkinsonism induced by bilateral infusion of MPTP into the SN, intragastrical administration of Madecassoside isolated from the *C. asiatica* [Ma ti cao (马蹄草)] to the rats improved motor function and significantly protected nigral dopaminergic neurons against toxicity and attenuated MPTP-induced reduction of dopamine in the striatum. Madecassoside treatment markedly decreased MDA contents while the GSH levels, Bcl-2/Bax ratio, and protein expression of BDNF were significantly increased compared with MPTP alone group (Xu, Qu, Zhang, Li, & Ma, 2013). This indicated that Madecassoside exerted its neuroprotective effect via its antioxidant activity, activating antiapoptotic pathway, increasing release of neurotrophic factor, and reversing the depletion of striatal DA.

In the rotenone-induced semiparkinsonian rats *F. forsythiae* [Lian qiao (连翘)] suspension extract FS8 dose dependently protected dopaminergic neurons against rotenone toxicity in the lesioned SN, markedly improved motor function, significantly decreased the expression of proinflammatory molecules (IL-6, TNF-α, iNOS, and COX-2), and downregulated the protein expression of proapoptotic molecules (p-PI3K, p-Akt, p-IkB, p-P65, cleaved caspase 8, and p-p38). This implied that FS8 protected dopamine neurons against rotenone toxicity via antioxidant and antiinflammatory effects (Zhang et al., 2016).

Studies summarized earlier showed that many compounds or extracts of Chinese herbs possess potential properties to act on different targets related to the causes of the death of dopamine neurones. Further, some compounds act on multitargets to protect dopamine neurons against different toxic effects and this makes them as ideal candidates of new drug development for PD.

## 3. MECHANISM OF ACTION OF CHINESE HERBAL FORMULA ON PARKINSON'S MODEL

Since the pathogenesis of Parkinson's appears to be multiple factor-derived, therapies based on single drug are unlikely to provide measurable disease modification or neuroprotection in Parkinson's. A combination of compounds, such as mixed herbs within Chinese herbal formulations that simultaneously interfere with different key events of PD pathology, seems to be a more promising strategy. Further, different compounds within the formulations are supposed to mutually enhance each other's effect more significantly than the simple sum of these ingredients (Ma et al., 2009). However, little has been researched about biological and molecular mechanisms of action of Chinese herbal formulas.

Zhichan tang (止颤汤) is composed of *Radix astragali* [Huang qi (黄芪)], *Radix paeoniae alba* [Bai shao (白芍)], *Radix salviae* [Dan shen (丹参)], prepared Rhubarb [Da huang (大黄)], *Uncaria rhynchophylla* [Guo teng (钩藤)], *Rhizoma anemarrhenae* [Zhi mu (知母)], *Rhizoma cimicifugae* [Sheng ma (升麻)], and *R. gastrodiae* [Tian ma (天麻)]. Zhichan decoction or its compounds called Zhichan powder has been used to effectively alleviate many symptoms of Parkinson's such as gait disturbance, tremor, speech, and writing disorder in China (Li et al., 2006; Yao, 2008). However, the mechanism of action of compounds is unclear. Recently, it has been shown that Zhichan powder, administered intragastrically twice a day to PD rat model induced by nigral infusion of 6-OHDA for 7 weeks significantly elevated mRNA expression of TH and antiapoptotic marker Bcl-2, and neutralized increased MAO-B, and proapoptotic markers such as caspase-3, caspase-8, Bax, and proinflammatory marker tumor necrosis factor receptor 1, Fas in the lesioned SN compared with controls (Chen et al., 2012; Zhou et al., 2012). In addition, Zhichan powder markedly increased levels of dopamine and its metabolites in the striatum and serum levels of SOD compared with controls (Zhou et al., 2012). This indicated that Zhichan powder exerted its therapeutic effect through regulate protein and gene expression of TH and signal transduction molecules, inhibit apoptosis and inflammation. In another study, coadministration of Zhichan decoction with neural stem cell transplantation into the 6-OHDA-lesioned SN improved motor function in semiparkinsonian rat model, and this was accompanied by the up to 10-fold increase in striatal dopamine metabolite, dihydroxyphenylacetic acid levels, compared with only threefold increase in transplantation only group (Shi, Song, & Yang, 2014). This suggested that Zhichan decoction significantly increased striatal dopamine content, indicating it may enhance differentiation and proliferation of cotransplantation and improve motor function.

Liuwei Dihuang decoction (LWDH) is a classic CHM formula consisting of *Radix rehmanniae* [Di huang (地黄)], *R. dioscoreae* [Shan yao (山药)], *Fructus corni* [Shan zhu yu (山茱萸)], *Cortex moutanradicis*

[Mu dan pi (牡丹皮)], *Rhizoma Alismatis* (Ze xie泽泻), and Poria [Fu ling (茯苓)]. LWDH has been prescribed as a primary or adjuvant treatment for many conditions, such as diabetics, neurosis, and dementia (van Wietmarschen, van der Greef, Schroen, & Wang, 2013; Zhou, Cheng, & Zhang, 2016). Recently, clinical studies were conducted to evaluate potential neuroprotective mechanisms of LWDH in Parkinson's model (Tseng, Chang, & Lo, 2014). It was shown that water extract of LWDH protected primary mesencephalic dopamine neurons against toxic substance $MPP^+$ via increasing antioxidant defense; reflected by the elevated expression of SOD and GSH; inhibited neuronal apoptosis pathway by blocking expression of Bax, cytochrome *c*, and caspase-3; and reduced production of ROS (Tseng et al., 2014). Further, in vivo studies showed that water exact of LWDH significantly reduced the loss of dopamine neurons in the SN of MPTP-treated parkinsonian mice and this was accompanied by the improved locomotor activity compared with control groups (Tseng et al., 2014).

Bushen Huoxue Yin (BSHXY, 补肾活血饮) is composed of *Herba cistanchis* [Rou Cong Rong (肉丛蓉)], *F. corni* [Shan zhu yu (山茱萸)], *Radix angelicae sinensis* [Dang Gui (当归)], *Radix polygoni multiflori* [He Shou Wu (何首乌)], *Radix paeoniae rubra* [Chi Shao (赤芍)], *R. chuanxiong* [Chuan Xiong (川芎)], Scolopendra [Wu Gong (蜈蚣)], and *Rhizoma acori graminei* [Shi Chang Pu (石菖蒲)]. BSHXY formula was used to study the mechanism of its antiparkinsonian action in Parkinson's models. In MPTP-treated mouse model, mice were intragastrically given BSHXY decoction 1 h before intraperitoneal MPTP injection, once a day for consecutive 7 days. It was found that in MPTP-treated alone group there was a significant increase in the nuclear transcription factor kappa B (NF-κB) and nitric oxide (NO) content and the levels of TNF-α and interferon-γ in the whole brain tissue; however, in BSHXY–MPTP-treated group the elevated brain NF-κB and NO content was markedly neutralized compared with MPTP alone group (Li, Li, et al., 2012; Li, Liu, & Yand, 2011, 2012). In other studies, intragastrical infusion of BSHXY to 6-OHDA-induced parkinsonian rats once a day for 8 consecutive weeks, significantly improved rotational behavioral changes induced by 6-OHDA and increased TH-positive cells and the level of Nurr1 mRNA expression in the lesioned SN (Yang, Wang, & Liu, 2011). Further, BSHXY administration increased dopamine D2 receptor positive neurons in the 6-OHDA-lesioned SN compared with 6-OHDA-lesion alone group (Wang et al., 2011). These results indicated that BSHXY decoction exerted its antiparkinsonian effect via a multi targets mechanism through promoting dopaminergic neuron survival, modulating dopamine receptor expression, and antiinflammatory activities.

Mi Fang Ding Zhen Wan (MFDZW, 定震丸) consists of *R. gastrodiae* [Tian ma (天麻)], *R. astragali* [Huang qi (黄芪)], *Radix rehmanniae praeparata* [Shu di huang (熟地黄)], and *Radix paeoniae alba* [Bai shao yao (白芍药)], and was used to assess its antidyskinetic effect on levodopa-induced AIMs in levodopa-treated parkinsonian rats. Intragastrical infusion of WFDZW to parkinsonian rats with levodopa-induced AIMs daily for 4 weeks significantly decreased levodopa-induced peak rotations compared with control groups (Wu, Yang, Song, Wei, & Liu, 2013). Moreover, WFDZW prevented the decreased expression of G protein-coupled receptor kinase 6 (GRK6) and β-arrestin1 in rat striatum compared with control group (Wu et al., 2013). As GRK6 and β-arrestin1 are closely associated with hyperactive expression of dopamine receptors which was believed to be one of the causes associated with development of AIMs following levodopa administration, WFDZW might indirectly modulate striatal levels of dopamine receptors leading to the reduction of the AIMs.

San Huang Xie Xin Tang (SHXXT, 三黄泻心汤) consists of *Rhizoma coptidis* [Huang lian (黄连)], *R. scutellariae* [Huang qin (黄芩)], and *Radix et rhizoma rhei* [Da huang (大黄)]. SHXXT has been reported to possess a variety of physiological and pharmacological properties, including anti-inflammatory, antioxidant, antiapoptotic, and neuroprotective effects (Hwang et al., 2015). Lo, Shih, Tseng, and Hsu (2012) investigated neuroprotective effects of SHXXT on PD models. In vitro study showed that SHXXT significantly increased the numbers of TH-positive neurons, via increasing the levels of GSH and SOD, reducing apoptotic signals markers such as cytochrome and caspase and ROS production in $MPP^+$-treated mesencephalic neuron cultures. Then, in MPTP-induced parkinsonian mouse model, SHXXT markedly increased TH-positive neurons in the SN pars compacta and improved motor activity of parkinsonian mice (Lo et al., 2012).

Taken together, reports summarized earlier showed that many Chinese herb formulas have potential to target the causes of dopamine neuronal degeneration. The studies shed the light on the mechanisms of action of those herb formulas.

## 4. CLINICAL STUDY OF CHM ON PARKINSON'S DISEASE

Many CHM formulas have been used to treat PD depending on their function in China for a very long time (Li et al., 2006). However, there were only a few reported clinical studies about the effectiveness and mechanism of action of the formulas on PD.

Di Huang Yin Zi (DHYZ, 地黄饮子) is composed of *Radix rehmanniae preparata* [Shu di huang (熟地黄)], *Placenta hominis* [Zi he che (紫河车)], *Radix salviae liguliobae* [Dan shen (丹参)], *Rhizoma acori talarinowii* [Shi chang pu (石菖蒲)], *Colla carapacis et plastri testudinis* [Gui ban jiao (龟板胶)], Poria [Fu shen (茯神)], and *Fructus alpiniae oxyphyllae* [Yi zhi ren (益智仁)]. DHYZ is a classical prescription for neurological disorders and was used to help recovery of patients with stroke and spinal cord injury (Li, Li, et al., 2012; Li, Liu, et al., 2012; Yu et al., 2015). Gu et al. (2015) conducted a randomized, double-blinded study to assess combined therapeutic effect of DHYZ plus Donepezil in patients with PD dementia. The patients with PD dementia were divided into DHYZ plus Donepezil group and donepezil alone group. Outcomes of combination of DHYZ plus Donepezil were measured before and after drug treatment by mini-mental state examination (MMSE) and Montreal cognitive assessment (MoCA) for cognition; Assessment Scale–Cognitive Subscale (ADAS–Cog) for orientation, memory, language, reasoning, and praxis; as well as the Barthel Index for activities of daily living (ADL). At the end of 6-month study, it was found that all premeters including MMSE, MoCA, ADAS–Cog, and ADL were improved in both DHYZ plus Donepezil and donepezil alone group compared with the baseline. Further, all outcome measurements showed more significant improvement in DHYZ plus Donepezil treatment group than donepezil alone group by the end of 6-month treatment (Gu et al., 2015). Although the underlying mechanism of the combined therapy is not clear it might be relevant to the synergic alleviation of brain cholinergic system or DHYZ may exert its neuroprotection via other unknown actions.

Jia Wei Liu Jun Zi Tang (JWLJZT, 加味六君子汤) is composed of *Radix codonopsis pilosulae* [Dang shen (党参)], *R. rehmanniae* [Sheng di huang (生地黄)], Poria [Fu ling (茯苓)], *Ramulus uncariae cum uncis* [Gou teng (钩藤)], *Rhizoma atractylodis macrocephalae* [Bai Zhu (白术)], *Radix angelicae sinensis* [Dang gui (当归)], *Rhizoma pinelliae preparata* [Fa ban xia (法半夏)], *R. chuanxiong* [Chuan xiong (川芎)], *Radix achyranthis bidentatae* [Huai niu xi (怀牛膝)], *Pericarpium citri reticulatae* [Chen pi (陈皮)], and *Radix glycyrrhizae* [Gan cao (甘草)]. JWLJZT is an ancient formula with the specific function of tonifying the energy (Qi) of spleen and stomach, and was used to treat some parkinsonian symptoms (Huang & Zhang, 2000). Recently, JWLJZT was used to assess its antiparkinsonian effect on patients with PD. Fifty-five patients with PD were randomly divided into JWLJZT treatment group and placebo group. Outcome measures included the 39-item Parkinson's Disease Questionnaire (PDQ-39), Unified

Parkinson's Disease Rating Scale (UPDRS), Short-Form-36 Health Survey (SF-36), and Geriatric Depression Scale (GDS). JWLJZT treatment significantly improved many nonmotor symptoms judged by perimeters such as UPDRS IVC, PDQ-39, SF-36, and GDS compared with placebo group; although no obvious improvement in motor function was observed following JWLJZT treatment compared with placebo group (Kum et al., 2011).

Bu shen Huo xue Yin (BSHXY, 补肾活血饮) was not only used to assess its mechanisms of action in PD models (Li, Liu, et al., 2011, 2012; Li, Li, et al., 2012; Wang et al., 2011) but also used to evaluate its antiparkinsonian effect in patients with PD (Wang, Yang, Liu, Li, & Li, 2014; Yang et al., 2010). A multicenter, randomized, double-blinded, and placebo-controlled clinical study was conducted to evaluate antiparkinsonian effect of BSHXY on PD motor symptoms. One-hundred and twenty PD patients were recruited in which 60 PD patients were treated with BSHXY plus levodopa for 3 months, while another 60 PD patients were administered with placebo plus madopar (Yang et al., 2010). Changes in symptoms were monitored at the baseline, 1, 2, and 3 months by measuring UPDRS III, rise time of 10-m back and forth exercise and resting muscle tension. Although improvement of motor symptoms was found within two groups at the end of 3-month treatment compared with the respective baseline, BSHXY treatment resulted in a significant improvement in all measurements at the end of treatment compared with placebo group (Yang et al., 2010). Recently a randomized study by Wang et al. (2014) investigated effect of BSHXY on nonmotor symptoms such as depression in patients with PD. Sixty-two PD patients treated with madopar were divided into two groups, one group was treated with BSHXY for 3 months and another group was given Fluoxetine hydrochloride, an antidepression medication, for 3 months. Hamilton depression rating scale (HAMD), content of brain neurotransmitters NE and 5-HT were measured before and after 3-month treatment. It was found that at the end of treatment both BSHXY and Fluoxetine hydrochloride improved depression symptoms judged by HAMD scores and content of NE and 5-HT compared with baseline (Wang et al., 2014). Further, BSHXY treatment markedly improved HAMD scores and content of NE and 5-HT at the end of treatment compared with Fluoxetine hydrochloride (Wang et al., 2014).

Zeng xiao An shen Zhi chan granule 2 (ZAZ2, 增效安神治颤丸2) is made up of Chinese herbs including *Ramulus uncariae cum uncis* [Gou teng (钩藤)], *R. rehmanniae* [Sheng di huang (生地黄)], *F. corni* [Shan zhu yu (山茱萸)], *Radix asparagi* [Tian men dong (天门冬)] Paeonia lactiflora

[Shao yao (芍药)], *H. cistanches* [Rou cong rong (肉苁蓉)], Radix puerariae [Ge gen (葛根)], Rhizoma arisaematis [Tian nan xing (天南星)], *Radix salviae liguliobae* [Dan shen (丹参)], *Rhizoma acori talarinowii* [Shi chang pu (石菖蒲)], *Rhizoma curcumae longae* [Jiang huang (姜黄)], *Radix morindae officinalis* [Ba ji tian (巴戟天)], *R. gastrodiae* [Tian ma (天麻)], and *R. chuanxiong* [Chuan xiong (川芎)]. ZAZ2 is commonly used to treat symptoms derived from "insufficiency of Kidney yang" which is believed to be a one of causes of PD according to theory of Chinese medicine. Pan et al. (2011) evaluated effect of ZAZ2 on PD in a single blinded clinical study in which 110 patients with PD were randomly allocated to ZAZ2 treatment group and placebo group. ZAZ2 or placebo treatment lasted for 11 weeks during which all patients were still treated with conventional antiparkinsonian medications. At the end of treatment, patients in AZA2-treated group showed significant improvements in their motor function, daily activities, and quality of night sleep judged by the UPDRS Part II, Part II + III, and Part IV scores compared with placebo group (Pan et al., 2011).

Yang Xue Qing Nao granules (YXQN 养血清脑丸), containing *Radix angelicae sinensis* [Dang Gui (当归)], *R. chuanxiong* [Chuan Xiong (川芎)], *Radix paeoniae alba* [Bai shao (白芍)], Ramulus uncariae cum uncis [Gou Teng (钩藤)], Caulis spatholobi [Ji Xue Teng (鸡血藤)], Spica prunellae [Xia Ku Cao (夏枯草)], Concha margaritifera usta [Zhen Zhu Mu (珍珠母)], *R. rehmanniae preparata* [Di Huang (地黄)], Semen cassiae [Jue Ming Zi (决明子)], Rhizoma corydalis yanhusuo [Yan Hu Suo (延胡素)], and Herba asari [Xi Xin (细辛)]. YXQN were shown to improve sleep disturbance in stroke-related condition (Xu et al., 2009). Recently in a randomized study, YXQN was used as adjuvant to conventional medication in PD patients with sleep difficulty and found to significantly improve the sleep dysfunction, judged by analyzing actigraphic measurements, in YXQN-treated PD patients compared with placebo-treated patients following 4-month treatment (Pan, Kwak, Li, Chen, & Cai, 2013).

Together, studies of CHM formulas in patients with PD presented evidence that within randomized or single- or double-blinded or multicenter setting, CHM formulas not only improved some motor symptoms but also nonmotor symptoms such as dementia, depression in patients with PD. Combination of CHM with dopaminergic drugs showed a synergistic improvement in the assessment of general symptoms and UPDRS scores and decreased in the dose of dopaminergic drugs, and reduced adverse effect rate.

Similarly, many studies by systematic review and metaanalysis suggested that CHM as an adjuvant therapy for PD, helped to improve both motor and nonmotor symptoms, prolonged therapeutic effect of levodopa, reduced the dose usage of levodopa, and reduced the occurrence and severity of levodopa-induced dyskinesia, although no conclusive evidence of efficacy of CHM on PD could be researched (Kim, Cho, Jung, & Lee, 2012; Wang, Xie, Lu, Fu, & Zheng, 2012; Zhang et al., 2015).

## 5. FUTURE RESEARCH OF CHM IN PARKINSON'S

As pathogenesis and causes of PD are believed to be multifactor derived, it is reasonable to use such a combined treatment as herbal formulation with multiple biologically active components to address a variety of pathogenesis aspects. The use of Chinese herbal extracts or formulas would be able to not only alleviate both motor and nonmotor symptoms, but might be able to protect the surviving dopaminergic neurons in the SN by its multiple actions on different pathogenic factors, which may lead to delay or even stop the progression of disease.

However, the lack of reliable quality control and resulting poor reproducibility of the biological effects of herbs have retarded the research and development of herbal extracts and formulations (Wang et al., 2008). Despite these limitations and difficulties, we should be optimistic about the future of herbal extracts and formulations for the prevention or treatment of PD. The use of multidiscipline, multimodule, and system biology approaches will help to evaluate the preclinical efficacy of herbal extracts and formulations, to examine the active components and clarify the mechanisms of action in future studies. Further, adverse effects relevant to herbal recipes also need to be investigated and described clearly. In order to establish the efficacy, strengthen the advantages and standardization of CHM extracts and formulas, more large sample, multicenter, and longitudinal studies should be carried out. Only by doing so can we have a deeper understanding of the scientific connotation of CHM treatment with Parkinson's and better approaches to prevent, treat, and cure the disease.

## ACKNOWLEDGMENT

This work was partly supported by ATCM (The Association of Traditional Chinese Medicine and Acupuncture UK). Author thanks for the IT support from King's College London.

## REFERENCES

Ahmad, M., Saleem, S., Ahmad, A. S., Ansari, M. A., Yousuf, S., & Hoda, M. N. (2005a). Neuroprotective effects of Withania somnifera on 6-hydroxydopamine induced parkinsonism in rats. *Human & Experimental Toxicology*, *24*, 137–147.

Ahmad, M., Saleem, S., Ahmad, A. S., Yousuf, S., Ansari, M. A., & Khan, M. B. (2005b). Ginkgo biloba affords dose-dependent protection against 6-hydroxydopamine-induced parkinsonism in rats: Neurobehavioural, neurochemical and immunohistochemical evidences. *Journal of Neurochemistry*, *93*, 94–104.

Bido, S., Marti, M., & Morari, M. (2011). Amantadine attenuates levodopa-induced dyskinesia in mice and rats preventing the accompanying rise in nigral GABA levels. *Journal of Neurochemistry*, *118*, 1043–1055.

Blanchet, J., Longpré, F., Bureau, G., Morissette, M., DiPaolo, T., & Bronchti, G. (2008). Resveratrol, a red wine polyphenol, protects dopaminergic neurons in MPTP-treated mice. *Progress in Neuro-Psychopharmacology & Biological Psychiatry*, *32*, 1243–1250.

Chaudhuri, K. R., Healy, D. G., & Schapira, A. H. (2006). Non-motor symptoms of Parkinson's disease: Diagnosis and management. *The Lancet Neurology*, *5*, 235e45.

Chaudhuri, K. R., Odin, P., Antonini, A., & Martinez-Martin, P. (2011). Parkinson's disease: The non-motor issues. *Parkinsonism & Related Disorders*, *17*, 717–723.

Chen, J., Ma, J., Qiu, Y., Yi, S., Liu, Y., & Zhou, Q. (2012). Effects of zhichan powder on signal transduction and apoptosis-associated gene expression in the substantia nigra of Parkinson's disease rats. *Neural Regeneration Research*, 7, 2115–2122.

Chen, X. C., Zhou, Y. C., Chen, Y., Zhu, Y. G., Fang, F., & Chen, L. M. (2005). Ginsenoside Rg1 reduces MPTP-induced substantia nigra neuron loss by suppressing oxidative stress. *Acta Pharmacologica Sinica*, *26*, 56–62.

Connolly, B. S., & Lang, A. E. (2014). Pharmacological treatment of Parkinson disease: A review. *Journal of the American Medical Association*, *311*, 1670–1683.

Doo, A. R., Kim, S. N., Hahm, D. H., Yoo, H. H., Park, J. Y., Lee, H., et al. (2014). Gastrodia elata blume alleviates L-DOPA-induced dyskinesia by normalizing FosB and ERK activation in a 6-OHDA-lesioned Parkinson's disease mouse model. *BMC Complementary and Alternative Medicine*, *14*, 107.

Ferrer, I. (2011). Neuropathology and neurochemistry of nonmotor symptoms in Parkinson's disease. *Parkinson's Disease*, *2011*, 708404.

Geng, X., Tian, X., Tu, P., & Pu, X. (2007). Neuroprotective effects of echinacoside in the mouse MPTP model of Parkinson's disease. *European Journal of Pharmacology*, *564*, 66–74.

Gu, C., Shen, T., An, H., Yuan, C., Zhou, J., Ye, Q., et al. (2015). Combined therapy of Di-Huang-YiZhi with donepezil in patients with Parkinson's disease dementia. *Neuroscience Letters*, *606*, 13–17.

Haleagrahara, N., & Ponnusamy, K. (2010). Neuroprotective effect of Centella asiatica extract (CAE) on experimentally induced parkinsonism in aged Sprague-Dawley rats. *The Journal of Toxicological Sciences*, *35*, 41–47.

Hirsch, E. C., Jenner, P., & Przedborski, S. (2013). Pathogenesis of Parkinson's disease. *Movement Disorders*, *28*, 24–30.

Hornykiewicz, O. (2001). Chemical neuroanatomy of the basal ganglia–normal and in Parkinson's disease. *The Journal of Chemical Neuroanatomy*, *22*, 3–12.

Huang, T. K., & Zhang, D. B. (2000). Extrapyramidal disease. In *Neuropathy and psychosis of traditional Chinese medicine* (pp. 426–427). Beijing: Chinese Medical Science Publishers.

Hwang, M. W., Ahn, T. S., Hong, N. R., Jeong, H. S., Jung, M. H., Ha, K. T., et al. (2015). Effects of traditional Chinese herbal medicine San-Huang-Xie-Xin-Tang on gastrointestinal motility in mice. *World Journal of Gastroenterology*, *21*, 1117–1124.

Jin, G. Z., & Sun, B. C. (1995). L-Stepholidine and its analogues act on brain dopamine receptors. *Progress in Natural Science*, *5*, 55–63.

Jin, F., Wu, Q., Lu, Y. F., Gong, Q. H., & Shi, J. S. (2008). Neuroprotective effect of resveratrol on 6-OHDA-induced Parkinson's disease in rats. *European Journal of Pharmacology, 600*, 78–82.

Kim, T. H., Cho, K. H., Jung, W. S., & Lee, M. S. (2012). Herbal medicines for Parkinson's disease: A systematic review of randomized controlled trials. *PloS One, 7,* e35695.

Kim, H. G., Ju, M. S., Shim, J. S., Kim, M. C., Lee, S. H., Huh, Y., et al. (2010). Mulberry fruit protects dopaminergic neurons in toxin-induced Parkinson's disease models. *British Journal of Nutrition, 104*, 8–16.

Kum, W. F., Durairajan, S. S., Bian, Z. X., Man, S. C., Lam, Y. C., Xie, L. X., et al. (2011). Treatment of idiopathic Parkinson's disease with traditional Chinese herbal medicine: A randomized placebo-controlled pilot clinical study. *Evidence-Based Complementary and Alternative Medicine, 2011,* 724353.

Li, Y. L., Li, L. T., Yu, M., Wang, Y. Z., Ge, H. Y., & Song, C. Q. (2012). Beneficial effects of the herbal medicine Di Huang Yin Zi in patients with spinal cord injury: A randomized, placebo-controlled clinical study. *The Journal of International Medical Research, 40*, 1715–1724.

Li, S. D., Liu, Y., & Yand, M. H. (2011). Effect of Bushenhuoxue Yin on cerebral levels of nitric oxide, tumor necrosis factor-α and interferon-γ in a mouse model of Parkinson disease. *Journal of Southern Medical University, 31*, 90–92.

Li, S. D., Liu, Y., & Yand, M. H. (2012). Effects of Bushen Huoxue Yin (补肾活血饮) on brain NF-κB and NO content in the Parkinson's disease model mouse. *Journal of Traditional Chinese Medicine, 32*, 67–70.

Li, X. Z., Zhang, S. N., Liu, S. M., & Lu, F. (2013). Recent advances in herbal medicines treating Parkinson's disease. *Fitoterapia, 84*, 273–285.

Li, Q., Zhao, D., & Bezard, E. (2006). Traditional Chinese medicine for Parkinson's disease: A review of Chinese literature. *Behavioural Pharmacology, 17*, 403–410.

Lo, Y. C., Shih, Y. T., Tseng, Y. T., & Hsu, H. T. (2012). Neuroprotective effects of San-Huang-Xie-Xin-Tang in the MPP(+)/MPTP models of Parkinson's disease in vitro and in vivo. *Evidence-Based Complementary and Alternative Medicine, 2012,* 501032.

Lu, K. T., Ko, M. C., Chen, B. Y., Huang, J. C., Hsieh, C. W., Lee, M. C., et al. (2008). Neuroprotective effects of resveratrol on MPTP-induced neuron loss mediated by free radical scavenging. *The Journal of Agricultural and Food Chemistry, 56*, 6910–6913.

Luo, F. C., Wang, S. D., Qi, L., Song, J. Y., Lv, T., & Bai, J. (2011). Protective effect of panaxatriol saponins extracted from Panax notoginseng against MPTP-induced neurotoxicity in vivo. *The Journal of Ethnopharmacology, 133*, 448–453.

Lv, C., Hong, T., Yang, Z., Zhang, Y., Wang, L., Dong, M., et al. (2012). Effect of quercetin in the 1-methyl-4-phenyl-1,2,3,6-tetrahydropyridine-induced mouse model of Parkinson's disease. *Evidence-Based Complementary and Alternative Medicine, 2012,* 928643.

Ma, X. H., Zheng, C. J., Han, L. Y., Xie, B., Jia, J., Cao, Z. W., et al. (2009). Synergistic therapeutic actions of herbal ingredients and their mechanisms from molecular interaction and network perspectives. *Drug Discovery Today, 14*, 579–588.

Manyam, B. V., & Sanchez-Ramos, J. R. (1999). Traditional and complementary therapies in Parkinson's disease. *Advances in Neurology, 80*, 565–574.

Mo, J., Zhang, H., Yu, L. P., Sun, P. H., Jin, G. Z., & Zhen, X. (2010). L-stepholidine reduced L-DOPA-induced dyskinesia in 6-OHDA-lesioned rat model of Parkinson's disease. *Neurobiology of Aging, 31*, 926–936.

Mohanasundari, M., & Sabesan, M. (2007). Modulating effect of Hypercom perforatum extract on astrocytes in MPTP induced Parkinson's disease in mice. *European Review for Medical and Pharmacological Sciences, 11*, 17–20.

Mu, X., He, G., Cheng, Y., Li, X., Xu, B., & Du, G. (2009). Baicalein exerts neuroprotective effects in 6-hydroxydopamine-induced experimental parkinsonism in vivo and in vitro. *Pharmacology Biochemistry & Behavior, 92*, 642–648.

Mu, X., He, G. R., Yuan, X., Li, X. X., & Du, G. H. (2011). Baicalein protects the brain against neuron impairments induced by MPTP in C57BL/6 mice. *Pharmacology Biochemistry & Behavior, 98*, 286–291.

Nash, K. M., & Shah, Z. A. (2015). Current perspectives on the beneficial role of Ginkgo biloba in neurological and cerebrovascular disorders. *Integrative Medicine Insights, 10*, 1–9.

Obeso, J. A., Marin, C., Rodriguez-Oroz, C., Blesa, J., Benitez-Temiño, B., Mena-Segovia, J., et al. (2008). The basal ganglia in Parkinson's disease: Current concepts and unexplained observations. *Annals of Neurology, 64*(Suppl. 2), S30–S46.

Olanow, C. W., & Schapira, A. H. (2013). Therapeutic prospects for Parkinson disease. *Annals of Neurology, 74*, 337–347.

Pan, W., Kwak, S., Li, G., Chen, Y., & Cai, D. (2013). Therapeutic effect of Yang-Xue-Qing-Nao granules on sleep dysfunction in Parkinson's disease. *Chinese Medicine, 8*, 14. http://dx.doi.org/10.1186/1749-8546-8-14.

Pan, W., Kwak, S., Liu, Y., Sun, Y., Fang, Z., Qin, B., et al. (2011). Traditional Chinese medicine improves activities of daily living in Parkinson's disease. *Parkinson's Disease, 2011*, 789506, 7 pages.

Rajeswari, A. (2006). Curcumin protects mouse brain from oxidative stress caused by 1-methyl-4-phenyl-1,2,3,6-tetrahydropyridine. *European Review for Medical and Pharmacological Sciences, 10*(4), 157–161.

Rojas, P., Garduño, B., Rojas, C., Vigueras, R. M., Rojas-Castañeda, J., Rios, C., et al. (2001). EGb761 blocks MPP+-induced lipid peroxidation in mouse corpus striatum. *Neurochemical Research, 26*, 1245–1251.

Rojas, P., Montes, P., Rojas, C., Serrano-García, N., & Rojas-Castañeda, J. C. (2012). Effect of a phytopharmaceutical medicine, Ginko biloba extract 761, in an animal model of Parkinson's disease: Therapeutic perspectives. *Nutrition, 28*, 1081–1088.

Rojas, P., Rojas, C., Ebadi, M., Montes, S., Monroy-Noyola, A., & Serrano-García, N. (2004). EGb761 pretreatment reduces monoamine oxidase activity in mouse corpus striatum during 1-methyl-4-phenylpyridinium neurotoxicity. *Neurochemical Research, 29*, 1417–1423.

Rojas, P., Ruiz-Sánchez, E., Rojas, C., & Ogren, S. O. (2012). Ginkgo biloba extract (EGb 761) modulates the expression of dopamine-related genes in 1-methyl-4-phenyl-1,2,3,6-tetrahydropyridine-induced Parkinsonism in mice. *Neuroscience, 223*, 246–257.

Rojas, P., Serrano-García, N., Mares-Sámano, J. J., Medina-Campos, O. N., Pedraza-Chaverri, J., & Ogren, S. O. (2008). EGb761 protects against nigrostriatal dopaminergic neurotoxicity in 1-methyl-4-phenyl-1,2,3,6-tetrahydropyridine-induced Parkinsonism in mice: Role of oxidative stress. *European Journal of Neuroscience, 28*, 41–50.

Sengupta, T., Vinayagam, J., Nagashayana, N., Gowda, B., Jaisankar, P., & Mohanakumar, K. P. (2011). Antiparkinsonian effects of aqueous methanolic extract of Hyoscyamus niger seeds result from its monoamine oxidase inhibitory and hydroxyl radical scavenging potency. *Neurochemical Research, 36*, 177–186.

Shi, H., Song, J., & Yang, X. (2014). Zhichan decoction induces differentiation of dopaminergic neurons in Parkinson's disease rats after neural stem cell transplantation. *Neural Regeneration Research, 9*, 931–936.

Song, J. X., Sze, S. C., Ng, T. B., Lee, C. K., Leung, G. P., Shaw, P. C., et al. (2012). Anti-Parkinsonian drug discovery from herbal medicines: What have we got from neurotoxic models? *Journal of Ethnopharmacology, 139*, 698–711.

Sriraksa, N., Wattanathorn, J., Muchimapura, S., Tiamkao, S., Brown, K., & Chaisiwamongkol, K. (2012). Cognitive-enhancing effect of quercetin in a rat model of Parkinson's disease induced by 6-hydroxydopamine. *Evidence-Based Complementary and Alternative Medicine, 2012*, 823206.

Stocchi, F., Jenner, P., & Obeso, J. A. (2010). When do levodopa motor fluctuations first appear in Parkinson's disease? *European Neurology, 63*, 257–266.

Tan, Z. (2009). Neural protection by naturopathic compounds-an example of tetramethylpyrazine from retina to brain. *Journal of ocular biology, diseases, and informatics*, *2*, 57–64.

Tanaka, K., Galduróz, R. F., Gobbi, L. T., & Galduróz, J. C. (2013). Ginkgo biloba extract in an animal model of Parkinson's disease: A systematic review. *Current Neuropharmacology*, *11*, 430–435.

Tseng, Y. T., Chang, F. R., & Lo, Y. C. (2014). The Chinese herbal formula Liuwei dihuang protects dopaminergic neurons against Parkinson's toxin through enhancing antioxidative defense and preventing apoptotic death. *Phytomedicine*, *21*, 724–733.

van Wietmarschen, H. A., van der Greef, J., Schroen, Y., & Wang, M. (2013). Evaluation of symptom, clinical chemistry and metabolomics profiles during Rehmannia six formula (R6) treatment: An integrated and personalized data analysis approach. *Journal of Ethnopharmacology*, *150*, 851–859.

Wang, X., Liangm, X. B., Li, F. Q., Zhou, H. F., Liu, X. Y., Wang, J. J., et al. (2008). Therapeutic strategies for Parkinson's disease: The ancient meets the future—Traditional Chinese herbal medicine, electroacupuncture, gene therapy and stem cells. *Neurochemistry Research*, *33*(10), 1956–1963.

Wang, Y., Xie, C. L., Lu, L., Fu, D. L., & Zheng, G. Q. (2012). Chinese herbal medicine paratherapy for Parkinson's disease: A meta-analysis of 19 randomized controlled trials. *Evidence-Based Complementary and Alternative Medicine*, *2012*, 534861.

Wang, H. M., Yang, M. H., Dou, Y. Q., Liu, Y., Li, S. D., & Li, M. (2011). Effect of Bushen Huoxue decoction on dopamine D2 receptor in the brain of rats with Parkinson's disease. *Journal of Southern Medical University*, *31*, 1879–1881.

Wang, H. M., Yang, M. H., Liu, Y., Li, S. D., & Li, M. (2014). Effectiveness of Bushen Huoxue granule (补肾活血颗粒) on 5-serotonin and norepinephrine in the brain of Parkinson's disease patients with depressive state. *Chinese Journal of Integrative Medicine*, *20*, 944–948.

Wu, N., Yang, X., Song, L., Wei, J., & Liu, Z. (2013). Effect of Tianqi antitremor granules on behavioral manifestations and expression of G protein-coupled receptor kinase 6 and β-arrestin1 in levodopa-induced dyskinesia in a rat model of Parkinson's disease. *Drug Design, Development and Therapy*, *7*, 1481–1489.

Xu, D., Duan, H., Zhang, Z., Cui, W., Wang, L., Sun, Y., et al. (2014). The novel tetramethylpyrazine bis-nitrone (TN-2) protects against MPTP/MPP+-induced neurotoxicity via inhibition of mitochondrial-dependent apoptosis. *Journal of Neuroimmune Pharmacology*, *9*, 245–258.

Xu, B. B., Liu, C. Q., Gao, X., Zhang, W. Q., Wang, S. W., & Cao, Y. L. (2005). Possible mechanisms of the protection of ginsenoside Re against MPTP-induced apoptosis in substantia nigra neurons of Parkinson's disease mouse model. *Journal of Asian Natural Products Research*, *7*, 215–224.

Xu, X. S., Ma, Z. Z., Wang, F., Hu, B. H., Wang, C. S., Liu, Y. Y., et al. (2009). The antioxidant cerebral care granule attenuates cerebral microcirculatory disturbance during ischemia-reperfusion injury. *Shock*, *32*, 201–209.

Xu, C. L., Qu, R., Zhang, J., Li, L. F., & Ma, S. P. (2013). Neuroprotective effects of madecassoside in early stage of Parkinson's disease induced by MPTP in rats. *Fitoterapia*, *90*, 112–118.

Xu, D. P., Zhang, K., Zhang, Z. J., Sun, Y. W., Guo, B. J., Wang, Y. Q., et al. (2014). A novel tetramethylpyrazine bis-nitrone (TN-2) protects against 6-hydroxyldopamine-induced neurotoxicity via modulation of the NF-κB and the PKCα/PI3-K/Akt pathways. *Neurochemistry International*, *78*, 76–85.

Yang, M. H., Li, M., Dou, Y. Q., Liu, Y., Luo, X. D., Chen, J. Z., et al. (2010). Effects of Bushen Huoxue granule on motor function in patients with Parkinson's disease: A multicenter, randomized, double-blind and placebo-controlled trial. *Journal of Chinese Integrative Medicine*, *8*, 231–237.

Yang, M. H., Wang, H. M., & Liu, Y. (2011). Effect of Bushen Huoxue decoction on the orphan receptor and tyrosine hydroxylase in the brain of rats with Parkinson's disease. *Chinese Journal of Integrative Medicine*, *17*, 43–47.

Yao, X. (2008). Clinical research of Zhi-chan capsule on the treatment of Parkinson's disease. *Jilin Medical Journal*, *29*, 653–654.

Yu, M., Sun, Z. J., Li, L. T., Ge, H. Y., Song, C. Q., & Wang, A. J. (2015). The beneficial effects of the herbal medicine Di-Huang-Yin-Zi (DHYZ) on patients with ischemic stroke: A randomized, placebo controlled clinical study. *Complementary Therapies in Medicine*, *23*, 591–597.

Zhang, L., & Liu, S. M. (2006). Recent developments of traditional Chinese medicine in treatment of Parkinson disease. *Chinese Journal of Clinical Rehabilitation*, *10*, 152–154.

Zhang, S., Shao, S. Y., Song, X. Y., Xia, C. Y., Yang, Y. N., Zhang, P. C., et al. (2016). Protective effects of Forsythia suspense extract with antioxidant and anti-inflammatory properties in a model of rotenone induced neurotoxicity. *Neurotoxicology*, *52*, 72–83.

Zhang, G., Xiong, N., Zhang, Z., Liu, L., Huang, J., Yang, J., et al. (2015). Effectiveness of traditional Chinese medicine as an adjunct therapy for Parkinson's disease: A systematic review and meta-analysis. *PloS One*, *10*, e0118498.

Zheng, G. Q. (2009). Therapeutic history of Parkinson's disease in Chinese medical treatises. *Journal of Alternative and Complementary Medicine*, *15*, 1223–1230.

Zhou, Q., Chen, J., Yi, S., Lou, Y., Tang, W., Liu, Y., et al. (2012). Zhichan powder regulates nigrostriatal dopamine synthesis and metabolism in Parkinson's disease rats. *Neural Regeneration Research*, *7*, 2107–2114.

Zhou, W., Cheng, X., & Zhang, Y. (2016). Effect of Liuwei Dihuang decoction, a traditional Chinese medicinal prescription, on the neuroendocrine immunomodulation network. *Pharmacology & Therapeutics*, *162*, 170–178.

CHAPTER FOUR

# Neurobiology of Chinese Herbal Medicine on Major Depressive Disorder

**Gang Chen[1], Xiaoyan Guo**

Center for Translational Systems Biology and Neuroscience, Key Laboratory of Integrative Biomedicine for Brain Diseases, Nanjing University of Chinese Medicine, Nanjing, Jiangsu, PR China

[1]Corresponding author: e-mail address: hdn_2001@yahoo.com

## Contents



## Abstract

Major depressive disorder, or depression, becomes a serious public health problem globally. As current mainstream conventional antidepressants have limitations in unsatisfied response and remission rate, late onset of efficacy and side effects, increasing attention has been drawn to traditional Chinese medicine (TCM), in which mixture of herbs is commonly prescribed after the clinicians make pattern identification for diagnosis. Here, the principle of herbal formulation for TCM (Chinese herbal medicine, CHM) on depression as well as some clinical outcome research is briefly introduced. The neurobiological mechanisms underlying antidepressant effects of CHM, including enhancement of monoamine, dehyperactivation of hypothalamic–pituitary–adrenal axis, neuroimmune response, and neurotrophic factors are discussed. Additionally, the results from emerging research into rapid antidepressant efficacy using CHM formula Yueju, as well as the associated neuroplastical mechanisms, are summarized. The in-depth study of the

*International Review of Neurobiology*, Volume 135
ISSN 0074-7742
http://dx.doi.org/10.1016/bs.irn.2017.02.005

antidepressant efficacy, and underlying chemical substrates, as well as the molecular and systems neurobiological mechanisms may shed new lights for better understanding the mechanisms of CHM and improving the therapy.

## 1. INTRODUCTION

Major depressive disorder (MDD), or depression, is a major public health challenge. The years lived with disability caused by depression already ranks second among all the diseases worldwide until 2013 (Vos et al., 2015). It is predicted that depression will become the first leading cause of disease burden in the west by 2030 (Ng et al., 2014). In China, the morbidity of depression is increasing year by year, bringing a heavy psychological and economic burden to patients, family, and society. As to the therapeutic efficacy to depression, limitations of the conventional antidepressants (CADs) are still remarkable. A significant portion of the depressives who seek the treatment are prone to relapse, or nonresponsive to various antidepressants (Trivedi et al., 2006). Additionally, it takes several weeks for 5-hydroxytryptamine (5-HT) reuptake inhibitors (SSRIs), the mainstream antidepressants, to manifest their effects. Furthermore, side effects associated with the acute and chronic administration of the CADs also have obvious impacts on the therapy. Side effects using SSRIs in the treatment of depression include nausea, vomiting, dizziness, and drowsiness. 30%–50% or more of depressed people who take SSRIs for treatment may experience sexual dysfunction (Keltner, Mcafee, & Taylor, 2002). Chinese herbal medicine (CHM) is the major form of prescription of traditional Chinese medicine (TCM), which has a long history of using a variety of herbs to treat depression-like syndromes with better safety profile. Therefore, CHM treatment of depression is getting more and more attentions. In recent years, some important progress has been made in the basic and clinical researches, with better understanding of the underlying neurobiological basis.

## 2. TCM VIEW ON DEPRESSION

Pattern differentiation is essential in TCM in the process of making a diagnosis from symptoms and signs, prior to prescription of formulae of herbs, acupuncture, or other approaches to treat the diseases. According to TCM theory, symptoms, and signs of diseases are interpreted as imbalanced internal organs (related to but not in the sense of anatomic structures), such as Excess (实) or Deficiency (虚) of Liver or Spleen. The pattern identification from complex signs and symptoms forms the basis of individual treatment, which is the selection of herb formulas tailored to each patient's diagnosis (Maciocia, 1989; Scheid, 2002). Identification of the diagnosis pattern associated with depression has been widely studied. As the diversity of symptoms associated with TCM patterns was displayed in depression, different TCM patterns for depression have been put forward. Based on TCM theory, it is widely held that depressive symptoms are attributable to Stagnation of Liver (肝郁), a form of Excess pattern. Correspondingly, formulas that essentially dredge the Liver Stagnation are among the mostly prescribed for treatment of depression, such as Chaihu-Shugan-San (CHSGS, 柴胡疏肝散), and Xiaoyao San (XYS, 逍遥散), two most widely studied CHM for antidepressant effects. These formulas contain the herb Bupleuri Radix (in Chinese "Chaihu," 柴胡), to clear Liver Stagnation, which is viewed to play a primary role in the treatment. In fact, many studies indicate that depression has both Excess and Deficiency patterns. Liver Stagnation is frequently accompanied with Deficiency patterns of other organs, and Liver Stagnation plus Spleen or Kidney Deficiency patterns have been commonly reported (Tang, Miao, & Bao, 2008; Xu et al., 2013). Consequently, Bupleuri Radix is frequently combined with other herbs that promote the deficient Spleen and/or Kidney. When patients show obvious signs of Heat (火热), such as high body temperature, reddish tongue, or yellowish urine, it should be dealt with preferentially using herbs cooling down the Heat. For example, Dan-Zhi Xiaoyao San (DZXYS, 丹栀逍遥散) derived from XYS by adding two cooldown herbs including Gardenia jasminoides Ellis (Zhizi, 栀子) (GJ). Additionally, GJ or some other herbs that treat different Excess patterns are also formulated for depression treatment, such as Yueju pill (越鞠丸) and Banxia Houpo decoction (半夏厚朴汤). On the other hand, a number of patients only display symptoms indicating Deficiency in both Heart and Spleen, without obvious Excess patterns such as Liver Stagnation. One of most famous formulae for this pattern is Gan-maidazao (GMDZ, 甘麦大枣汤) decoction. Depression is a heterogeneous disease with different etiological and pathological factors, and thus identification of specific patterns based on TCM theory may facilitate a precise and individualized treatment. However, a uniform standard for TCM pattern identification for depression remains to be further established operationally.

## 3. CLINICAL FINDINGS DEMONSTRATED THE ANTIDEPRESSANT EFFICACY OF CHM

Besides personalized medicine with treatment based on individual's pattern, CHM has certain advantages and potentials in the prevention and treatment of depression. For example, people with subthreshold depression can be treated with CHM, acupuncture, etc., instead of taking conventional antidepressants. Previous studies indicate that under the circumstance of treatment based on pattern identification, the combination of CHM and mainstream antidepressants not only significantly improve the response and remission rate of patients with severe depression but also remarkably reduce the side effects of antidepressants (Liu et al., 2014). A meta-analysis demonstrates that under the rule of treatment based on pattern identification, the combination of Chinese and mainstream antidepressants lead to better treatment efficacy than that of using the antidepressants alone (Kou & Chen, 2012). The first formula that showed clinical antidepressant efficacy following rigorous procedure and was reported in the international journal is Dan-Zhi Xiaoyao san (DZXYS), and a meta-analysis supports its antidepressant effects (Qin et al., 2011; Zhang, Kang, Li, & Tan, 2007). Luo, Qian, and Zhao (2006) compared DZXYS with maprotiline in the treatment of depression. In the study, the therapy effects were accessed with Hamilton's depressive scale, self-rating depression scale, self-rating anxiety scale, and the scale for TCM syndrome and symptom differentiation, and the side effect was evaluated by Asberg side effect scale. According to these scores, there was no remarkable difference in therapeutic effects between these two groups, indicating a comparable antidepressant efficacy. Importantly, DZXYS showed fewer side effects than maprotiline. Another randomized and double-blinded study examined the effects of DZXYS on neuroimmuno-endocrine system in depressive patients. The therapeutic effects were also verified, which were suggested to link to its regulation of the levels of neurotransmitter 5-HT, brain-derived neurotrophic factor (BDNF), stress hormone corticosterone (CORT), and cytokine interleukin-6 (IL-6) (Li, Luo, & Qian, 2007).

A systematic review of randomized controlled trials demonstrated the combination of XYS and CADs was more effective than using CADs alone, and as the monotherapy, XYS was even superior to CADs (Zhang et al., 2012). Similarly, patients taking XYS alone or in combination with CADs experienced fewer adverse events of dizziness, headache, dry mouth, nausea, and constipation compared to CADs alone, suggesting XYS may have a higher safety profile compared to CADs (Qin et al., 2011). Another study result also supported the antidepressant effect of XYS and found that it reduced side effects such as insomnia and constipation associated with CADs (Man, Li, Gong, Xu, & Yu, 2014). Collectively, XYS appears very effective in reducing constipation. XYS can also be used to treat poststroke depression (PSD). In an 8-week clinical trial, the response rate between XYS and fluoxetine, an SSRI, was comparable in PSD patients (Li et al., 2008). In addition, compared to fluoxetine, XYS made greater improvement at week 2 and in the activities of daily living at the end of this trial. Similarly, a systematic review on the efficacy and safety of GMDZ decoction for depression suggests GMDZ has antidepressant effects, without any obvious side effects; and adjunct administration of GMDZ to CADs enhances efficacy and reduces side effects (El-Alfy, Abourashed, & Matsumoto, 2012). In a clinical research, CHSGS showed similar antidepressant effects to fluoxetine, with fewer side effects. Regional cerebral blood flow perfusion defects were reduced significantly or even disappeared and new lesions were not detectable following CHSGS in the depressive patients (Qiu, Hu, Shi, & Wang, 2014). There are certainly more studies concerning the efficacy CHM treatment of depression since 1990s, which support the efficacy of CHM. However, more studies with rigorous double-blind, parallel-controlled design are still needed (Butler & Pilkington, 2013; Zhang et al., 2007).

## 4. NEUROBIOLOGICAL MECHANISMS OF ANTIDEPRESSANT EFFECTS OF CHM

Depression is a heterogeneous disease, and thus several etiopathological hypotheses have been proposed on the basis of the complexity of symptoms. Similarly, different mechanisms have been found to be involved in antidepressant effects. Here only a few major mechanisms underling CHM antidepressant effects are reviewed.

### 4.1 Enhancement of Monoamine

Monoamine neurotransmitters, such as 5-HT, norepinephrine (NE), and dopamine (DA), are the important bioactive substances in central nervous system (CNS), which participate in many physiological activities of the body, including emotion, learning, and memory (Wu, Yan, et al., 2016). In the past years, the monoamine hypothesis or its derivatives have dominated the field. It is believed that the dysfunction of monoamine neurotransmitters, their metabolites, as well as the maladaptive alternations of their receptors in the

CNS are involved in the pathogenesis of depression (Li et al., 2012). A major support for the hypothesis comes from the facts that until recently, most of the CADs promote CNS monoamines, particularly, serotonin. It is thus not surprising that for a long time, the focus of the pharmacological study on CHM treatment of depression has been targeting on NE, 5-HT, and DA in the brain or serum. For example, after the treatment of GMDZ, the serum levels of 5-HT and NE increased (Ma et al., 2014). Oral administration of the polysaccharides from Banxia Houpo decoction at a low dose for 4 weeks resulted in a significant increase in the monoamine neurotransmitter 5-HT and DA levels in whole mouse brain (Guo, Kong, Wang, & Huang, 2004). XYS′ antidepressant effects were also been suggested to be mediated through the increase in NE, DA, and 5-HT, and the decrease in the metabolite of 5-HT, 5-hydroxyindoleacetic acid (5-HIAA) (Tao & Tao, 2010).

DZXYS could significantly increase the 5-HT and DA contents and decrease the NE content in the hippocampus, meanwhile reverse the ratio of 5-HIAA/5-HT in a rat model of depression (Wu, Yan, et al., 2016). Baihe Dihuang (百合地黄汤) decoction enhanced 5-HT and NA levels in mouse cerebrum as well as decreased the ratio of 5-HT compared to its metabolite, 5-HIAA (Chen, Gao, He, & Chen, 2012). In contrast to fluoxetine, Baisong (白松片) tablet could largely improve NE and 5-HT levels, which was proposed to underlie its antidepressant effects (Song, Sui-Yu, Huang, Chen, & Wang, 2005). The total triterpenes of herb Centella asiatica (积雪草) also exhibit antidepressant effects by increasing the contents of 5-HT, NE, DA, and their metabolites including 5-HIAA in rat brain (Zhang, Hou, & Xu, 2005). Doron et al. (2014) identified a novel herbal treatment (NHT) for depression and found it significantly lowered serotonin transporter (SERT) level, which increased serotonin in the synapse of the prefrontal cortex in the NHT group in comparison to the CAD escitalopram. This may link to NHT's antidepressant effect. In contrast, NHT induced a higher level of SERT in the hypothalamus than escitalopram, in parallel with the much less impairment in sexual function by NHT than escitalopram.

## 4.2 Dehyperactivation of HPA Axis

In recent years, some pharmacological studies on TCM treatment of depression started to shift focus from the pharmacological action of NE and 5-HT to other targets. One of the major attentions is put on hypothalamic–pituitary–adrenal (HPA) axis, which plays an essential role in stress response.

The alterations of HPA axis are observed in some depressed patients, with the elevation of basal cortisol, dexamethasone-mediated negative-feedback resistance, increased cerebrospinal fluid levels of corticotropin-releasing factor (CRF), and a blunted adrenocorticotropic hormone (ACTH) response to challenge with exogenous CRF (Plotsky, Owens, & Nemeroff, 1998). Therefore, it is predicted that the recovery of the HPA axis function may participate in normalizing behaviors after antidepressant treatment.

DZXYS significantly decreased the CORT, CRH, and ACTH contents in plasma and hypothalamus, suggesting that antidepressant effects exerted by DZXYS were partly due to inhibiting the hyperactivity of the HPA axis (Wu, Yan, et al., 2016). CHM Shuyusan (舒郁散) displayed antidepressant activity and decreases glucocorticoid levels activity, by regulating the function of HPA axis, and inhibiting glucocorticoid receptor expression in hippocampus (Chen, Chen, et al., 2012). Chronic administration of the total flavonoids isolated from the extract of CHM Xiaobuxin (小补心汤) decoction modulated the HPA axis by significantly decreasing serum CORT level and its upstream stress hormone ACTH level and upregulating hippocampal glucocorticoid receptor expression (An et al., 2011). 3,60-Disinapoyl sucrose, the active oligosaccharide ester component from roots of the herb Polygala tenuifolia (远志), a herb suggested primarily responsible for antidepressant effects of CHM Kai-Xin-San (开心散), reversed the abnormality in the HPA axis by significantly reducing serum CORT, ACTH, and CRH levels, and enhancing the expression of glucocorticoid receptor and mineralocorticoid receptor mRNA in the chronic mild-stressed (CMS; Hu, Liao, Liu, Guo, & Rahman, 2009) rats. Oral administrations of CHM Xiangsu (香苏散) powder remarkably increased levels of corticotrophin-releasing hormone mRNA expression in the hypothalamus and proopiomelanocortin mRNA expression in the pituitary, and down-regulated glucocorticoid receptor protein expression in the hypothalamus paraventricular nucleus, suggesting that Xiangsu powder confers antidepressant effect through suppression of the hyperactivity of the HPA axis (Ito et al., 2006).

## 4.3 Neuroimmune Response

Increasing evidence suggests that immune system including the cytokine network and the inflammatory response is involved in the pathophysiology of MDD (Qu et al., 2015). Pan et al. discovered that 12-week-long chronic stress resulted in the increased expression of IL-1β in prefrontal cortex and the activation of nuclear factor-κB (NF-κB) inflammatory pathway, which were closely related to the increased expression of NLRP3 inflammasome in the nervous system (Pan, Chen, Zhang, & Kong, 2014). Icariin, a flavonoid extracted from the traditional Chinese herb Herba Epimedii (淫羊藿), induces antidepressant-like effects by enhancing antioxidant status and antiinflammatory effects on the brain tissue via the inhibition of NF-κB signaling activation and the NLRP3-inflammasome/caspase-1/IL-1β axis (Liu, Xu, et al., 2015). Treatment with St. John's wort increases the expression of adiponectin, an antiinflammatory adipokine, decreases the TNF-α-induced increase in the mRNA levels of proinflammatory adipokines, IL-6, and monocyte chemoattractant protein-1 (Hatano et al., 2014). The antiinflammatory activity of St. John's wort was demonstrated by its inhibition of the activation of NF-κB (Hatano et al., 2014). Li et al. discovered detrimental effects of chronic stress may be associated with inhibition in NK cell activity and a decrease in proinflammatory cytokines, and that Banxia Houpu decoction modulates the immune dysregulation by raising the NK cell activity and IL-2 levels (Li et al., 2003). Shenqi Jieyu (参芪解郁) formula exerts an antidepressant effects by regulating the organs of the immune system and T lymphocyte subsets in rat model of postpartum depression (PPD) (Qu et al., 2015).

## 4.4 Metabolic Alternation

Studies also demonstrated that the antidepressant effects of some CHM may have connection with the improvement of the metabolic function. Liu et al. discovered that the plasma levels of trimethylamine oxide, glutamine, and lactate in depressed patients increased significantly and the levels of phenylalanine, valine, alanine, glycine, leucine, citrate, choline, lipids, and glucose decreased significantly. After XYS treatment, they returned to normal levels (Liu, Wu, et al., 2015). Genipin, the extract from Gardenia jasminoides Ellis, showed optimal antidepressant effects via downregulation of inositol, glutamate, lactate, and alanine, and upregulation of glycine and *N*-acetylaspartate (Peng, Shi, Tian, Gao, & Qin, 2014). The Total Alkaloid of Corydalis Rhizoma (延胡索) plays an antidepressant role in chronically stressed rats via regulating the flowing endogenous plasma metabolites: myo-inositol, glycerol, glycine, creatine, glutamine, glutamate, β-glucose, α-glucose, acetoacetate, 3-hydroxybutyrate, leucine, and unsaturated lipids (L7, L9) (Wu et al., 2015). Fufang Chaigui (复方>柴胡) prescription increased contents of ACTH, CORT, and T3; decreased contents of

TSH and TESTO; and normalized the levels of lactate, α-glucose, choline, *N*-acetyl glycoprotein, trimethylamine oxide, and leucine (Chen et al., 2015). Ginkgo biloba (银杏) extract also demonstrated antidepressant effects, which may link to reversing the alterations in the animal model of depression in the level of glutamate biosynthesis, aspartate metabolism, oxidative stress, neuroinflammation, and lipid and energy metabolism, and enhances GABA, inosine, and kynurenic acid levels in the hippocampus (Liang et al., 2016).

## 4.5 Neurotrophic Factors

Accumulating evidence indicates that neurotrophic factors (NTs), which are a family of proteins responsible for the growth and survival of neurons during development and for maintenance of adult neurons, for example, nerve growth factor (NGF), BDNF, glial cell-derived neurotrophic factor (GDNF), NT3, and NT-4/5. They play important roles in the treatment of depression (Zhu et al., 2013). Shuyu Ningxin (舒郁宁心) Recipe could exert antidepressant effects by improving the expression of hippocampal BDNF and its receptor TrkB (Sun, Huang, Xu, & Chen, 2013). Shuyu Formula and XYS exerted depression-like activity by upregulating BDNF in specific encephalic regions (Jing, Zhu, Lv, & Sun, 2015; Wang, Qiao, Xue, & Wei, 2015). Kai-Xin-San was also found to significantly increase the expression and secretion of NTs, including NGF, BDNF, and GDNF (Zhu et al., 2013). BDNF levels also increased in the prefrontal cortex of mice treated with the NHT, comparably with the CAD escitalopram (Doron et al., 2014).

In conclusion, studies suggest that different CHM formula confer antidepressant effects likely through multiple neurobiological mechanisms. For example, Banxia Houpo could induce antidepressant effect by the increase of monoamine neurotransmitter 5-HT and DA levels, as well as regulation of the immune dysregulation. Shuyu Formula decreases glucocorticoid levels activity, regulates the function of HPA axis, and upregulates BDNF in the treatment of depression. XYS exerts antidepressant effects via the modulation of monoamine neurotransmitters, improvement of the metabolic disorder systematically, and upregulation of BDNF levels. As depression is a complex disease that involves multiple mechanisms, it is an advantage for acting on multiple mechanisms simultaneously by using multiple herbs, and each herb may contain many bioactive compounds that work in a synergic manner. More details remain to be detected regarding the interactions of different mechanisms underlying the antidepressant effects of CHM.

## 5. NEUROBIOLOGICAL MECHANISMS OF RAPID AND PERSISTENT ANTIDEPRESSANT EFFECTS OF CHM

Patients with severe depression have a tendency to commit suicide. The onset of efficacy of the CADs such as SSRIs is slow, typically about 2–4 weeks after continuous administration, which is ineffective to reduce risk of suicide of those patients and also reduces the treatment compliance. With the treatment course extended, the economic burden of the patients has also risen (Trivedi et al., 2006). The delayed onset time, the low remission rate of CADs and side effects that may be associated with long-term CAD administration remain major challenges for the current MDD treatment (Machado-Vieira, Salvadore, Diazgranados, & Zarate, 2009). Consequently, it is urgently needed nowadays for new medications that have rapid onset of antidepressant effects (Zarate et al., 2006).

In recent years, the rapid and lasting antidepressant effects of low-dose ketamine, a noncompetitive *N*-methyl-D-aspartate (NMDA) receptor antagonist, has drawn wide attention in terms of both clinical and experimental fields. After a single injection of ketamine, it takes only 2 h to initiate the antidepressant effects and the efficacy of the administration can sustain for 1 week (Zarate et al., 2006), in sharp contrast to the conventional monoamine-based antidepressants. This rapid and lasting antidepressant effects were also recapitulated experimentally using a variety of animal models of depression. Even though esketamine has been granted in United States as a breakthrough designation, essentially to prevent suicide with a quick action, concerns regarding its potential addictive and neurotoxic effects may restrict its clinical utility. Yueju pill (越鞠丸), a traditional Chinese herbal formula invented by one of the most famous Chinese Medicine Doctor, 800 years ago, is popularly prescribed in China for the treatment of depression. In a double-blinded, placebo-controlled, and randomized pilot clinical trial (Wu et al., 2015), daily treatment of Yueju plus fluoxetine, a SSRI known with slow onset, induced a remarkable alleviation of depression symptoms starting from 3 days after drug administration, compared with herbal placebo plus fluoxetine, which induced unaltered clinical outcomes in patients on the first 10 days. Moreover, the response rate of the experimental group is much higher than the control group. Additionally, there was a significant correlation between the serum BDNF levels on day 1 and day 7 in experimental group but not the control group. This is the first clinical study designated to test the rapid antidepressant effects of a CHM and supported that Yueju may play a leading role in the fast antidepressant effect of treatment on depression.

The first study that specifically tested the ketamine-like rapid antidepressant effects of CHM was carried out by Xue et al. (2013). They used two animal behavioral paradigms (learned helplessness and tail suspension test, TST) that model some aspects of depression to assess the rapid antidepressant efficacy of Yueju, and disclosed some underlying neuromolecular mechanism. Similar to ketamine, 30 min after a single dose of Yueju significantly attenuated depressive-like symptoms in TST, which could last for 1 day in the ICR mouse strain, indicating that Yueju is able to quickly act as an antidepressant. Moreover, the antidepressant effect of Yueju could sustain for 24 h after one dose of administration and relieved depression-like behavior in the learned helplessness and novelty suppressed feeding tests (Xue et al., 2013; Zhang et al., 2015), which require chronic administration of CADs. The fast antidepressant of Yueju was linked to quick modulation of the BDNF expressions. Yueju acutely upregulated the expression of BDNF and reduced the phosphorylation of eukaryotic elongation factor 2, leading to recovery of BDNF protein synthesis, without gene expression change (Xue et al., 2013). Taken together, Yueju exhibited fast antidepressant effect through the instant enhancement of BDNF expression, indicating Yueju's upregulation of neuroplasticity. Xia et al. carried out a study to investigate the association of acute and lasting antidepressant effect of Yueju with NMDA signaling in the hippocampus of Kunming strain mice (Xia, Zhang, et al., 2016). The study showed Yueju reduced immobility time in the forced swimming test from 30 min to 5 days post a single administration; Yueju rapidly attenuated the protein expression of the obligatory subunit NR1 and nonobligatory NR2B, with NR2A expression remained unchanged in the hippocampus. However, NR1, but not NR2B, expression remained downregulated 5 days post-Yueju administration. Unlike ketamine, the expression of another glutamate receptor α-amino-3-hydroxy-5-methyl-4-isoxazolepropionic acid (AMPA) subunit GluR1 remained unaltered, neither did the blockade of AMPA signaling have a impact on antidepressant effects of Yueju (Xia, Zhang, et al., 2016), suggesting the relative independence of AMPA singling. Yueju improved the depressive symptoms at least for 72 h in learned helplessness mice and reversed the upregulated expression of NR1 in the learned helpless paradigms 1 or 3 days postadministration. Collectively, this finding supports the modulation of NMDA signaling may be critical for Yueju's immediate and last antidepressant effects.

Tang et al. carried out a study to examine the antidepressant of Yueju and ketamine in chronically stressed Kunming strain mice and the involvement with the changes in glutamatergic receptors and mammalian target of rapamycin (mTOR) signaling in the prefrontal cortex (Tang et al., 2015). Chronic stress paradigm is viewed to closely mimic the clinical depression. The study revealed both Yueju and ketamine, but not a CAD fluoxetine, could alleviate the depression-like symptoms in sucrose preference test which assesses anhedonia, the core symptoms of depression, 2 h after a single administration. Interestingly, a single dose of Yueju demonstrated antidepressant efficacy for 6 days, longer than ketamine. At the neuromolecular level, Yueju, like ketamine, reversed the decreased phosphorylation of the mTOR effectors 4E-BP1 and p70S6K, their upstream regulators ERK and Akt, and downstream targets including synaptic protein GluR1, and increased level of NMDA receptor subunit NR1 expression, which was induced by CMS. Only Yueju, but not ketamine, reversed the abnormity of phosphor-Akt and NR1 expression at 6 days postadministration, suggesting they are likely the key molecular substrates underlying Yueju's persistent antidepressant efficacy. The results suggest that the alterations in NMDA receptors and associated mTOR signaling may be involved in the mechanism of Yueju's antidepressant effect. In an animal model of PPD, in which the dams showed long-term PPD-like behaviors, a single dose of Yueju also induced a rapid antidepressant effects (Xia, Chen, et al., 2016). However, the following study provided further evidence that unlike ketamine, mTOR signaling itself may not be required for the action of Yueju. By comparison of the persistency of antidepressant response between ICR and Kunming strain mice, Xue et al. confirmed the strain-dependent difference in the antidepressant response to Yueju or ketamine (Xue et al., 2016). The longer antidepressant effect in Kunming strain mice required the activation of PKA-CREB signaling that regulates the transcription of BDNF, whereas there was no activation of the signaling in ICR strain mice. This study offered the first evidence that activation of specific molecular signaling pathway may lead to strain-dependent behavioral difference in response to the same CHM, which is implicated in the personalized medicine of TCM. Additionally, the study revealed a major signaling pathway that delineated the rapid antidepressant effects of Yueju from ketamine (Xue et al., 2016).

Yueju consists of five herbs, in which only Gardenia jasminoides Ellis displayed rapid antidepressant potential (Tao et al., 2014; Zhang et al., 2015), using learned helpless paradigm and NSF paradigms. The antidepressant effects of GJ began at 2 h post-GJ administration and lasted for over 20 h, with higher concentration than that in Yueju. Furthermore, the BDNF level increased in line with the antidepressant effect of GJ from 2 to 24 h after GJ administration. A study further identified the fractions of GJ responsible for rapid antidepressant efficacy and found two fractions both showed the effects (Ren, Tao, Xue, Wu, & Chen, 2015; Ren et al., 2016). However, only one fraction increased expression of BDNF and its receptor TrkB. The yellow pigment of GJ (GYP) is a major ingredient in the GJ fraction that displayed rapid antidepressant effects without association with upregulation of BDNF. Particularly, the immediate antidepressant activity of GYP is dependent on the synthesis of proteins including some synaptic proteins, but not BDNF, whereas the persistent antidepressant effects of GYP may link to the activation of PKA-CREB-BDNF signaling (Wu, Tao, et al., 2016).

Radix Polygala is widely prescribed in China for the treatment of depression. Shin et al. found that only minutes after a single oral administration of Radix Polygala extract, the depressive symptoms were alleviated in the forced swim test, TST, and female urine sniffing test (Shin et al., 2014). After two oral administration of Radix Polygala extract, the latency to feeding is reduced in the novelty-suppressed feeding test, and the number of failed escapes in the learned helplessness paradigm was also reduced, indicating the rapid antidepressant potential of Radix Polygala. Additionally, the fast antidepressant effect of Radix Polygala is associated with the decrease of phosphorylation of GluR1 serine-845 in the hippocampus. Radix Polygala extract could exert fast antidepressant effect by the regulation of glutamatergic synapses. Some other herb extracts or ingredients may not show antidepressant effects as fast as ketamine, but still faster than CADs. Timosaponin is an important ingredient isolated from the Chinese herb Rhizoma Anemarrhenae (知母). Timosaponin displayed antidepressant effect 1 week earlier than fluoxetine, and at week 2, according to the TST, the immobility time of the timosaponin group returned to normal, whereas the fluoxetine group did not show much improvement in the chronically stressed mice (Zhang et al., 2016). Fuzi polysaccharide, a new water-soluble polysaccharide extracted from Fuzi (附子), alleviated depressive symptoms on chronic depression animal model after a 2-week long administration, compared to imipramine, which needs 4 weeks to take effect (Li et al., 2016). Moreover, total alkaloid, another important component of

Fuzi, improves the ratio of phospho-CREB/CREB and BDNF protein level in the frontal cortex and hippocampus of ovariectomized mice may link to its relatively fast antidepressant-like effect, compared to CADs (Liu et al., 2012). It is worth to note that the effects of these herb ingredients or extracts were demonstrated in animal models. Up to now, the only clinical study that specifically tested the rapid onset of herbal medicine or natural products was conducted by Wu et al. using Yueju (2015), and more studies are yet to investigate this important profile for antidepressant efficacy of CHM. As the studies on Yueju demonstrated, different CHM may preferentially target different signaling pathways or neural pathways to elicit antidepressant effects.

## 6. CONCLUSION AND OUTLOOK

Some significant progress has been made on research into CHM treatment of depression, which is important for guiding and improving the clinical practice. Clinical research revealed the efficacy of treatment of the disease based on pattern identification, and it also reduced the side effects when used as the monotherapy or combined with CADs. Liver Stagnation pattern is relatively common, and thus Chaihu-like herbs that dredge Stagnation are most frequently prescribed. Some CHM components or compounds with antidepressant effects have potential to be used for development of new drugs. Yueju displayed fast onset of the antidepressant efficacy, which may be mediated by rapid promotion of neural plasticity via specific neuromolecular mechanisms. The neurobiological studies on the antidepressant effects of CHM reflect the multiple targets characteristic of CHM treatment of depression. In the future, research should focus on certain effective antidepressant CHM therapies. In order to establish the efficacy and strengthen the advantages of CHM including rapid antidepressant effects, more large-sample, multicenter, and longitudinal studies should be carried out. CHM treatment based on pattern identification for depression should be standardized. In addition, evidence of the efficacy of the suitable formula should be valued. The genome-wide association study methods could be used to reveal its genetic biological foundation. In experimental studies, it is important to establish animal models that are more consistent with clinical observations. The use of multidiscipline, multimodule, and systems biology approaches will help to discover new targets and mechanisms. Moreover, more attention should be paid to the verification of the role of different neural and molecular pathways and targets for the actions of CHM. In this way, we can have a deeper understanding of the scientific connotation of CHM treatment of depression and better approaches to prevent, treat, and cure the disease.

## ACKNOWLEDGMENTS

The study was supported by the National Natural Science Foundation (No. 81673625), Natural Science Foundation of Jiangsu Province (BK20151568), and the Priority Academic Program Development of Jiangsu Higher Education Institutions (PAPD).

## REFERENCES

An, L., Zhang, Y. Z., Liu, X. M., Yu, N. J., Chen, H. X., Zhao, N., et al. (2011). Total flavonoids extracted from xiaobuxin-tang on the hyperactivity of hypothalamic-pituitary-adrenal axis in chronically stressed rats. *Evidence-Based Complementary and Alternative Medicine*, *2011*(3), 1–7.

Butler, L., & Pilkington, K. (2013). Chinese herbal medicine and depression: The research evidence. *Evidence-Based Complementary and Alternative Medicine: eCAM*, *2013*(1), 221–229.

Chen, L., Chen, M., Wang, F., Sun, Z., Quanzhi, H., Geng, M., et al. (2012). Antidepressant-like effects of shuyusan in rats exposed to chronic stress: Effects on hypothalamic-pituitary-adrenal function. *Evidence-Based Complementary and Alternative Medicine*, *2012*(6), 155–237.

Chen, M. L., Gao, J., He, X. R., & Chen, Q. (2012). Involvement of the cerebral monoamine neurotransmitters system in antidepressant-like effects of a Chinese herbal decoction, baihe dihuang tang, in mice model. *Evidence-Based Complementary and Alternative Medicine*, *2012*(4), 214–215.

Chen, L., Liu, H., Chen, J. L., Gao, X. X., Zhou, Y. Z., Tian, J. S., et al. (2015). Anti-depressive mechanism of Fufang Chaigui prescription based on neuroendocrine hormone and metabolomic correlation analysis. *Zhongguo Zhong yao za zhi = Zhongguo zhongyao zazhi = China Journal of Chinese Materia Medica*, *40*(20), 4080–4087.

Doron, R., Lotan, D., Einat, N., Yaffe, R., Winer, A., Marom, I., et al. (2014). A novel herbal treatment reduces depressive-like behaviors and increases BDNF levels in the brain of stressed mice. *Life Sciences*, *94*(2), 151–157.

El-Alfy, A. T., Abourashed, E. A., & Matsumoto, R. R. (2012). Nature against depression. *Current Medicinal Chemistry*, *19*(14), 2229–2241.

Guo, Y., Kong, L., Wang, Y., & Huang, Z. (2004). Antidepressant evaluation of polysaccharides from a Chinese herbal medicine Banxia-houpu decoction. *Phytotherapy Research*, *18*(3), 204–207.

Hatano, T., Sameshima, Y., Kawabata, M., Yamada, S., Shinozuka, K., Nakabayashi, T., et al. (2014). St. john's wort promotes adipocyte differentiation and modulates NF-κB activation in 3t3-l1 cells. *Biological & Pharmaceutical Bulletin*, *37*(7), 1132–1138.

Hu, Y., Liao, H. B., Liu, P., Guo, D. H., & Rahman, K. (2009). A bioactive compound from Polygala tenuifolia regulates efficiency of chronic stress on hypothalamic-pituitary-adrenal axis. *Pharmazie*, *64*(9), 605–608.

Ito, N., Nagai, T., Yabe, T., Nunome, S., Hanawa, T., & Yamada, H. (2006). Antidepressant-like activity of a Kampo (Japanese herbal) medicine, Koso-san (Xiang-su-san), and its mode of action via the hypothalamic-pituitary-adrenal axis. *Phytomedicine*, *13*(9–10), 658–667.

Jing, L. L., Zhu, X. X., Lv, Z. P., & Sun, X. G. (2015). Effect of Xiaoyaosan on major depressive disorder. *Chinese Medicine*, *10*(1), 1.

Keltner, N. L., Mcafee, K. M., & Taylor, C. L. (2002). Mechanisms and treatments of SSRI-induced sexual dysfunction. *Perspectives in Psychiatric Care*, *38*(3), 111–116.

Kou, M. J., & Chen, J. X. (2012). Integrated traditional and Western medicine for treatment of depression based on syndrome differentiation: A meta-analysis of randomized controlled trials based on the Hamilton depression scale. *Journal of Traditional Chinese Medicine*, *32*(1), 1–5.

Li, J. M., Kong, L. D., Wang, Y. M., Cheng, C. H. K., Zhang, W. Y., & Tan, W. Z. (2003). Behavioral and biochemical studies on chronic mild stress models in rats treated with a Chinese traditional prescription Banxia-houpu decoction. *Life Sciences*, *74*(1), 55–73.

Li, J., Liu, Q. T., Chen, Y., Liu, J., Shi, J. L., Liu, Y., et al. (2016). Involvement of 5-HT1A receptors in the anxiolytic-like effects of quercitrin and evidence of the involvement of the monoaminergic system. *Evidence-Based Complementary and Alternative Medicine: eCAM*, *2016*, 6530364.

Li, L. F., Lu, J., Li, X. M., Xu, C. L., Yang, J., Qu, R., et al. (2012). Antidepressant-like effects of the saponins extracted from Chaihu-jia-longgu-muli-tang, in a rat unpredictable chronic mild stress model. *Fitoterapia*, *83*(1), 93–103.

Li, Y. J., Luo, H. C., & Qian, R. Q. (2007). Effect of Danzhi xiaoyao powder on neuro-immuno-endocrine system in patients with depression. *Zhongguo Zhong xi yi jie he za zhi Zhongguo Zhongxiyi jiehe zazhi = Chinese journal of integrated traditional and Western medicine/Zhongguo Zhong xi yi jie he xue hui, Zhongguo Zhong yi yan jiu yuan zhu ban*, *27*(3), 197–200.

Li, L. T., Wang, S. H., Ge, H. Y., Chen, J., Yue, S. W., & Yu, M. (2008). The beneficial effects of the herbal medicine free and easy wanderer plus (FEWP) and fluoxetine on post-stroke depression. *Journal of Alternative & Complementary Medicine (New York, N.Y.)*, *14*(7), 841–846.

Liang, Z., Bai, S., Peng, S., Hu, Q., Wang, X., Dong, M., et al. (2016). Gc-ms-based metabolomic study on the antidepressant-like effects of diterpene ginkgolides in mouse hippocampus. *Behavioural Brain Research*, *314*, 116–124.

Liu, L. Y., Feng, B., Chen, J., Tan, Q. R., Chen, Z. X., Chen, W. S., et al. (2014). Herbal medicine for hospitalized patients with severe depressive episode: A retrospective controlled study. *Journal of Affective Disorders*, *170*, 71–77.

Liu, L., Li, B., Zhou, Y., Wang, L., Tang, F., Shao, D., et al. (2012). Antidepressant-like effect of Fuzi total alkaloid on ovariectomized mice. *Journal of Pharmacological Sciences*, *120*(4), 280–287.

Liu, C. C., Wu, Y. F., Feng, G. M., Gao, X. X., Zhou, Y. Z., Hou, W. J., et al. (2015). Plasma-metabolite-biomarkers for the therapeutic response in depressed patients by the traditional Chinese medicine formula Xiaoyaosan: A 1H NMR-based metabolomics approach. *Journal of Affective Disorders*, *185*, 156–163.

Liu, B., Xu, C., Wu, X., Liu, F., Du, Y., Sun, J., et al. (2015). Icariin exerts an antidepressant effect in an unpredictable chronic mild stress model of depression in rats and is associated with the regulation of hippocampal neuroinflammation. *Neuroscience*, *294*, 193–205.

Luo, H. C., Qian, R. Q., & Zhao, X. Y. (2006). Clinical observation on effect of danzhi xiaoyao powder in treating depression. *Zhongguo Zhong xi yi jie he za zhi Zhongguo Zhongxiyi jiehe zazhi = Chinese journal of integrated traditional and Western medicine/Zhongguo Zhong xi yi jie he xue hui, Zhongguo Zhong yi yan jiu yuan zhu ban*, *26*(3), 212–214.

Ma, X. J., Zhao, J., Feng, Z. Y., Chang, J. M., Meng, S., Liu, H. Z., et al. (2014). Effects of modified ganmai dazao decoction on neuroendocrine system in patients with climacteric depression. *Zhongguo Zhong yao za zhi = Zhongguo zhongyao zazhi = China journal of Chinese materia medica*, *39*(23), 4680–4684.

Machado-Vieira, R., Salvadore, G., Diazgranados, N., & Zarate, C. A. (2009). Ketamine and the next generation of antidepressants with a rapid onset of action. *Pharmacology & Therapeutics*, *123*(2), 143–150.

Maciocia, G. (1989). The foundations of Chinese medicine: A comprehensive text for acupuncturists and herbalists. *Focus on Alternative & Complementary Therapies*, *11*(1), 112.

Man, C., Li, C., Gong, D., Xu, J., & Yu, F. (2014). Meta-analysis of Chinese herbal Xiaoyao formula as an adjuvant treatment in relieving depression in Chinese patients. *Complementary Therapies in Medicine*, *22*(2), 362–370.

Ng, M., Fleming, T., Robinson, M., Thomson, B., Graetz, N., Margono, C., et al. (2014). Global, regional, and national prevalence of overweight and obesity in children and adults during 1980–2013: A systematic analysis for the global burden of disease study 2013. *Lancet*, *384*(9945), 766–781.

Pan, Y., Chen, X. Y., Zhang, Q. Y., & Kong, L. D. (2014). Microglial NLRP3 inflammasome activation mediates IL-1β-related inflammation in prefrontal cortex of depressive rats. *Brain Behavior & Immunity*, *41*(1), 90–100.

Peng, G. J., Shi, B. Y., Tian, J. S., Gao, S., & Qin, X. M. (2014). 1H NMR based metabonomics study on the antidepressant effect of genipin in rat hippocampus. *Acta Pharmaceutica Sinica*, *49*(49), 209–216.

Plotsky, P. M., Owens, M. J., & Nemeroff, C. B. (1998). Psychoneuroendocrinology of depression: Hypothalamic-pituitary-adrenal axis. *Psychiatric Clinics of North America*, *21*(2), 293–307.

Qin, F., Wu, X. A., Tang, Y., Huang, Q., Zhang, Z. J., & Yuan, J. H. (2011). Meta-analysis of randomized controlled trials to assess the effectiveness and safety of free and easy wanderer plus, a polyherbal preparation for depressive disorders. *Journal of Psychiatric Research*, *45*(11), 1518–1524.

Qiu, J., Hu, S. Y., Shi, G. Q., & Wang, S. E. (2014). Changes in regional cerebral blood flow with ChaihuShugan-san in the treatment of major depression. *Pharmacognosy Magazine*, *10*(40), 503–508.

Qu, M., Tang, Q., Li, X., Zhao, R., Li, J., Xu, H., et al. (2015). Shen-Qi-Jie-Yu-Fang has antidepressant effects in a rodent model of postpartum depression by regulating the immune organs and subsets of t lymphocytes. *Neuropsychiatric Disease & Treatment*, *11*, 1523–1540.

Ren, L., Tao, W. W., Xue, W. D., Wu, H. X., & Chen, G. (2015). Potent rapid antidepressant effects of petroleum ether fraction of Yueju pill associated with BDNF and TrkB up-regulation. *Chinese Pharmacological Bulletin*, *31*(12), 1754–1759.

Ren, L., Tao, W., Zhang, H., Xue, W., Tang, J., Wu, R., et al. (2016). Two standardized fractions of Gardenia jasminoides Ellis with rapid antidepressant effects are differentially associated with BDNF up-regulation in the hippocampus. *Journal of Ethnopharmacology*, *187*, 66–73.

Scheid, V. (2002). *Chinese medicine in contemporary China: Plurality and synthesis*. Durham, NC: Duke University Press.

Shin, I. J., Son, S. U., Park, H., Kim, Y., Park, S. H., Swanberg, K., et al. (2014). Preclinical evidence of rapid-onset antidepressant-like effect in Radix Polygalae extract. *PloS One*, *9*(2), e88617.

Song, W. X., Sui-Yu, H. U., Huang, C. Y., Chen, C. H., & Wang, Z. (2005). Effects of baisong tablets on the behavior and levels of norepinephrine and dopamine in the brain of stress rats. *Zhong nan da xue xue bao. Yi xue ban = Journal of Central South University Medical sciences*, *30*(5), 574–578.

Sun, Z. G., Huang, Q. Z., Xu, C. Y., & Chen, L. P. (2013). Effects of shuyu ningxln recipe on the praxiology and the expressions of hippocampal BDNF and trkb of model rats with chronic stress-induced depression. *Zhongguo Zhong xi yi jie he za zhi Zhongguo Zhongxiyi jiehe zazhi = Chinese journal of integrated traditional and Western medicine/Zhongguo Zhong xi yi jie he xue hui, Zhongguo Zhong yi yan jiu yuan zhu ban*, *33*(3), 370–375.

Tang, Q. S., Miao, Q. U., & Bao, Z. X. (2008). Bayesian network research on tcm syndromes of depression. *Journal of Traditional Chinese Medicine*, *49*(11), 1013–1015.

Tang, J., Xue, W., Xia, B., Li, R., Tao, W., Chen, C., et al. (2015). Involvement of normalized NMDA receptor and mTOR-related signaling in rapid antidepressant effects of Yueju and ketamine on chronically stressed mice. *Scientific Reports*, *5*(5), 338–341.

Tao, W., & Tao, W. (2010). Effects of Chinese herbal medicine Xiaoyao powder on monoamine neurotransmitters in hippocampus of rats with postpartum depression. *Journal of Chinese Integrative Medicine*, *8*(11), 1075–1079.

Tao, W., Zhang, H., Xue, W., Ren, L., Xia, B., Zhou, X., et al. (2014). Optimization of supercritical fluid extraction of oil from the fruit of Gardenia jasminoides and its antidepressant activity. *Molecules*, *19*(12), 19350–19360.

Trivedi, M. H., Rush, A. J., Wisniewski, S. R., Nierenberg, A. A., Warden, D., Ritz, L., et al. (2006). Evaluation of outcomes with citalopram for depression using measurement-based care in star*d: Implications for clinical practice. *American Journal of Psychiatry*, *163*(1), 28–40.

Vos, T., Barber, R. M., Bell, B., Bertozzi-Villa, A., Biryukov, S., Bolliger, I., et al. (2015). Global, regional, and national incidence, prevalence, and years lived with disability for 301 acute and chronic diseases and injuries in 188 countries, 1990–2013: A systematic analysis for the Global Burden of Disease Study 2013. *The Lancet*, *386*(9995), 743–800.

Wang, F. R., Qiao, M. Q., Xue, L., & Wei, S. (2015). Possible involvement of μ opioid receptor in the antidepressant-like effect of Shuyu formula in restraint stress-induced depression-like rats. *Evidence-Based Complementary and Alternative Medicine*, *2015*, 452412.

Wu, R., Tao, W., Zhang, H., Xue, W., Zou, Z., Wu, H., et al. (2016). Instant and persistent antidepressant response of Gardenia yellow pigment is associated with acute protein synthesis and delayed up-regulation of BDNF expression in the hippocampus. *ACS Chemical Neuroscience*, 7(8), 1068–1076.

Wu, H., Wang, P., Liu, M., Tang, L., Fang, J., Zhao, Y., et al. (2015). A 1H-NMR-based metabonomic study on the anti-depressive effect of the total alkaloid of Corydalis Rhizoma. *Molecules*, *20*(6), 10047–10064.

Wu, L. L., Yan, L., Yan, C., Yi, P., Su, J. F., & Wu, W. K. (2016). Antidepressant-like effects of fractions prepared from Danzhi-xiaoyao-san decoction in rats with chronic unpredictable mild stress: Effects on hypothalamic-pituitary-adrenal axis, arginine vasopressin, and neurotransmitters. *Evidence-Based Complementary and Alternative Medicine*, *2016*(3), 1–11.

Wu, R., Zhu, D., Xia, Y., Wang, H., Tao, W., Xue, W., et al. (2015). A role of Yueju in fast-onset antidepressant action on major depressive disorder and serum BDNF expression: A randomly double-blind, fluoxetine-adjunct, placebo-controlled, pilot clinical study. *Neuropsychiatric Disease & Treatment*, *11*, 2013–2021.

Xia, B., Chen, C., Zhang, H., Xue, W., Tang, J., Tao, W., et al. (2016). Chronic stress prior to pregnancy potentiated long-lasting postpartum depressive-like behavior, regulated by Akt-mTOR signaling in the hippocampus. *Science Reports*, *6*, 35042.

Xia, B., Zhang, H., Xue, W., Tao, W., Chen, C., Wu, R., et al. (2016). Instant and lasting down-regulation of NR1 expression in the hippocampus is associated temporally with antidepressant activity after acute Yueju. *Cellular & Molecular Neurobiology*, *36*, 1189–1196.

Xu, C. Y., Tian, J. Z., Shi, J., et al. (2013). TCM characteristics of the sign of depression. *Chinese Archives of Traditional Chinese Medicine*, *31*(4), 810–813.

Xue, W., Wang, W., Gong, T., Zhang, H., Tao, W., Xue, L., et al. (2016). PKA-CREB-BDNF signaling regulated long lasting antidepressant activities of Yueju but not ketamine. *Scientific Reports*, *6*, 26331.

Xue, W., Zhou, X., Yi, N., Jiang, L., Tao, W., Wu, R., et al. (2013). Yueju pill rapidly induces antidepressant-like effects and acutely enhances bdnf expression in mouse brain. *Evidence-Based Complementary and Alternative Medicine, 2013*(1), 70–75.

Zarate, C. A., Jr., Singh, J. B., Carlson, P. J., Brutsche, N. E., Ameli, R., Luckenbaugh, D. A., et al. (2006). A randomized trial of an N-methyl-d-aspartate antagonist in treatment-resistant major depression. *Archives of General Psychiatry, 63*(8), 856–864.

Zhang, Q., Guo, F., Fu, Z. W., Zhang, B., Huang, C. G., & Li, Y. (2016). Timosaponin derivative YY-23 acts as a non-competitive NMDA receptor antagonist and exerts a rapid antidepressant-like effect in mice. *Acta Pharmacologica Sinica, 37*(2), 166–176.

Zhang, Y., Han, M., Liu, Z., Wang, J., He, Q., & Liu, J. (2012). Chinese herbal formula Xiao yao san for treatment of depression: A systematic review of randomized controlled trials. *Evidence-Based Complementary and Alternative Medicine, 2012*, 931636.

Zhang, W., Hou, T., & Xu, X. (2005). Effects of total triterpenes of Centella asiatica on the corticosterone levels in serum and contents of monoamine in depression rat brain. *Journal of Chinese Medicinal Materials, 28*(6), 492–496.

Zhang, Z. J., Kang, W. H., Li, Q., & Tan, Q. R. (2007). The beneficial effects of the herbal medicine free and easy wanderer plus (FEWP) for mood disorders: Double-blind, placebo-controlled studies. *Journal of Psychiatric Research, 41*(10), 828–836.

Zhang, H., Xue, W., Wu, R., Gong, T., Tao, W., Zhou, X., et al. (2015). Rapid antidepressant activity of ethanol extract of Gardenia jasminoides Ellis is associated with upregulation of bdnf expression in the hippocampus. *Evidence-Based Complementary and Alternative Medicine: eCAM, 2015*(15–16), 13–19.

Zhu, K. Y., Xu, S. L., Choi, R. C., Yan, A. L., Dong, T. T., & Tsim, K. W. (2013). Kai-xin-san, a Chinese herbal decoction containing Ginseng Radix et Rhizoma, Polygalae Radix, Acori Tatarinowii Rhizoma, and Poria, stimulates the expression and secretion of neurotrophic factors in cultured astrocytes. *Evidence-Based Complementary & Alternative Medicine: eCAM, 2013*(8), 731385.

CHAPTER FIVE

# Treatment of Insomnia With Traditional Chinese Herbal Medicine

**Amrinder Singh, Kaicun Zhao[1]**
Traditional Chinese Herbal Medicine Programme, Middlesex University, The Borough, Hendon, London, United Kingdom
[1]Corresponding author: e-mail address: k.zhao@mdx.ac.uk

## Contents



## Abstract

Insomnia is a condition with sleep problems and many people suffered from it. Chronic insomnia can last for long time and it will severely affect people's health and the quality of life. In conventional medicine, the most commonly used the medicine is benzodiazepine. It is effective but also has significant side effects. Patients try to use some kinds of alternative medicines.

Chinese medicinal herbs and formulas have been used in the treatment of insomnia for more than 2000 years in China. In recent decades, Chinese herbal medicine has been widely used in the Western countries. Many clinical studies including randomized controlled clinical trials and research on pharmacological action mechanisms of the herbs for treatment of insomnia have been conducted. It is very important and very helpful to review the published research papers to gather the available information for a critical analysis.

*International Review of Neurobiology*, Volume 135
ISSN 0074-7742
http://dx.doi.org/10.1016/bs.irn.2017.02.006

This chapter evaluated the data from both of clinical studies and pharmacological researches on the therapeutic formulas and on some key herbs used in the treatment of insomnia. Clinical studies showed a very wide spectrum of herbs that were used in clinical treatment of insomnia. This was due to different syndrome patterns happened with insomnia. This brought complexity and difficulties to identify which are the essential key herbs or formulas. It was found Suanzaoren decoction (*Ziziphus spinose* decoction 酸枣仁汤) is the most frequently used formula for the treatment of insomnia. Based on the clinical data, several herbs were identified as most frequently used sedative and hypnotic herbs in Chinese herbal medicine including Suanzaoren (*Ziziphus spinose* 酸枣仁), Fuling (*Poria cocos* 茯苓), and Gancao (*Glycyrrhiza uralensis* 甘草).

The underlying pharmacological action mechanisms discovered in the studies on some key herbs used in the treatment of insomnia were evaluated. The major pharmacological action mechanisms shared by most of the sedative herbs are to act through the neurotransmitter gamma-aminobutyric acid (GABA) or via stimulation of $GABAA_A$ receptor. Some herbs exert sedative activities via inhibition of 5-hydroxytryptamine 1A receptor. Another mechanism shown by some herbs is to upregulate the expression of orexin-A, leptin, orexin receptor-1, and leptin receptor in the brain, reducing insomnia-induced negative consequences, and thus indirectly help improvement of insomnia.

## 1. INTRODUCTION

Humans spend their one-third of life span in sleeping. Enough amount of sleep with a good quality is important for people to maintain efficient daytime activities and psychological health. However, there are around 10%–30% of population suffering from a sleep disorder insomnia (Erden et al., 2016; Grewal & Doghramji, 2010; Hu, Zhang, Liu, Li, & Zhang, 2015). Insomnia is the most ubiquitous form of sleep disturbance characterized by difficulty to fall asleep or to maintain staying asleep appropriately. Poor quality of sleep results in negative consequences to the daytime activities manifested as mood swings, lethargy, and forgetfulness (Abd-Allah, Abdel-Aziz, & Abo El-Seoud, 2014; Pigeon, 2010). Insomnia is the second most common disorder reported after pain (Grewal & Doghramji, 2010). The incidences of insomnia were found even higher in aged population. It was reported that 20%–40% of the elderly people show sleep problems and 25% of them actually suffer from insomnia (Kim et al., 2017; Mai & Buysse, 2008). It also found that women are most prevalent than men (Leger et al., 2012; Zhang & Wing, 2006).

Insomnia can be divided into two stages according to the length of the insomnia condition onset, acute (less than 4 weeks) and chronic (more than

4 weeks, with experience of symptoms for 3 or more nights per week) (Reeve & Bailes, 2010). Considering the causes, insomnia can be classified as primary and secondary insomnia. While secondary insomnia is clearly associated with certain pathological conditions such as asthma and pain, there is no clearly contributable medical, psychiatric, or environmental causes for primary insomnia. Studies have found that the concurrent diseases with insomnia are most likely related to disorders in cardio-cerebrovascular or metabolic systems and anxiety–depression conditions (Baker, Baldwin, & Garner, 2015; Cui et al., 2014).

For treatment of insomnia particularly chronic insomnia, pharmacological therapy is the major option in relieving symptoms. It is reported that about 60% of the insomnia patients are prescribed with sedative and hypnotic medications and 4%–11% of the general population use the medication for their sleep (Hermes & Rosenheck, 2014). Benzodiazepine (BDZ) is widely prescribed for insomnia. Investigation among the insomnia patients in China has found the most frequently used hypnotic is alprazolam (Cui et al., 2014). However, long-term use of these medicines is potentially causing dependence and clear adverse effects, such as cognitive decline and psychiatric issues (Doghramji & Jangro, 2016; Murphy, Wilson, Goldner, & Fischer, 2016). More and more insomnia patients are looking for alternative treatment. Traditional Chinese herbal medicine is one of the most sophisticated and integrated traditional medicine system. It has been commonly used in Asian countries and more and more widely recognized in Western countries (Lu & Lu, 2014; Pearson, Johnson, & Nahin, 2006). There is a long history in treatment of insomnia in traditional Chinese herbal medicine. A number of Chinese medicinal herbs and formulas are used in the treatment of insomnia (Chen et al., 2011). Much research in clinical effectiveness of Chinese herbal medicine in the treatment of insomnia is also reported. This paper will evaluate the potential effects of Chinese herbal medicines on insomnia by reviewing the published reports in both of clinical studies and pharmacological research in this area.

## 2. CLINICAL APPLICATIONS OF CHINESE MEDICINAL HERBS AND FORMULAS FOR TREATMENT OF INSOMNIA

A specific feature in the clinical practice of traditional Chinese medicine is syndrome differentiation. This means that during a diagnosis process, any symptoms must be considered in conjunction with the condition context of individual patient. When conducting or evaluating researches in traditional Chinese medicine, this should be taken into account (Jiang et al., 2012). For primary insomnia, according to traditional Chinese medicine principles, it can be potentially differentiated into 10 syndromes patterns. In terms of traditional Chinese medicine, they are deficiency in both the heart and spleen, hyperactivity of fire due to yin deficiency, liver qi stagnation transforming into fire, heart–kidney incoordination, internal disturbance of phlegm heat, qi deficiency of the heart and gallbladder, liver fire flaming upward, heart deficiency with timidity, disharmony between spleen and stomach, and stomach disharmony (Yeung et al., 2012a). Recent studies reported that among various different patterns, deficient conditions and excessive fire (overactive) are more likely happen with insomnia (Kung, Yang, Chiu, & Kuo, 2016; Tian & Liu, 2015). Most recently, it has been pointed out that in the Chinese medicine diagnosis of insomnia, the "spleen deficient and liver heat syndrome" is the main subtype diagnosed from the viewpoint of TCM, where spleen deficiency and liver heat syndrome are the two different syndromes in insomnia (Liu et al., 2016). These different syndrome patterns have led to applications of different herbs and formulas varied very widely. In addition, there are clear inconsistencies in diagnosis for differentiation of the syndrome patterns (Yeung et al., 2012b). This has made it more complicated and difficult to identify which herbs or formulas are most appropriate for treatment of insomnia.

Suanzaoren Tang (*Ziziphus spinose* decoction 酸枣仁汤), containing Chao Suanzaoren (*Ziziphus spinose* 炒酸枣仁), Fuling (*Poria cocos* 茯苓), Chuanxiong (*Ligusticum chuanxiong* 川芎), Zhimu (*Anemarrhena aspodeloidea* 知母), and Zhigancao (*Glycyrrhiza glabra* 炙甘草), is used traditionally as a sedative and hypnotic in Chinese medicine. Previous reviews also found it is a most frequently used Chinese herbal formula for treatment of insomnia (Ni et al., 2015; Yeung et al., 2012a). Recently, Hu et al. (2015) investigated the therapeutic efficacy of Suanzaoren Tang (SZRT) combined with Zhizi Chi Tang (*Gardenia jasminoides* and *Glycine max* decoction, 栀子豉汤) containing Zhizi (*Gardenia jasminoides* 栀子), and Dandouchi (*Glycine max* 淡豆豉) (SZR-ZZC) in patients of insomnia with a Yin deficiency and excessive heat pattern. In this randomized controlled clinical trial, Polysomnography (PSG), Insomnia Severity Index (ISI), Pittsburgh Sleep Quality Index (PSQI), and Self-Rating Anxiety Scale (SAS) were monitored to evaluate the efficacy of combined formula SZR-ZZC. The patients were randomly assigned to the SZR-ZZC treatment group ($n=60$) and lorazepam control group ($n=59$). SZR-ZZC exhibited better results than lorazepam in improvement of insomnia as indicated by significant reduction of ISI, PSQI scores, and improved PSG parameters.

In another randomized controlled clinical trial, a group of heart and gallbladder Qi-deficiency type of insomnia patients were treated with Zhenjingdingzhi decoction (镇惊定志汤) which is composed of Tiannanxing (*Arisaematis Erubescentis* 天麻*i*), Shichangpu (*Acori Tatarinowii* 石菖蒲), Zhenzhumu (*Concha Margaritifera* 珍珠母), Yuanzhi (*Polygala tenuifolia* 远志), Danshen (*Salviae miltiorrhizae* 丹参), Fuxiaomai (*Tritici Levis* 浮小麦), Dazao (*Ziziphus jujuba* 大枣), and Gancao (*Glycyrrhiza uralensis* 甘草) (Ye, Zhou, Yuan, Yuan, & Yang, 2015). After 8 weeks of treatment, PSQI scores were significantly reduced and the total effective rate (PSQI reduction >25%) was 84%. Interestingly, this study used Suanzaoren Tang (酸枣仁汤) as a control and it also showed a total effective rate of 59.6%. In comparison of the two herbal formulas, their compositions are completely different. For the clinical effectiveness, although both of the formulas showed positive results but there was a significant difference between the treatment and control groups, particularly in the improvement of sleep quality and sleep duration. The treatment formula Zhenjingdingzhi decoction (镇惊定志汤) demonstrated significant better therapeutic effects than Suanzaoren Tang (酸枣仁汤) (Ye et al., 2015).

The above studies suggested that there are herbal formulas with completely different herbs but they may all be able to achieve a satisfied clinical effectiveness. Furthermore, the result also suggested that using appropriate herbal formula based on the syndrome differentiation of traditional Chinese medicine may achieve better clinical outcomes. This is special feature of clinical practice of traditional Chinese medicine. However, there is lack of scientific evidence to support this practice and need more research on this aspect.

## 3. POTENTIAL EFFECTIVE HERBS AND PHARMACOLOGICAL ACTION MECHANISMS

As a normal practice, traditional Chinese medicine is using herbal formula rather than a single herb for therapeutic treatment. Additionally, diverse categories of herbal formulas are used for different syndrome patterns. In a latest literature review, 66 formulas were reviewed for their clinical effectiveness in the treatment of insomnia (Ni et al., 2015) and include more than 100 individual herbs. This makes it is very complicated and

**Table 1** Herbs With Higher Frequency in Clinical Use for Treatment of Insomnia

| Herb | Latin | Frequency (%)[a] |
|---|---|---|
| Fuling 茯苓 | *Poria cocos* | 61.8 |
| Suanzaoren 酸枣仁 | *Ziziphi Spinosae* | 58.8 |
| Gancao 甘草 | *Glycyrrhiza uralensis* | 44.1 |
| Danshen 丹参 | *Salvia miltiorrhiza* | 38.2 |
| Yuanzhi 远志 | *Polygala tenuifolia* | 35.3 |
| Chuanxiong 川芎 | *Ligusticum chuanxiong* | 32.4 |
| Danggui 当归 | *Angelica sinensis* | 29.4 |
| Chaihu 柴胡 | *Bupleurum chinensis* | 29.4 |
| Chenpi 陈皮 | *Citrus reticulata* | 20.6 |
| Shengdihuang 生地黄 | *Rehmannia glutinosa* | 20.6 |
| Yejiaoteng 夜交藤 | *Polygonum multiflorum* | 20.6 |
| Baiziren 柏子仁 | *Platycladus orientalis* | 17.6 |
| Banxia 半夏 | *Pinellia ternata* | 14.7 |
| Wuweizi 五味子 | *Schisandra chinensis* | 14.7 |

[a]The percentage is the frequency of the herbs used in total of 34 formulas reported in the studies of Ye, Yuan, and Dai (2011), Li, Gao, and Liu (2009), Liu et al. (2016), Ye et al. (2015), Li et al. (2013), Gan, Tian, and Qin (2013), Zou, Zhang, Jia, and Cao (2011), Li, Xu, and Xiao (2009), Zhang et al. (2007), Zhou et al. (2007), Yu, Liang, and Sun (2007), Ushiroyama, Ikeda, Sakuma, and Ueki (2005), Yang et al. (2005), Ni et al. (2015), and Xia, Xia, Deng, and Zhu (2009).

difficult to identify the potential key herbs which are pharmacologically effective in the treatment of insomnia. Analysis of the formulas reported in recent years for treatment of insomnia has found several herbs that are most frequently used in clinical practice. Table 1 shows the 15 herbs that appeared most frequently in clinical treatment of insomnia with Suanzaoren (*Ziziphus spinose* 酸枣仁), Fuling (*Poria cocos* 茯苓), Gancao (*Glycyrrhiza uralensis* 甘草), Yuanzhi (*Polygala tenuifolia* 远志), and Danshen (*Salviae miltiorrhizae* 丹参) as the top popular ones.

Researches on the underlying pharmacological action mechanisms are also developing. A number of herbs have been found to be able to regulate sleep via inhibitory action on the central nervous system (CNS). Most of these herbs are also appeared on the list of the most frequently used herbs for treatment of insomnia. The active constituents in regulation of sleep are also identified in some commonly used herbs.

Regulation of sleep may be a complex process with different neurotransmitters involved. In general, catecholamines, orexin, and histamine are known as suppressor of sleep, and gamma-aminobutyric acid (GABA), adenosine, serotonin, melatonin, prostaglandin D2 are sleep-promoting factors (Levenson, Kay, & Buysse, 2015). GABA is in general an inhibitory neurotransmitter in CNS. It inhibits arousal systems to promote sleep by binding to the $GABA_A$ receptor, which is the main target for many sedative and hypnotic medicines to act on for the treatment of insomnia (Chen, 2009; Krystal, Richelson, & Roth, 2013).

## 3.1 Suanzaoren (*Ziziphi spinosae* 酸枣仁)

Suanzaoren is a typical herb used in treatment of insomnia in traditional Chinese medicine for more than 2000 years. Oral administration of Suanzaoren decoction to rat increased nonrapid eye movement (NREM) sleep, but had no significant effect on rapid eye movement (REM) sleep. Evidence implicated that the effects were mediated through the GABA(A) receptors, but not GABA(B) receptors (Yi, Lin, Tsai, Lin, & Chang, 2007; Yi, Tsai, Chen, & Chang, 2007). Study in mice also showed that the water extract of Suanzaoren can shorten sleep latency and increase sleeping time significantly (Fang et al., 2010).

Saponin is one of the major components of *Ziziphi spinosae* (Suanzaoren 酸枣仁). The hypnotic effect of jujubosides was investigated in rat. The results showed that the Saponin significantly increased the total sleep time and also increased the REM sleep or NREM depending on the time. REM increased during daytime but NREM increased during the night time, suggesting the hypnotic effect of jujubosides on normal rats may be influenced by circadian rhythm. The study also found that the serotonergic system may involve in the hypnotic effect of jujubosides (Cao et al., 2010). Jujubosides may need to be hydrolyzed before it can pass the blood–brain barrier. It has been suggested that the hydrolyzed product jujubogenin is the compound in fact distributed into the brain (Chen, Chen, Wu, & Tsai, 2008). From the molecular decoking analysis, it has been confirmed that jujubogenin is the constitute of Suanzaoren decoction paying key role in the modulation of $GABA_A$ receptor (Chen, 2009). This established the effectiveness of jujubogenin as the $GABA_A$ receptor agonist.

Sanjoinine A is one of the major alkaloid compounds in Suanzaoren (*Ziziphi spinosae* 酸枣仁). Coadministration of sanjoinine A with pentobarbital to mouse showed increase in sleeping rate and time but decrease in sleeping latency. Sanjoinine A worked in a manner similar to GABA(A) receptor agonist but did not induce sleep directly. It increased the glutamic acid decarboxylase (GAD) activities to ultimately increase the synthesis of GABA (Ma et al., 2007). With an alcohol–water extract of Suanzaoren, a similar result was obtained. The content of GABA in the hypothalamus was significantly increased following administration of the extract to rat (Gao, Ji, Jiang, & Chen, 2013).

Spinosin is a C-glycoside of flavonoid identified in *Ziziphi spinosae* (酸枣仁). In the rat model, spinosin potentiated pentobarbital-induced sleep by reducing sleep latency and increasing the total, NREM and REM sleep time. Evidence suggested that the potentiating effects of spinosin on pentobarbital-induced sleep may be via its activity as an antagonist at post-synaptic 5-hydroxytryptamine 1A (5-$HT_{1A}$) receptor (Wang et al., 2010).

## 3.2 Danshen (*Salviae miltiorrhizae* 丹参)

Danshen is also traditionally used for treatment of insomnia in Chinese herbal medicine. Study showed that the ether extract of Danshen can shorten sleep latency significantly and increase sleeping time in mice (Fang et al., 2010). Evidence also showed that the water extract of Danshen is able to act on the BDZ binding sites of $GABA_A$ receptors (Liao et al., 1995). Furthermore, combined use with Suanzaoren (*Ziziphi spinosae* 酸枣仁) showed significant synergistic effects in promoting sleep (Fang et al., 2010). In Chinese herbal medicine clinical practice, it is indeed these two herbs are frequently used together for treatment of insomnia. This study provided strong evidence to suggest that combined use of Suanzaoren and Danshen should be recommended for the treatment of insomnia.

## 3.3 Yuanzhi (*Polygala tenuifolia* 远志)

Yuanzhi is traditionally recorded as a tranquilizer promoting sleep in Chinese herbal medicine. It is widely used in the treatment of insomnia and palpitation (Chen, Lee, Huang, Kuan, & Chen, 2015). Polygalasaponins are one of the major active constituents discovered in the herb (Yao et al., 2010). Tenuifolin is the main polygalasaponin compound extracted from Yuanzhi. Most recently in an in vivo study with mice, it has been found that oral administration of tenuifolin at doses of 40 and 80 mg/kg significantly prolonged the total sleep time. It increased the amount of both of NREM and REM sleep (Cao et al., 2016). The author found that tenuifolin significantly reduced the noradrenaline (NA) levels in the area where either the sleep-promoting (NREM and REM) or the wake-promoting neurons are located. At the same time, it significantly increased GABA levels in this area. It also increased the acetylcholine (Ach) levels in the area where REM sleep-promoting neurons are accumulated. The evidence is clearly indicated that the sleep-promoting activity of Yuanzhi is mediated by activation of GABAergic systems and/or inhibition of noradrenergic systems.

Another active compound identified from Yuanzhi (*Polygala tenuifolia* 远志) is 3,4,5-trimethoxycinnamic acid (TMCA). This is a chemically different compound from polygalasaponin. It has been reported that TMCA potentiate pentobarbital-induced sleep by increasing the total sleep time and the NREM sleep but reduced the sleep latency and the number of sleep/wake cycles (Lee, Han, Hong, & Oh, 2013). It was also found that TMCA increased the activity of GAD and the expressions of γ-subunit of $GABA_A$ receptors, suggesting that TMCA may enhance sleep by modulation of the GABAAergic systems. In the mice model, TMCA decreased the locomotor activity but tenuifolin did not (Cao et al., 2016; Lee et al., 2013).

In another study, a methyl derivative of TMCA which is also identified in Yuanzhi was found effective to protect the heart from arrhythmias via inhibition of the calcium channel in rabbit myocites (Zhao et al., 2013). In addition, studies have shown Yuanzhi has cognitive improving and cerebral protective activity (Ikeya et al., 2004). The polygalasaponins were also found having good anxiolytic and antidepressant activities (Yao et al., 2010). These additional activities on heart and CNS suggest Yuanzhi is a very useful herb for treatment of insomnia accompanied with anxiety, palpitations, and depression. The evidence provided a scientific basis for the traditional use of this herb. Yuanzhi is ideally used in Chinese herbal medicine for the treatment of insomnia with the heart–kidney incoordination pattern which is characterized with insomnia, palpitation, anxiety, dreaminess, and forgetfulness.

## 3.4 Baizhu (*Atractylodes macrocephala* 白术)

Another Chinese herb Baizhu was highlighted for its positive modulation of $GABA_A$ receptor by an in vitro study. Baizhu is recorded in traditional

Chinese medicine as a Qi-tonifying herb with major applications for deficient conditions in the digestion system. According to the principles of traditional Chinese medicine theories, it is thought a harmonizing condition in the stomach is a key to maintain a good quality sleep. The main chemical components of Baizhu, atractylenolide II and III, showed positive modulation to GABA-induced chloride currents (Singhuber, Baburin, Kahlig, et al., 2012; Singhuber, Baburin, Khom, et al., 2012). This result suggested that Baizhu may be an effective herb for clinical use as a sedative and hypnotic medicine.

## 3.5 Sinisan (四逆散)

Sinisan is a traditional herbal formula recorded in the ancient traditional Chinese medicine book "Shang Han Lun." It is a basic prescription of TCM for coordinating the liver and spleen functions in terms of Chinese medicine to alleviate some digestion related symptoms including some emotional feelings. The formula is composed of four herbs including Chaihu (*Bupleuri chinensis* 柴胡), Baishao (*Paeoniae alba* 白芍), Zhishi (*Aurantii immaturus* 枳实), and Gancao (*Glycyrrhiza uralensis* 甘草). It has been reported in an in vivo study with mice that the remedy significantly potentiated the pentobarbital-induced sleeping effects. The sleep latency of mice was significantly decreased and the sleeping time was markedly prolonged (Li et al., 2014). It has been demonstrated that Baishao played a key role for the formula to promote sleep. Although the in vivo study showed that only Baishao was able to produce significant sedative effects, but among the ingredients of the formula, Chaihu and Gancao are also shown being active in modulating the GABA or $GABA_A$ receptor (Cho et al., 2010; Liao et al., 1995). In addition, studies in sleep deprivation rat have shown that Sinisan may relieve insomnia by regulating the level of cation-chloride cotransporters, Na-K-Cl-cotransporter (NKCC1), and K(+)/Cl(−) cotransporter (KCC2), in the hippocampus (Liu et al., 2015). Theoretically, Sinisan should be a useful formula for treatment of insomnia. However, it seems there are not many clinical reports on the application of this formula for improving sleeping.

## 3.6 Wen-Dan Decoction (温胆汤)

Wen-Dan decoction is another traditional formula with a long history in use for treatment of insomnia. The composition of Wen-Dan decoction includes Banxia (*Pinellia ternate* 半夏), Zhuru (*Bambusa tuldoides* 竹茹), Zhishi

(*Citrus aurantium* 枳实), Chenpi (*Citrus reticulate* 陈皮), Gancao (*Glycyrrhiza uralensis* 甘草), Fuling (*Poriae cocos* 茯苓). Wu et al. (2014) investigated the potential action mechanisms of the formula in improvement of insomnia. It has been found that Wen-Dan decoction improves insomnia caused negative emotion conditions through upregulation of orexin-A, leptin, orexin receptor-1, and leptin receptor in the brain. Both orexin and leptin are involved in regulating the sleep and arousal states. Insomnia can cause significant decrease in the levels of orexin and leptin in the brain (Ebrahim, Howard, Kopelman, & Williams, 2002; Wu et al., 2014). Wen-Dan decoction exhibits clear effects on preventing the development of negative conditions caused by insomnia. Another study conducted by Song et al. (2014) supported the findings related to Wen-Dan decoction. Evidence has suggested that Wen-Dan decoction does not function as a sedative–hypnotic but regulates the emotional disorders caused by insomnia, therefore, improve sleep quality (Wu et al., 2014).

## 3.7 Gancao (*Glycyrrhiza uralensis* 甘草) and Hehuanpi (*Albizzia julibrissin* 合欢皮)

Cho et al. (2010) suggested that both liquorice (*Glycyrrhiza uralensis* 甘草, Gancao) and Hehuanpi (*Albizzia julibrissin* 合欢皮, Hehuanpi) modulate the $GABA_A$ and serotonin 5-HT2C receptors in mice. Pentobarbital-induced sleep was potentiated by both of the herbs as manifested with reduced the sleep latency and increased sleep duration. Kim, Kim, Lee, and Jang (2007) augmented the results by confirming Hehuanpi acted on the $5\text{-}HT_{1A}$ receptor after oral administration in mice. These findings regarding to the herb Hehuanpi were further supported by studies from Jung et al. (2005).

## 3.8 Other Herbs

A number of herbs have been studied on their potential sedative and hypnotic activities as well as the underlying action mechanisms. Most of the studied herbs showed actions on CNS with a major influence on the inhibitory GABA neurological pathways. Some commonly used Chinese medicinal herbs which have modulation activities on GABA, 5-HT neurological systems are listed in Table 2. Theoretically these herbs potentially can be used for sedative and hypnotic purposes to help improvement of sleep.

**Table 2** Commonly Used Traditional Chinese Medicinal Herbs With Modulate Activities to GABA or $GABA_A$ Receptors

| Herbs | Binding Activities | References |
|---|---|---|
| Baizhu (*Atractylodes macrocephala* 白术)—atractylenolide II and III | Positive allosteric modulation of the $GABA_A$ receptor; but independent from the benzodiazepine (BDZ) binding site | Singhuber, Baburin, Kahlig, et al. (2012) |
| Chaihu (*Bupleurum chinensis,* 柴胡) | Modulate Dopamine $D_2$, $5\text{-}HT_{1A}$, GABA | Liao et al. (1995) and Sugiyama, Muteki, and Kano (1996) |
| Chuanxiong (*Ligusticum chuanxiong* 川芎) | Modulate GABA, $5\text{-}HT_{1A}$ | Liao et al. (1995) |
| Danggui (*Angelicae sinensis* 当归) | Modulate GABA, $5\text{-}HT_{1A}$ | Liao et al. (1995) and Xin, Zhang, Yang, Deng, and Xie (2013) |
| Danshen (*Salviae miltiorrhizae* 丹参) | Modulate $GABA_A$–BDZ | Lee et al. (1991), Chang et al. (1991), Liao et al. (1995), and Cho et al. (2010) |
| Dengxincao (*Juncus effuses* 灯心草)—effusol and dehydroeffusol | Modulate $GABA_A$ receptor but independent of the benzodiazepine binding site | Singhuber, Baburin, Khom, et al. (2012) |
| Duhuo (*Angelicae pubescentis* 独活) | Modulate GABA, $5\text{-}HT_{1A}$, $D_2$, $D_1$ | Liao et al. (1995) and Zaugg, Eickmeier, Rueda, Hering, and Hamburger (2011) |
| Gancao (*Glycyrrhiza glabra* 甘草) | Modifies $GABA_A$–BZD receptors; modulating $5\text{-}HT_{2C}$ receptors | Cho et al. (2010) |
| Hehuanpi/ Hehuanhua (*Albizzia julibrissin* 合欢皮/合欢花) | Modulating $GABA_A$ and $5\text{-}HT_{2C}$ receptors | Cho et al. (2010) |
| Houpo (*Magnolia officinalis* 厚朴)—honokiol | Modifies the benzodiazepine site of the $GABA_A$ receptor | Qu et al. (2012) |
| Huangqin (*Scutellaria baicalensis* 黄芩) | Modifies the $GABA_A$ receptor | Liao et al. (1995), De Carvalho, Duarte, and De Lima (2011), and Chang et al. (2011) |

**Table 2** Commonly Used Traditional Chinese Medicinal Herbs With Modulate Activities to GABA or $GABA_A$ Receptors—cont'd

| Herbs | Binding Activities | References |
|---|---|---|
| Juhua (*Chrysanthemum morifolium* 菊花) | Increase expression levels of GAD | Zhang and Zhang (2007) |
| Longyanrou (*Dimocarpus longan* 龙眼肉) | Modifies the $GABA_A$ receptor | Ma et al. (2009) |
| Nan Wuweizi (*Kadsura longipedunculata* 南五味子) | Modulate $5\text{-}HT_{1A}$ receptor | Zheng et al. (2015) |
| Qinjiao (*Gentianae macrophyllae* 秦艽) | Modulate GABA–BDZ, $5\text{-}HT_{1A}$, $D_2$ | Liao et al. (1995) |
| Suanzaoren (*Ziziphi spinosae* 酸枣仁) and Wuweizi (*Schisandrae chinensis* 北五味子) mixed extract | Regulate the content of GABA in the hypothalamus | Gao et al. (2013) |
| Suanzaoren (*Zizphi spinose* 酸枣仁) | $5\text{-}HT_{1A}$, $5\text{-}HT_2$, GABA | Liao et al. (1995), Yi, Lin, et al. (2007), and Yi, Tsai, et al. (2007) |
| Tianwang Buxin decoction (天王补心丹) | Modulate 5-HT | Li, Xu, et al. (2009) |
| Yuanzhi (*Polygala tenuifolia*, 远志) | Modulate GABA, GAD, Ach; $GABA_A$ receptor | Cao et al. (2016) and Lee et al. (2013) |

*Ach*, acetylcholine; *BZD*, benzodiazepine; *D1, D2*, Dopamine D1, D2 receptors; *GABA*, γ-aminobutyric acid; *GAD*, glutamic acid decarboxylase; *5-HT1A*, 5-hydroxytryptamine.

## 4. SUMMARY

Chinese herbal medicine is traditionally used for treatment of insomnia conditions. As a special feature of the medicine system, syndrome differentiation leads to clinical application of herbs varied very widely, causing complexity, and difficulties to identify the key formula, and key herbs that may be mainly responsible for the therapeutic effectiveness. It is still a not very clear how to evaluate the importance of syndrome differentiation for achieving good clinical therapeutic effects. Based on previously published research papers, it can be found some formulas or herbs may be used more frequently. However, the formula appeared through this way may not be the most effective one. According to traditional Chinese medicine principles, syndrome differentiation must be taken into account before a formula can be selected. Due to the quality problems in methodology, it is still unable to give a conclusive answer about the clinical effectiveness of Chinese herbal medicine in treatment of insomnia, but most of the clinical trials and case reports provided a clearly positive trend.

Several most frequently used herbs are identified. This is the same as those reported previously. Some of these herbs are traditionally used as sedative and hypnotic herbs. More scientific researchers are on these herbs. While the clinical data are still not conclusive, studies on the action mechanisms of the herbs may be valuable. Although some data are come from animals or in vitro models, the positive information is encouraging. These finding may be useful not only for us to understand how the herbs work but also potentially very useful for practitioners to make selection when they prescript a herbal remedy in their clinical practice.

In the future, it is warranted that more high quality clinical trials are needed to obtain more conclusive answers to the clinical effectiveness of the herbal medicine treatment of insomnia. At the same time, more pharmacological research on the herbs is also required. It will help to understand the action mechanisms of the relevant herbs. This knowledge may mutually promote the development with clinical studies.

## REFERENCES

Abd-Allah, E., Abdel-Aziz, H., & Abo El-Seoud, A. (2014). Insomnia: Prevalence, risk factors, and its effect on quality of life among elderly in Zagazig city, Egypt. *Journal of Nursing Education and Practice*, *4*(8), 52–69.

Baker, L., Baldwin, D., & Garner, M. (2015). Daytime intrusive thoughts and subjective insomnia symptoms. *Psychiatry Research*, *229*(3), 1038–1042.

Cao, Q., Jiang, Y., Cui, S. Y., Tu, P. F., Chen, Y. M., Ma, X. L., et al. (2016). Tenuifolin, a saponin derived from Radix Polygalae, exhibits sleep-enhancing effects in mice. *Phytomedicine*, *23*(14), 1797–1805.

Cao, J., Zhang, Q., Cui, S., Cui, X., Zhang, J., Zhang, Y., et al. (2010). Hypnotic effect of jujubosides from Semen Ziziphi Spinosae. *Journal of Ethnopharmacology*, *6*(1), 163–166.

Chang, H., Chui, K., Tan, F., Yang, Y., Zhong, Z., Lee, C., et al. (1991). Compounds from Danshen. Part 4. Structure-activity relationship of miltirone, an active central benzodiazepine receptor ligand isolated from Salvia miltiorrhiza Bunge (Danshen). *Journal of Medicinal Chemistry*, *34*(5), 1675–1692.

Chang, H. H., Yi, P. L., Cheng, C. H., Lu, C. Y., Hsiao, Y. T., Tsai, Y. F., et al. (2011). Biphasic effects of baicalin, an active constituent of Scutellaria baicalensis Georgi, in the spontaneous sleep-wake regulation. *Journal of Ethnopharmacology*, *135*(2), 359–368.

Chen, C. (2009). Chemoinformatics and pharmacoinformatics approach for exploring the GABA-A agonist from Chinese herb suanzaoren. *Journal of the Taiwan Institute of Chemical Engineers*, *40*(1), 36–47.
Chen, C., Chen, Y., Wu, C., & Tsai, H. (2008). What is the effective component in suanzaoren decoction for curing insomnia? Discovery by virtual screening and molecular dynamic simulation. *Journal of Biomolecular Structure and Dynamics*, *26*(1), 57–64.
Chen, F., Jong, M., Chen, Y., Kung, Y., Chen, T., Chen, F., et al. (2011). Prescriptions of Chinese herbal medicines for insomnia in Taiwan during 2002. *Evidence-Based Complementary and Alternative Medicine*, *2011*, 1–9.
Chen, Y. L., Lee, C. Y., Huang, K. H., Kuan, Y. H., & Chen, M. (2015). Prescription patterns of Chinese herbal products for patients with sleep disorder and major depressive disorder in Taiwan. *Journal of Ethnopharmacology*, *171*, 307–316.
Cho, S., Shimizu, M., Lee, C., Han, D., Jung, C., Jo, J., et al. (2010). Hypnotic effects and binding studies for GABAA and 5-HT2C receptors of traditional medicinal plants used in Asia for insomnia. *Journal of Ethnopharmacology*, *132*(1), 225–232.
Cui, J., Yang, W., Xie, Y., Sun, Y., Zhuang, Y., & Wang, Y. (2014). Real-world analysis of concurrent diseases and medicine use among patients with insomnia. *Zhongguo Zhong Yao Za Zhi*, *39*(18), 3519–3526.
De Carvalho, R., Duarte, F., & De Lima, T. (2011). Involvement of GABAergic non-benzodiazepine sites in the anxiolytic-like and sedative effects of the flavonoid baicalein in mice. *Behavioural Brain Research*, *221*(1), 75–82.
Doghramji, K., & Jangro, W. (2016). Adverse effects of psychotropic medications on sleep. *Sleep Medicine Clinics*, *11*(4), 503–514.
Ebrahim, I., Howard, R., Kopelman, M., & Williams, A. (2002). Hypocretin/orexin system. *Journal of Royal Society of Medicine*, *95*, 227–230.
Erden, V., Abitagaoglu, S., Guler, C., Dogan, Z., Kırgezen, S., & Abut, Y. (2016). Insomnia may increase anesthetic requirement. *Journal of Clinical Anesthesia*, *34*, 367–372.
Fang, X., Hao, J., Zhou, H., Zhu, L., Wang, J., & Song, F. (2010). Pharmacological studies on the sedative-hypnotic effect of Semen Ziziphi spinosae (Suanzaoren) and Radix et Rhizoma Salviae miltiorrhizae (Danshen) extracts and the synergistic effect of their combinations. *Phytomedicine*, *17*(1), 75–80.
Gan, J., Tian, G., & Qin, G. (2013). Study on efficacy of zaoren anshen capsules in treating senile insomnia and changes in its hemorheology. *Zhongguo Zhong Yao Za Zhi*, *38*(2), 273–275.
Gao, J., Ji, W., Jiang, H., & Chen, J. (2013). Effects of extracts from ziziphi spinosae semen and schisandrae chinensis fructus on amino acid neurotransmitter in rats with insomnia induced by PCPA. *Zhong Yao Cai*, *36*(10), 1635–1639.
Grewal, R., & Doghramji, K. (2010). Epidemiology of insomnia. In H. P. Attarian & Catherine Schuman (Eds.), *Clinical handbook of insomnia* (pp. 13–22). NY: Humana Press.
Hermes, E., & Rosenheck, R. (2014). Prevalence, pharmacotherapy and clinical correlates of diagnosed insomnia among veterans' health administration service users nationally. *Sleep Medicine*, *15*(5), 508–514.
Hu, L., Zhang, X., Liu, W., Li, M., & Zhang, Y. (2015). Suan Zao Ren tangin combination with Zhi Zi Chi tangas a treatment protocol for insomniacs with anxiety: A randomized parallel-controlled trial. *Evidence-Based Complementary and Alternative Medicine*, *2015*, 1–7.
Ikeya, Y., Takeda, S., Tunakawa, M., Karakida, H., Toda, K., Yamamaguchi, T., et al. (2004). Cognitive improving and cerebral protective effects of acylated oligosaccharides in *Polygala tenuifolia*. *Biological & Pharmaceutical Bulletin*, *27*(7), 1081–1085.
Jiang, M., Lu, C., Zhang, C., Yang, J., Tan, Y., Lu, A., et al. (2012). Syndrome differentiation in modern research of traditional Chinese medicine. *Journal of Ethnopharmacology*, *140*(3), 634–642.

Jung, J., Cho, J., Ahn, N., Oh, H., Kim, S., Jang, C., et al. (2005). Effect of chronic treatment on 5-hydroxytryptamine receptors in rat brain. *Pharmacology, Biochemistry, and Behavior, 81*(1), 205–210.
Kim, K., Kang, S., Yoon, I., Lee, S., Ju, G., Han, J., et al. (2017). Prevalence and clinical characteristics of insomnia and its subtypes in the Korean elderly. *Archives of Gerontology and Geriatrics, 68*, 68–75.
Kim, J., Kim, S., Lee, S., & Jang, C. (2007). Antidepressant-like effects of Albizzia julibrissin in mice: Involvement of the 5-HT1A receptor system. *Pharmacology, Biochemistry, and Behavior, 87*(1), 41–47.
Krystal, A., Richelson, E., & Roth, T. (2013). Review of the histamine system and the clinical effects of H1 antagonists: Basis for a new model for understanding the effects of insomnia medications. *Sleep Medicine Reviews, 17*(4), 263–272.
Kung, Y., Yang, C., Chiu, J., & Kuo, T. (2016). The application of yin-yang concept on heart rate variability patterns in menopausal women with insomnia. *Chinese Journal of Integrative Medicine*, 1–7. http://dx.doi.org/10.1007/s11655-016-2590-2.
Lee, C. I., Han, J. Y., Hong, J. T., & Oh, K. W. (2013). 3,4,5-Trimethoxycinnamic acid (TMCA), one of the constituents of Polygalae Radix enhances pentobarbital-induced sleeping behaviors via GABAAergic systems in mice. *Archives of Pharmacal Research, 36*(10), 1244–1251.
Lee, C., Wong, H., Chui, K., Choang, T., Hon, P., & Chang, H. (1991). Miltirone, a central benzodiazepine receptor partial agonist from a Chinese medicinal herb Salvia miltiorrhiza. *Neuroscience Letters, 127*(2), 237–241.
Leger, D., Morin, C., Uchiyama, M., Hakimi, Z., Cure, S., & Walsh, J. (2012). Chronic insomnia, quality-of-life, and utility scores: Comparison with good sleepers in a cross-sectional international survey. *Sleep Medicine, 13*(1), 43–51.
Levenson, J., Kay, D., & Buysse, D. (2015). The pathophysiology of insomnia. *Chest, 147*(4), 1179–1192.
Li, H., Gao, Y., & Liu, P. (2009). Effect of Ttianwang Buxin decoction or Tianwang Buxin without Radix platycodi decoction on brain neurotransmitter of rats hyposomnia model. *Zhongguo Zhong Yao Za Zhi, 34*(2), 217–223.
Li, Y., Wu, P., Ning, Y., Yan, X., Zhu, T., Ma, C., et al. (2014). Sedative and hypnotic effect of freeze-dried paeoniflorin and Sini San freeze-dried powder in pentobarbital sodium-induced mice. *Journal of Traditional Chinese Medicine, 34*(2), 184–187.
Li, Y., Xu, B., & Xiao, F. (2009). Effect of modified xiaoyao powder for improving sleep in patients with psychological stress insomnia. *Zhongguo Zhong Xi Yi Jie He Za Zhi, 29*(3), 208–211.
Li, H., Yan, X., Li, T., Xu, J., Wang, Q., & Xu, Y. (2013). Efficacy of Huadananshen mistura on insomnia: A randomized, double-blind, placebo-controlled, and multi-center clinical trial. *Journal of Traditional Chinese Medicine, 33*(4), 423–427.
Liao, J., Jan, Y., Huang, S., Wang, H., Yu, L., & Chen, C. (1995). Evaluation with receptor binding assay on the water extracts of ten CNS-active Chinese herbal drugs. *Proceedings of the National Science Council, Republic of China. Part B, 19*(3), 151–158.
Liu, J., Li, F., Tang, X. D., Ma, J., Mao, M., Wu, F. Z., et al. (2015). Sini powder (四逆散) decoction alleviates mood disorder of insomnia by regulating cation-chloride cotransporters in hippocampus. *Chinese Journal of Integrative Medicine, 2015*, 1–7. http://dx.doi.org/10.1007/s11655-015-2308-x.
Liu, Q. Q., Zhang, J., Guo, R. J., Xie, Y. Z., Fu, Q. N., He, T., et al. (2016). Efficacy and safety of the Chaihuguizhiganjiang-suanzaoren granule on primary insomnia: Study protocol for a randomised controlled trial. *BMJ Open, 6*(2), 1–7.
Lu, W., & Lu, D. (2014). Impact of Chinese herbal medicine on American society and health care system: Perspective and concern. *Evidence-Based Complementary and Alternative Medicine, 2014*, 1–6.

Ma, Y., Han, H., Eun, J., Kim, H., Hong, J., & Oh, K. (2007). Sanjoinine a isolated from Zizyphi Spinosi semen augments pentobarbital-induced sleeping behaviors through the modification of GABA-ergic systems. *Biological & Pharmaceutical Bulletin*, *30*(9), 1748–1753.

Ma, Y., Ma, H., Eun, J., Nam, S., Kim, Y., Hong, J., et al. (2009). Methanol extract of Longanae Arillus augments pentobarbital-induced sleep behaviors through the modification of GABAergic systems. *Journal of Ethnopharmacology*, *122*, 245–250.

Mai, E., & Buysse, D. (2008). Insomnia: Prevalence, impact, pathogenesis, differential diagnosis, and evaluation. *Sleep Medicine Clinics*, *3*(2), 167–174.

Murphy, Y., Wilson, E., Goldner, E., & Fischer, B. (2016). Benzodiazepine use, misuse, and harm at the population level in Canada: A comprehensive narrative review of data and developments since 1995. *Clinical Drug Investigation*, *36*(7), 519–530.

Ni, X., Shergis, J., Guo, X., Zhang, A., Li, Y., Lu, C., et al. (2015). Updated clinical evidence of Chinese herbal medicine for insomnia: A systematic review and meta-analysis of randomized controlled trials. *Sleep Medicine*, *16*(12), 1462–1481.

Pearson, N., Johnson, L., & Nahin, R. (2006). Insomnia, trouble sleeping, and complementary and alternative medicine: Analysis of the 2002 national health interview survey data. *Archives of Internal Medicine*, *166*, 1775–1782.

Pigeon, W. (2010). Diagnosis, prevalence, pathways, consequences & treatment of insomnia. *The Indian Journal of Medical Research*, *131*, 321–332.

Qu, W. M., Yue, X. F., Sun, Y., Fan, K., Chen, C. R., Hou, Y. P., et al. (2012). Honokiol promotes non-rapid eye movement sleep via the benzodiazepine site of the GABA(A) receptor in mice. *British Journal of Pharmacology*, *167*(3), 587–598.

Reeve, K., & Bailes, B. (2010). Insomnia in adults: Etiology and management. *The Journal for Nurse Practitioners*, *6*(1), 53–60.

Singhuber, J., Baburin, I., Kahlig, H., Urban, E., Kopp, B., & Hering, S. (2012a). GABAA receptor modulators from Chinese herbal medicines traditionally applied against insomnia and anxiety. *Phytomedicine*, *19*(3-4), 334–340.

Singhuber, J., Baburin, I., Khom, S., Zehl, M., Urban, E., Hering, S., et al. (2012b). GABA (A) receptor modulators from the Chinese herbal drug Junci Medulla—The pith of Juncus effuses. *Planta Medica*, *78*(5), 455–458.

Song, Y., Li, F., Liu, Y., Ma, J., Mao, M., Wu, F., et al. (2014). Effects of Wen Dan Tang on insomnia-related anxiety and levels of the brain-gut peptide Ghrelin. *Neural Regeneration Research*, *9*(2), 205–212.

Sugiyama, K., Muteki, T., & Kano, T. (1996). The Japanese herbal medicine 'saiko-keishi-to' activates GABAA receptors of rat sensory neurons in culture. *Neuroscience Letters*, *216*(3), 147–150.

Tian, J., & Liu, R. (2015). Mutual information-based correlation analysis of herbs against insomnia. *Zhongguo Zhong Yao Za Zhi*, *40*(19), 3873–3877.

Ushiroyama, T., Ikeda, A., Sakuma, K., & Ueki, M. (2005). Chai-hu-gui-zhi-gan-jiang-tang regulates plasma interleukin-6 and soluble interleukin-6 receptor concentrations and improves depressed mood in climacteric women with insomnia. *The American Journal of Chinese Medicine*, *33*(5), 703–711.

Wang, L., Cui, X., Cui, S., Cao, J., Zhang, J., Zhang, Y., et al. (2010). Potentiating effect of spinosin, a C-glycoside flavonoid of Semen Ziziphi spinosae, on pentobarbital-induced sleep may be related to postsynaptic 5-HT(1A) receptors. *Phytomedicine*, *17*(6), 404–409.

Wu, F., Song, Y., Li, F., He, X., Ma, J., Feng, T., et al. (2014). Wen-Dan decoction improves negative emotions in sleep-deprived rats by regulating orexin-A and leptin expression. *Evidence-Based Complementary and Alternative Medicine*, *2014*, 1–10.

Xia, C., Xia, C., Deng, S., & Zhu, P. (2009). The XIA's No. 1 sleeping prescription for the treatment of insomnia of the deficiency type: A clinical observation of 60 cases. *Journal of Traditional Chinese Medicine*, *29*(3), 211–215.

Xin, J., Zhang, J., Yang, Y., Deng, M., & Xie, X. (2013). Radix Angelica Sinensis that contains the component Z-ligustilide promotes adult neurogenesis to mediate recovery from cognitive impairment. *Current Neurovascular Research*, *10*(4), 304–315.

Yang, Y., Li, H., Zhang, S., Li, Q., Yang, X., Chen, X., et al. (2005). TCM treatment for 63 cases of senile dyssomnia. *Journal of Traditional Chinese Medicine*, *25*(1), 45–49.

Yao, Y., Jia, M., Wu, J. G., Zhang, H., Sun, L. N., Chen, W. S., et al. (2010). Anxiolytic and sedative-hypnotic activities of polygalasaponins from Polygala tenuifolia in mice. *Pharmaceutical Biology*, *48*(7), 801–807.

Ye, R., Yuan, Z., & Dai, C. (2011). Intervention of tianwang buxin decoction combined with dormancy hygiene education for treatment of sub-healthy insomnia patients of yin deficiency fire excess syndrome. *Zhongguo Zhong Xi Yi Jie He Za Zhi*, *31*(5), 618–621.

Ye, Q., Zhou, J., Yuan, X., Yuan, C., & Yang, X. (2015). Efficacy of Zhenjingdingzhi decoction in treating insomnia with Qi-deficiency of heart and gallbladder: A randomized, double-blind, controlled trial. *Journal of Traditional Chinese Medicine*, *35*(4), 381–388.

Yeung, W., Chung, K., Poon, M., Ho, F., Zhang, S., Zhang, Z., et al. (2012a). Chinese herbal medicine for insomnia: A systematic review of randomized controlled trials. *Sleep Medicine Reviews*, *16*(6), 497–507.

Yeung, W., Chung, K., Poon, M., Ho, F., Zhang, S., Zhang, Z., et al. (2012b). Prescription of Chinese herbal medicine and selection of acupoints in pattern-based traditional Chinese medicine treatment for insomnia: A systematic review. *Evidence-Based Complementary and Alternative Medicine*, *2012*, 1–16.

Yi, P. L., Lin, C. P., Tsai, C. H., Lin, J. G., & Chang, F. C. (2007). The involvement of serotonin receptors in suanzaorentang-induced sleep alteration. *Journal of Biomedical Science*, *14*(6), 829–840. Epub 2007 Jul 27.

Yi, P., Tsai, C., Chen, Y., & Chang, F. (2007). Gamma-aminobutyric acid (GABA) receptor mediates suanzaorentang, a traditional Chinese herb remedy—Induced sleep alteration. *Journal of Biomedical Science*, *14*(2), 285–297.

Yu, G., Liang, S., & Sun, Q. (2007). Study on modified xiaoyao decoction combining clomipramine treating depression. *Zhongguo Zhong Xi Yi Jie He Za Zhi*, *27*(4), 318–320.

Zaugg, J., Eickmeier, E., Rueda, D. C., Hering, S., & Hamburger, M. (2011). HPLC-based activity profiling of Angelica pubescens roots for new positive GABAA receptor modulators in Xenopus oocytes. *Fitoterapia*, *82*(3), 434–440.

Zhang, S., Chen, Z., Lin, Y., Cheng, Y., Liu, S., & Wang, C. (2007). Clinical observation of modified Suan Zao Ren decoction on insomnia of chronic hepatitis B patients. *Zhong Yao Cai*, *30*(11), 1482–1484.

Zhang, B., & Wing, Y. (2006). Sex differences in insomnia: A meta-analysis. *Sleep*, *29*(1), 85–93.

Zhang, Q. H., & Zhang, L. (2007). Research advance in chemical composition and pharmacological action of *Chrysanthemum morifolium*. *Food Drug*, *2*, 21–23.

Zhao, Z., Fang, M., Xiao, D., Liu, M., Fefelova, N., Huang, C., et al. (2013). Potential antiarrhythmic effect of methyl 3,4,5-trimethoxycinnamate, a bioactive substance from roots of polygalae radix: Suppression of triggered activities in rabbit myocytes. *Biological & Pharmaceutical Bulletin*, *36*(2), 238–244. Epub 2012 Nov 29.

Zheng, Y., Wu, J., Feng, X., Jia, Y., Huang, J., Hao, Z., et al. (2015). In silico analysis and experimental validation of lignan extracts from Kadsura longipedunculata for potential 5-HT1AR agonists. *PloS One*, *10*(6), e0130055.

Zhou, Y., Wei, Y., Zhang, P., Gao, S., Ning, G., Zhang, Z., et al. (2007). The short-term therapeutic effect of the three-part massotherapy for insomnia due to deficiency of both the heart and the spleen—A report of 100 cases. *Journal of Traditional Chinese Medicine*, *27*(4), 261–264.

Zou, J., Zhang, J., Jia, Z., & Cao, K. (2011). Evaluation of the traditional Chinese medicine Shensongyangxin capsule on treating premature ventricular contractions: A randomized, double-blind, controlled multicenter trial. *Chinese Medical Journal*, *124*(1), 76–83.

CHAPTER SIX

# Metabolic Factors and Adult Neurogenesis: Impacts of Chinese Herbal Medicine on Brain Repair in Neurological Diseases

**Chong Gao, Jiangang Shen[1]**
School of Chinese Medicine, The University of Hong Kong, Hong Kong SAR, China
[1]Corresponding author: e-mail address: shenjg@hkucc.hku.hk

## Contents



## Abstract

Adult neurogenesis plays the important roles in animal cognitive and emotional behaviors. Abnormal proliferation and differentiation of neural stem cells (NSCs) usually associate with the neural dysfunctions induced by different brain disorders. Therefore, targeting neurogenic factors could be a promoting strategy for neural regeneration and brain repair. Importantly, epidemiological studies suggest metabolism disorders like diabetes and obesity significantly increase the risk of neurological and psychiatric diseases. A large number of studies indicate that metabolic factors could serve as the modulators to adult neurogenesis, providing the potentials of metabolic factors to regulate NSCs growth and neural regeneration therapy. This chapter reviews the current studies

*International Review of Neurobiology*, Volume 135
ISSN 0074-7742
http://dx.doi.org/10.1016/bs.irn.2017.02.007

on the roles of metabolic factors in modulating adult neurogenesis and evaluates the potentials of Chinese Herbal Medicine (CHM) for the treatment of neurological or psychiatric disorders by targeting the metabolic factors.

Traditional Chinese Medicine (TCM) including CHM and acupuncture is now widely applied for the treatment of metabolic diseases, and neurological diseases in Asia, because its' therapeutic principles meet the multiple targets and complexity characteristics of most neurological disorders. Different studies indicate that there are many active compounds perform the regulations to metabolic factors and promoting neurogenesis. This chapter systematically summarizes the current progress and understanding of the active compounds and their underlying mechanisms of CHM formulas for promoting neurogenesis. Many CHM formulas and their active ingredients that originally used for metabolic disorders show the promising effects on mediating neurogenesis and brain repair for the treatments of neurodegenerative diseases. Therefore, further investigations about the relationship between neurogenesis and metabolic regulations of CHM will bring new insights into understanding the mechanisms of adult neurogenesis and provide great opportunities to develop new therapeutic strategies for neurological diseases. Those studies will provide scientific guidance to develop the drugs from TCM resource.

## 1. INTRODUCTION

Neurological and psychiatric diseases are major disease burdens in the world. Epidemiological studies indicate that metabolic disorders like obesity and diabetes can increase the risks of neurological or psychiatric diseases, especially stroke, Parkinson's disease (PD), and Alzheimer's disease (AD) (Li, Song, & Leng, 2015; Mitchell et al., 2015; Saaksjarvi, Knekt, Mannisto, Lyytinen, & Heliovaara, 2015). Understanding how metabolic factors affect cognitive and neurological functions in neurological diseases is important for developing effective treatments for different brain disorders.

Discovery of neural stem cells/neural progenitor cells (NSPCs) in adult mammalian brains breaks the dogma that adult brain is a fixed unchanging structure. Adult neurogenesis, a process that NSPCs give rise to new neurons/glia cells, occurs in several brain regions that require high neural plasticity such as olfactory bulb, subventricular zone (SVZ), and hippocampal dentate gyrus (DG) (Zhao, Deng, & Gage, 2008). Impaired neurogenesis usually coordinates with different types of brain disorders (Liu & Song, 2016). For decades, adult neurogenesis has attracted many investigations for its critical roles in maintaining the cognitive and emotional capacities of animals (Aimone et al., 2014; Kim et al., 2012). Large efforts have been made to explore the factors that regulate growth and differentiations of

NSPCs. Recent progress indicates that metabolism-modulating factors actively regulate the growth of NSPCs and targeting metabolic factors could promote the recovery of neurological deficits in many neurological or psychiatric diseases. Antidiabetic or antiobesity drugs also appear to have their therapeutic effectiveness for neurological diseases. Those studies indicate that the values of modulating metabolism factors for neural regenerative practice.

Traditional Chinese Medicine (TCM) has been utilized as the therapeutic practice in metabolism disorders for a very long history. Many herbal products showed multiple effects of antidiabetic/antiobesity and neurogenic promotion. Because of the relationship between metabolism factors and neurogenesis process, many Chinese Herbal Medicines (CHM) showed the potential in treating both metabolism disorders and neurological diseases. This notion sheds the light for developing new drugs for treating the complex neurological and psychiatric disorders. This chapter first introduces the roles of neurogenesis in regulating neurological behaviors, and how the factors produced from neurogenic niches regulate the process of the proliferation, differentiation, and maturation of NSCs. Then the relationship between metabolism disorders and neurological diseases, as well as the functions of metabolism factors on neurogenic regulations were summarized. Finally, and most importantly, this chapter introduces the updated progress about the therapeutic values of TCM formulas and their active components in the treatments of neurological and metabolism disorders and their underlying mechanisms for regulations of metabolism factors and neurogenesis.

## 2. OVERVIEW OF NEUROGENESIS

### 2.1 Adult Neurogenesis and its Relationship With Neurological Diseases

The glial fibrillary acidic protein (GFAP) positive radial-glial-like stem cells (RGLs) in adult neurogenic niches can develop into multiple cell lineage including neurons, astrocytes, and oligodendrocytes (Kempermann, 2011). In SVZ region, NSCs produce neuronal progenitors and migrate to OB through the rostral migratory stream. The emergence of neurogenesis in SVZ is associated with the changes of olfactory behaviors (Whitman & Greer, 2009). Continuous generation of neurons especially in hippocampus engages in the cognitive or emotional regulations (Fig. 1). For instance, neurogenesis in hippocampus affects animal pattern separation behaviors (Clelland et al., 2009). Neurogenesis occurs when the output firing condition

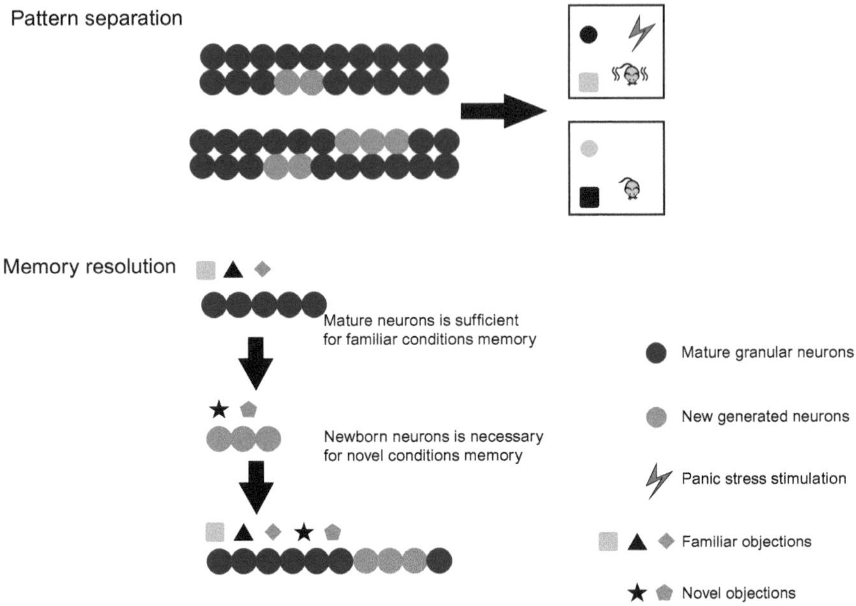


**Fig. 1** Neurogenesis is related with different types behaviors. Pattern separation is type of capability that makes the animal separate the safety condition from the endangered one based on the tiny different environmental cues. The so-called pattern separating ability is essential for animal to get rid of anxiety mood. Memory resolution is another neurogenesis-dependent brain function. This hypothesis declares that typical mature granular neurons are sufficient for memory of familiar objections. On the other hand, new generated neurons integrated into the neural circuit can help to encode the novel information. Therefore, combination of mature and new generated neurons could elevate the level of memory resolution.

of a network is less like the input firing patterns. Sufficient neurogenesis is necessary for pattern separation-dependent fear context discrimination task (Sahay et al., 2011). Conversely, focal ablation of hippocampal neurogenesis with low-dosage X-irradiation impaired mouse relative behavioral performance in spatial navigation radial arm maze task as well as the spatial, but non-navigable task in the mouse touch screen (Clelland et al., 2009). In addition, neurogenesis is also associated with other types of behaviors such as temporal context memory encoding and improvement of memory resolution (Morris, Curtis, Churchwell, Maasberg, & Kesner, 2013). All these studies indicate the important roles of NSCs in animal behavioral modulating.

Impaired neurogenesis is highly involved with the development of neurological and psychiatric disorders. As early in 1990s, the reduction of neuronal progenitor proliferation was detected in aging rodents (Kuhn,

Dickinson-Anson, & Gage, 1996), indicating that the involvement of neurogenesis impairment with neurodegenerative process. Impaired neurogenesis also occurs in many neurological diseases like stroke, AD, PD, and Huntington's diseases (HD) (Winner, Kohl, & Gage, 2011). Interestingly, increased neurogenesis has been found at the early stage of stroke and epilepsy (Cho et al., 2015; Jin et al., 2006). The possible reason might be that NSCs serve as supplementary mechanisms for neurological functions at the early stage of these disorders (Jin et al., 2006). Same appearance was also shown in other types of brain disorders. For example, in AD patients, the brain tissue showed increased expression levels of immature neuron markers doublecortin (DCX), polysialylated nerve cell adhesion molecule (PSA-NCAM), neurogenic differentiation factor (NeuroD), and Dihydropyrimidinase-like 3 (Drp-3) (Jin et al., 2004). Enhanced neurogenesis was in ischemic brain zones can be also detected in middle cerebral artery occlusion (MCAO) animal model (Zhang et al., 2006). Increased neurogenesis in the early stage of brain disorders suggests the important roles of NSCs in endogenous regenerative process. Therefore, clarifying the mechanisms of adult neurogenesis as well as its relationship with different neurological diseases would be beneficial for developing new therapeutic practice to brain disorders.

## 2.2 Cellular Signaling Pathways for Neurogenesis Regulation

In neurogenic niches, the proliferation and differentiations of RGLs are regulated by complex systems involving a series of extrinsic and intrinsic cellular signals. Crosstalk among different cellular signaling pathways triggers variable cellular events and affects the process of neurogenesis. Exploring the neurogenic regulatory effects of those factors could provide the guidance to ascertain the effective targets for neurogenesis. In this part, we briefly introduce several representative signaling pathways identified as neurogenic niche regulators (Table 1).

Wnt and Notch are two classic signaling pathways to regulate adult neurogenesis. The downstream signals of these factors, which comprise with different intrinsic molecules, are activated and eventually result in multiple events including the changes of proliferation, differentiation, and survival of the NSCs. Wnt pathway plays an important role in regulating NSCs niche (Lie et al., 2005). The signaling is activated by binding Wnt-protein ligand to a frizzled family receptor, which passes the signal to other signaling proteins inside the cells (Nusse & Varmus, 1992). Wnt signaling triggers the

**Table 1** Classical Regulation Factors on Neurogenesis

| Name of the Factor | Type | Function in Neurogenesis Regulation | References |
|---|---|---|---|
| Epidermal growth factor (EGF) | Growth factor | Gave rise to the olfactory neurons;<br>Enhance the generation of astrocytes in the olfactory bulb (OB) but resulted a reduction of total number of newborn neurons reaching the OB. | Mahanthappa and Schwarting (1993);<br>Kuhn, Winkler, Kempermann, Thal, and Gage (1997) |
| Vascular endothelial growth factor (VEGF) | Growth factor | Gene-transfer of VEGF resulted the increased neurogenesis and animal cognition. | Cao et al. (2004) |
| Fibroblast growth factor (FGF) | Growth factor | Increased the number of proliferating cells in SVZ. | Jin et al. (2005) |
| Transforming growth factor-beta (TGF-β) | Growth factor | Blockade of TGF-β activity could decrease the percentage newborn neurons in vivo. | Battista, Ferrari, Gage, and Pitossi (2006) |
| Brain-derived neurotrophic factor (BDNF) | Neurotrophic factors | Intrahippocampal infusion of the brain-derived neurotrophic factor (BDNF) resulted an increased production of new neurons locally. | Scharfman et al. (2005) |
| Ciliary neurotrophic factor (CNTF) | Neurotrophic factors | Increased the number of newborn cells in SVZ and SGZ regions by forebrain injection;<br>Increased the produce of the neuronal progenitor cells without affect the astrocytic differentiation and a slight increased gliogenesis in SVZ. | Emsley and Hagg (2003) |
| Glial cell-derived neurotrophic factor (GDNF) | Neurotrophic factors | Increased cell proliferation in the ipsilateral SVZ and promoted the recruitment of new neuroblasts into the striatum and improved survival of new mature neurons in ischemic stroke model. | Kobayashi et al. (2006) |
| Glutamate | Neurotransmitters | Increase the neurogenesis and synaptogenesis after ischemia, which contribute to memory recovery after global ischemia-induced hippocampal injury. | Bernabeu and Sharp (2000) |

| | | | |
|---|---|---|---|
| γ-Aminobutyric acid (GABA) | Neurotransmitters | Rapid increased quiescent stage and enhanced symmetrical self-renewal after conditional deletion of γ2 in RGLs. | Song et al. (2012) |
| 5-Hydroxytryptamine (serotonin) | Neurotransmitters | Both inhibition of serotonin synthesis and selective depletion of serotonin neurons could induce the decreasing number of newly generated cells in DG and SVZ region. | Brezun and Daszuta (1999) |
| Acetylcholine | Neurotransmitters | Increased the newly generated neurons in DG region. | Kotani, Yamauchi, Teramoto, and Ogura (2008) |
| Dopamine | Neurotransmitters | Promote precursor cell proliferation in the SVZ concomitant with a reduction in local epidermal growth factor (EGF) production. | Mahanthappa and Schwarting (1993) |
| β-Catenin | Transcriptional factors | Important downstream of Wnt signaling to promote neurogenesis. | Lie et al. (2005) |
| Glucocorticoid receptor (GR) | Transcriptional factors | Normalized the stress-induced reductions in neurogenesis. | Oomen, Mayer, de Kloet, Joels, and Lucassen (2007); Anacker et al. (2011) |
| cAMP response element-binding protein (CREB) | Transcriptional factors | Increased the proliferation of newborn cells in adult hippocampus;<br>Inhibition of endogenous CREB blocked ischemia-induced neurogenesis in the DG region of adult rats. | Nakagawa et al. (2002); Zhu, Lau, Liu, Wei, and Lu (2004) |
| Pax6 | Transcriptional factors | Pax6 mutant mice have a reduced neurogenic potential, but the neurogenic potential of nonradial glial precursors is not affected. | Heins et al. (2002) |

gene transcription by activating β-catenin protein and affecting calcium signaling through controlling endoplasmic reticulum (Komiya & Habas, 2008; MacDonald, Tamai, & He, 2009). Blockade of Wnt signaling reduced neurogenesis in vitro and almost completely abolished neurogenesis in vivo (Lie et al., 2005). Secreted frizzled-related protein 3 (sFRP3), a naturally secreted Wnt inhibitor, is highly expressed in adult DG granule neurons and performs as the inhibitor to RGLs proliferation and young neuronal development (Jang et al., 2013). Prospero-related homeodomain transcription factor (Prox1), a downstream singling of Wnt, is necessary for the maintenance of intermediate progenitors and the maturation of granule cells in the DG during NSCs development (Lavado, Lagutin, Chow, Baker, & Oliver, 2010). Notch pathway is another important regulator to adult neurogenesis. The ligands of Notch expressed in neuroblasts may act to instruct a cell-heritable to gliogenesis in neighboring stem cells (Morrison et al., 2000). Activation of Notch signaling induces the expression of transcriptional repressor genes such as Hes1, leading to repression of certain gene expression and maintenance of neural stem/progenitor cells (Imayoshi, Sakamoto, Yamaguchi, Mori, & Kageyama, 2010). Furthermore, a molecular perturbation experiment suggests that Notch1 signaling in cortical neurons could promote dendritic branching and inhibit dendritic growth (Redmond, Oh, Hicks, Weinmaster, & Ghosh, 2000).

Neurogenesis is also controlled by network regulation involving multiple signaling pathways. Neurotrophins like brain-derived neurotrophic factor (BDNF), as well as the growth factors including epidermal growth factor (EGF), vascular endothelial growth factor (VEGF), and fibroblast growth factors (FGFs) showed the proneurogenic effects in adult brain (Jin et al., 2002; Mahanthappa & Schwarting, 1993; Yoshimura et al., 2001). Additionally, neurotransmitters are also important factors affecting neurogenesis and act as the linkage between behavioral performance and NSCs in adulthood. One of the obvious evidence is the deficiency in monoamine neurotransmission (serotonin, norepinephrine, and dopamine) in the central nervous system underlies the development of depression (Kaufman, DeLorenzo, Choudhury, & Parsey, 2016). Serotonin (5-HT) attracts intensive attentions for its roles in modulating neurogenesis to affect depressive behaviors (Santarelli et al., 2003). Selective serotonin reuptake inhibitors are used as antidepressants revealed to stimulate NSPCs growth and promote neurogenesis, indicating the important role of serotonin in neurogenic regulations (Malberg, Eisch, Nestler, & Duman, 2000; Santarelli et al., 2003). Other neurotransmitters like γ-aminobutyric acid (GABA), glutamate, and acetylcholine (ACh) can perform as the neurogenic regulator in different ways. The abnormal interaction between GABA-induced depolarization and Disrupted-in-Schizophrenia 1 (DISC1) contributes to the impairment of neurogenesis and the development of the schizophrenia (Kim et al., 2012). Conditional knockout of GABA receptor increased proliferation, promoted gliogenesis, and self-renewing symmetric division of the RGLs, the pattern of which coincidently matched with the RGLs fate choice changes when animal expose to social isolation (Song et al., 2012). Glutamate is another important neurotransmitter regulating calcium signaling and multiple types of neural plasticity including synaptic plasticity and synaptogenesis (Collingridge & Bliss, 1995; Cottrell, Dube, Egles, & Liu, 2000). Glutamate acts the important roles in maintaining neurogenesis because of the critical role of calcium signaling in modulating cell proliferation (Sugiura, Imai, Moriyama, Horio, & Hori, 2000). Acute treatment with *N*-methyl-D-aspartate (NMDA) receptor antagonists increased the birth of neuronal progenitors and the overall density of neurons in the granule cell layer (Cameron, McEwen, & Gould, 1995). Conversely, glutamate-induced calcium overload could induce neuronal death and affect neurogenesis in various brain injuries like cerebral ischemia, epilepsy, and stress (Schlett, 2006). These lines of evidence suggest that neurotransmitters could be important therapeutic targets for neurogenesis therapy and brain repair in neurological diseases and psychiatric disorders.

## 2.3 Environmental Factors and Neurogenesis

Recent attentions have been paid on the neurogenic promotion effects of environment such as environmental enrichment (EE) and voluntary exercise (Farmer et al., 2004; Olson, Eadie, Ernst, & Christie, 2006). EE consisting of an enlarged enclosure along with tunnels, huts, toys and running wheels provides the multiple life innocuous experiences. Rodents raised in the EE for 4 weeks showed the elevated expressions of VEGF and BDNF as well as the increased production of new neurons in hippocampus, indicating the proneurogenic effects of EE (Cao et al., 2004). Recent study suggests that voluntary wheel running could also promote neurogenesis (Aimone et al., 2014). Collectively, variable neurogenic effects induced by different environmental factors are mediated by mix of different regulators from NSC niches as well as the factors from other systems. Particularly, the metabolism factors recently attracted many researches as its critical roles in NSPCs regulation.

## 3. METABOLIC FACTORS FOR NEUROGENIC REGULATIONS

Metabolic disorders and neurological diseases usually share same risk factors such as age, unhealthy diet habits, and sedentary lifestyle, indicating the possibility that metabolic factors might also be involved in regulations of neural functions. Importantly, metabolism disorders result in the impairment of neurogenesis and contribute to the development of the neurological diseases. The alterative characteristics of NSPCs in the metabolism disorders provide the hints about the functions of the relative metabolic factors in regulating neurogenesis. Therefore, this part will focus on discussing the relationship between metabolic disorders and neurogenesis, especially the factors that regulate both metabolic regulations and neurogenic mediations (Table 2).

### 3.1 Diabetes and Neurogenesis

Diabetes is a highly prevalent metabolism disorder with a long-term hyperglycemia due to insufficient insulin production. Resistance to insulin-stimulated glucose uptake is present in most patients with impaired glucose tolerance (IGT) and ultimately result in the diabetes. In brain tissue, activation of insulin signaling mediates the cognitive behaviors through mediating presynaptic neurotransmitters release (Zhao & Alkon, 2001). Insulin-like growth factors (IGFs) are well-documented molecules modulating both insulin sensitivity and neurogenesis (Clemmons, 2004). Systemic infusion of IGF-I increased NSCs proliferation and neurogenesis in the adult rat hippocampus (Anderson, Aberg, Nilsson, & Eriksson, 2002). As a type of growth factor, IGFs could regulate glucose metabolism as well as neurogenesis. Mutant mice with low serum IGF-I level (LID mice) revealed the reduced hippocampal neurogenesis and impaired spatial learning (Trejo, Llorens-Martin, & Torres-Aleman, 2008). Treatment of IGF-I prevented cognitive deficit in diabetic rats, independently of ongoing hyperglycemia (Lupien, Bluhm, & Ishii, 2003).

Incretins are a number of hormones functioning as the regulator the glucose metabolism as well as neurogenesis. Incretins stimulate the release of insulin from pancreatic beta-cells (Kahn, Hull, & Utzschneider, 2006). Glucagon-like peptide-1 (GLP-1) is recognized as a neuropeptide and a family member of incretins derived from the transcription product of proglucagon gene. GLP-1 (7–36 amide) regulates glucose-mediated insulin

**Table 2** Cell Metabolism Factors That Works on Neurogenesis

| Name of the Factor | Roles in Cell Metabolism and Targeted Therapy | References | Roles in Neurogenesis Regulation | References |
|---|---|---|---|---|
| Insulin/insulin-like growth factor (IGF) | Promoting the absorption of glucose from the blood to skeletal muscles and fat tissue. *Insulin/Ibutamoren* | Sonksen and Sonksen (2000) | ICV treatment of IGF-1 could ameliorate the age-related decline of hippocampal neurogenesis; Peripheral administration of IGF-I to rats increased the number of the newly generated neurons in granular cell layer of DG region. | Lopez-Lopez, LeRoith, and Torres-Aleman (2004); Aberg, Aberg, Hedbacker, Oscarsson, and Eriksson (2000) |
| Incretins (including glycagon-like peptide 1 (GLP-1) and gastric inhibitory polypeptide (GIP)) | Decrease blood glucose levels via stimulating insulin and suppresses glucagon secretion, inhibits gastric emptying, and reduces appetite and food intake. *Exenatide/Liraglutide* | Drucker and Nauck (2006) | Analogs of GLP-1, like liraglutide and lexisenatide, could get through the BBB and promote the neurogenesis in hippocampal DG region; Depletion of GIP signaling could result a decreased number of neuronal progenitor cells in the DG region, and the impaired synaptic plasticity in paired-pulse facilitation and a deficit of long-term potentiation in hippocampal CA1 region. | Hamilton, Patterson, Porter, Gault, and Holscher (2011) and Hunter and Holscher (2012); Faivre, Gault, Thorens, and Holscher (2011) |

*Continued*

**Table 2** Cell Metabolism Factors That Works on Neurogenesis—cont'd

| Name of the Factor | Roles in Cell Metabolism and Targeted Therapy | References | Roles in Neurogenesis Regulation | References |
|---|---|---|---|---|
| Leptin | Adipocyte-derived hormone that regulate energy balance by inhibiting hunger. *Metreleptin/energy balance* | Zhang et al. (1997) | Chronic i.p. injection of leptin increased cell proliferation without significant effects on the differentiation and the survival of newly proliferated cells in the dentate gyrus; Inhibit NSCs expansion via ERK/cyclin D1-triggered apoptosis, which might involved with the adiposity-dependent neurotoxicity. | Garza, Guo, Zhang, and Lu (2008); Segura et al. (2015) |
| Adiponectin | Deserve the insulin resistance in *db/db* and KKAy mice obesity model and result a decreased tissue triglyceride content. *Recombinant adiponectin* | Yamauchi et al. (2001) | Increased cell proliferation and decreased depression-like behaviors could be triggered by elevating the AND level in the brain. | Yau et al. (2014) |
| PPAR-γ | Promote adipogenesis in vitro and the differentiation of ES cells into fat. *Pioglitazone/Rosiglitazone* | Rosen et al. (1999) | Activation of PPAR-γ could also prevent the impairments in spatial memory and neurogenesis following transient illness induced by LPS. | Ormerod et al. (2013) |

secretion, insulin gene expression, as well as the growth and differentiation of beta-cells (Kjems, Holst, Volund, & Madsbad, 2003). GLP-1 plays important roles in neuronal plasticity and animal behavioral performance. Mice with knockout GLP-1 receptor failed establishing postsynaptic plasticity LTP (long-term plasticity) in hippocampal CA1 region and presented a pronounced deficit in object recognition task (Abbas, Faivre, & Holscher, 2009). Stimulation GLP-1 signaling reduced the brain damage and improved functional outcome in a transient MCAO stroke model (Li et al., 2009). Other incretins like gastric inhibitory polypeptide (GIP) have also the neuroprotective and neurogenic effects.

## 3.2 Lipid Metabolism and Neurogenesis

Besides glucose metabolism, lipid metabolism also regulates the activity of adult NSCs. Seven weeks of high-fat diet (HFD) treatment significantly decreased the numbers of newly generated neural cells in the DG of the hippocampus. HFD also exacerbated malondialdehyde (MDA) mediated toxicity in NPCs (Park et al., 2010). This evidence indicates the potential role of lipid metabolism factors in adult neurogenesis.

Leptin is the first identified adipokine as opposing to the "hunger hormone" for its inhibitory effects on hunger feeling. Deficient leptin signaling can usually induce obesity and diabetes. Mice with depleted leptin or leptin receptor (*ob/ob* or *db/db* mice) usually present profound metabolic symptomatology-like obese, insulin resistance and exhibit reduced activity, metabolism, and body temperature (Pelleymounter et al., 1995). Systemic infusion of low dosage of recombinant leptin overcame insulin resistance in lipodystrophic mice whose effects could not be mimicked by chronic food restriction (Shimomura, Hammer, Ikemoto, Brown, & Goldstein, 1999). Abundant expression of leptin receptor in neural tissues indicates the activity of leptin to regulate neural functions (Hakansson, Brown, Ghilardi, Skoda, & Meister, 1998). Actually, administration of leptin to adult mice increased NSCs proliferation but had no effect on the differentiation and the survival to newly proliferated cells in the hippocampal DG region (Garza et al., 2008). In experimental stroke animal model, treatment of leptin not only increased the number of NSCs in the cortex abutting the lesion but also stimulated angiogenesis and increased the number of blood vessels in the perilesional cortex (Avraham et al., 2011). In the rat depression model of chronic unpredictable stress (CUS), leptin reversed the CUS-induced reduction of hippocampal neurogenesis and evoked antidepressant behaviors through activating GSK-3β/β-catenin signaling (Garza, Guo, Zhang, & Lu, 2012).

Adiponectin is specifically and abundantly expressed in adipose tissue and directly sensitizes the body to insulin. The receptors, AdipoR1 and AdipoR2, are cloned and involved in obesity-linked insulin resistance (Kadowaki et al., 2006). Adiponectin decreases insulin resistance by down-regulating triglyceride content in the muscle and liver of obese mice (Yamauchi et al., 2001). Activation of adiponectin caused the exercise-induced hippocampal neurogenesis and antidepressant behavior (Yau et al., 2014). However, the effects of adiponectin on neurogenesis and neural behaviors are complicated and the underlying mechanisms are unclear yet. Adiponectin is also essential in EE-induced antidepressant effects; but depletion of adiponectin did not affect the hippocampal neurogenesis induced by EE (Nicolas et al., 2015), which suggests the knowledge gap between behavioral effects and neurogenesis.

Activation of peroxisome proliferator-activated receptor gamma (PPAR-γ or PPARG) could prevent the inflammation in white adipocyte and ameliorate obesity-induced insulin resistance (Tsuchida et al., 2005). As a group of transcriptional factors in adipocytes, PPARs regulate fatty acid storage and glucose metabolism. Stimulation of PPARG triggers the transcription of genes that regulates lipid uptake and adipogenesis. PPARG knockout mice failed to generate adipose tissue when fed a HFD (Jones et al., 2005), indicating the important roles of PPARG in lipid metabolism. Fenofibrate, agonist of PPARG, improved hippocampal neurogenesis and inhibited microglial activation after whole-brain irradiation (Ramanan et al., 2009), suggesting the promotion functions of PPARG in neurogenesis. PPARs activation prevented neuronal death by inhibiting of oxidative and inflammatory response in cerebral injury (Kapadia, Yi, & Vemuganti, 2008). Furthermore, activation of PPARalpha induced a vascular protection and prevented postischemic endothelial dysfunction (Bordet et al., 2006).

Collectively, interactions among different intercellular signalings and extrinsic factors establish a complex network to regulate adult neurogenesis. Current knowledge about the roles of metabolic factors in modulating adult neurogenesis provide the new rationale that metabolic and neural systems can affect each other and regulated by multiple factors. Targeting cellular metabolic factors would create new therapeutic opportunities for regeneration therapies in neurological diseases and psychiatric disorders.

# 4. TARGETING METABOLIC FACTORS INCREASE NEUROGENESIS AND APPLICATION FOR CHINESE MEDICINE TREATMENT

Cell metabolic factors impact the process of neurogenesis and could be an important target in developing new therapeutic strategies for neural regeneration. Many antidiabetic or antiobesity drugs like insulin, IGF, the agonists of GLP-1, and PPARG have been confirmed as the potential candidates for neural regenerations therapies in different neurological diseases and psychiatric disorders. However, it is a long journey that full of challenges to develop those drugs for clinical applications with reliable evidence to induce adult neurogenesis for neurodegenerative diseases and other neurological diseases.

TCM has been used for the treatments of neurological diseases and psychiatric disorders in China for centuries. Many active compounds or components isolated from medicinal herbs or Chinese Herbal Medicine (CHM) formulas could promote adult neurogenesis, such as curcumin, salvianolic acid B, baicalin, polysaccharides, and cornel iridoid glycoside (Zhang, Shen, & So, 2014). Recent studies have reported that many CHM formulas and their active components originally used for metabolic disorder have the neurogenesis-promoting effects. In this part, we conduct intensive reviews on the pure compounds, active extractions, and CHM formulas with the functions of regulating cellular metabolisms and promoting neurogenesis promotion. Furthermore, we discuss the potentials of those therapeutic strategies for treatments of neurological diseases and psychiatric disorders.

## 4.1 Effects of CHM on Neuroprotection/Neurogenesis and Metablic Modulations

According to Chinese Medicine Theory, *Yin-Yang* balance is essential for maintaining physiological functions of human body. Diseases are considered as the pathological statuses of *Yin-Yang* imbalance due to the dysfunctions of internal regulation systems and the dysfunctions of visceral systems as well as abnormal statues of channels and collaterals. The pathological statuses could be cataloged as different patterns of *Syndrome Differentiation*. The purpose of TCM treatment is to restore the Balance of *Yin-Yang* by using different therapeutic strategies. Based on the principles of "*Jun-Chen-Zuo-Shi*" (君-臣-佐-使) in the Chinese Medicinal theory, CHM formulas are made for the purpose by using different herbs based to their different functions. Each medicinal plant plays specific role and the different herbal components in the CHM formulas synergistically work for restoring the balance of *Yin-Yang*. With its long history, TCM accumulates huge number of herbs and CHM formulas with the therapeutic values. With the globalization and modernization of Chinese Medicine, large efforts have been made to explore the therapeutic principles and identify the active compounds. Many medicinal herbal plants and their effective components have revealed the neuroprotection/neurogenesis and metabolism-regulating effects, providing great opportunities for drug discovery.

Medicinal plants contain large number of chemical ingredients and active compounds through their synergic and independent bioactivities for regulating cellular metabolic factors and neurogenic factors. Bioactivity-guided studies explore the therapeutic principles and active ingredients contributing to their pharmacological functions. Wolfberry (*Lycium barbarum*, 枸杞 Gou Qi) is a representative medicinal herb with multiple functions in TCM practices. The fruit enriches bioactive components, including diester forms of lutein and zeaxanthin, phenolics, polysaccharides, betaine, and taurine (Tang et al., 2011). The dietary of wolfberry prevented the HFD-induced mitochondrial dysfunction and attenuated hepatic steatosis in an obese mouse model (Lin et al., 2014). Our recent study revealed that wolfberry polysaccharides (LBP) prevented the corticosterone-induced sexual inhibition via promoting neurogenesis in adult rat brains (Lau et al., 2012), providing the indirect evidence about the antidepressant effects of wolfberry. The metabolic modulation and neurogenic promotion effects of wolfberry could originate from different active compounds or same active compounds with multiple bioactivities. Further studies not only bring novel insight into the therapeutic principles of wolfberry for metabolic modulation as well as neurogenic promotion but also provide important sources for screening drug candidates for the purposes. By identifying the active compounds with different bioactivities from those medicinal plants, we could found valuable drug candidates with the properties of promoting neurogenesis, anti-inflammation and modulating cellular mechanisms, and others.

Many chemical components extracted from natural products present their therapeutic values for both metabolism disorders and neurological disorders (summarized in Table 3). For example, Salvianolic acid B (Sal B) is one of the major water-soluble components isolated from *Salvia miltiorrhiza* (丹参 Danshen). The compound can be potentially used for neurodegenerative diseases, cardiovascular, and cerebrovascular disorders (Durairajan et al., 2008). Sal B also showed the inhibitive effects on Aβ fibril formation and had protective effects against Aβ-induced cytotoxicity in brain tissue

**Table 3** Components Extracted From TCM Medications and Their Functions in Neuroprotection and Metabolism Regulations

| Name of Chemical | Herbal Plant Extracted | Neuroprotection | Metabolism Regulation |
|---|---|---|---|
| Salvianolic acid B | *Salvia miltiorrhiza* (Danshen) | Inhibit Aβ fibril formation and protecting against Aβ-induced cytotoxicity (Durairajan et al., 2008).<br>Promote neurogenesis in DG region (Zhuang et al., 2012). | Prevent diabetic nephropathy (Luo, Tan, Zhang, Li, & Mo, 2008).<br>Protected pancreatic beta-cells against cytotoxicity (Cheng et al., 2013). |
| Curcumin | *Crucuma longa* | Reserved the stress-induced decrease in 5-HT (1A) mRNA and BDNF protein levels in the hippocampal subfields (Xu et al., 2007).<br>Reduced the infarct volume in middle cerebral artery occlusion (MCAO) stroke model (Thiyagarajan & Sharma, 2004). | Ameliorated diabetes in HFD-induced obese and leptin-deficient *ob/ob* male mice (Weisberg, Leibel, & Tortoriello, 2008).<br>Activated PPAR-γ, Nrf2 in adipocytes (Shehzad, Ha, Subhan, & Lee, 2011). |
| Baicalin | *Scutellariae Radix* | Increased the neuronal differentiation of NPCs (Li, Zhuang, Shen, Zhang, & Shen, 2012).<br>Ameliorated memory impairment in AD model (Chen et al., 2015). | Combination with metformin treatment attenuated the STZ-induced plasma leptin reduction and hepatic triglycerides and cholesterol elevation (Waisundara, Hsu, Tan, & Huang, 2009). |
| Berberine | *Berberis vulgaris*, *Hydrastis canadensis*, *Coptis chinensis*, and *Arcangelisia flava* | Treatment enhanced motor balance by preventing dopaminergic neuronal damage (Kim et al., 2014).<br>Improved the hippocampal synaptic plasticity and electrophysiology impairments in STZ-induced AD model (Kalalian-Moghaddam, Baluchnejadmojarad, Roghani, Goshadrou, & Ronaghi, 2013). | Reduced liver weight, hepatic and plasma triglyceride, and cholesterol contents in *ob/ob* mice (Kim et al., 2009).<br>Reduced body weight and plasma triglycerides and improved insulin action in HFD Wistar rats (Lee et al., 2006). |

(Durairajan et al., 2008). In metabolism disorder, Sal B could be used for treatment of diabetic nephropathy (Luo et al., 2008). Moreover, Sal B inhibited the amyloid formation of human islet amyloid polypeptide and prevented the loss of pancreatic beta-cells, a major causative factor of T2DM (Cheng et al., 2013). In addition, Sal B increased the emergence of BrdU incorporation into NPCs in vitro and brain tissues in vivo, indicating the proneurogenic effect of Sal (Zhuang et al., 2012). Those studies suggest that Sal B has multiple bioactivities for modulating metabolic factors as well as neurogenic factors. Further investigations are warranted for understanding the underlying mechanisms.

Curcumin is a polyphenolic compound extracted from *Carcuma longa* (姜黄 Jiang Huang). Dietary curcumin could be used for diabetes treatment. Curcumin improved glucose and insulin tolerance and reduced hemoglobin A1c percentages in male mouse diabetes models of HFD-induced obesity and leptin-deficient *ob/ob* mice (Weisberg et al., 2008). Curcumin administration inhibited the expression of NF-kB, STAT-3, and activate PPAR-γ, Nrf2 in different cells including adipocytes, pancreatic cells, hepatic stellate cells, macrophages, and muscle cells (Bradford, 2013; Tiwari et al., 2014). Curcumin could downregulate the expressions of inflammatory cytokines, resistin, and leptin, and upregulate adiponectin and other associated proteins for lipid and glucose metabolism, indicating the capacity of curcumin in regulations of lipid metabolism and antiobesity effects (Sergent, Vanderstraeten, Winand, Beguin, & Schneider, 2012; Shehzad et al., 2011). Importantly, curcumin treatment increased the number of hippocampal neonatal neurons in chronic stress rats, and significantly reserved the expression of 5-HT(1A) mRNA and BDNF protein level in the hippocampal subfields and promoted hippocampal neurogenesis in the chronic stress-induced decrease in the chronic stress rats (Xu et al., 2007). The neuroprotection of curcumin also reflects on its therapeutic effects against cerebral ischemia injury. Treatment of curcumin significantly reduced the infarct volume in a MCAO stroke animal model (Thiyagarajan & Sharma, 2004). These studies indicate the curcumin has multiple bioactivities of modulating cellular metabolisms, improving neurogenesis, and having neuroprotective effects.

Baicalin is a bioactive agent extracted from *Scutellariae Radix* (黄芩 Huang Qin), which showed significant antioxidant and antidiabetic effects (Waisundara, Hsu, Huang, & Tan, 2008). Both baicalin and the combination of baicalin and metformin increased plasma leptin level and decreased the plasma levels of hepatic triglycerides and cholesterol in the streptozotocin (STZ)-induced diabetic animal model (Waisundara et al.,

2009). On the other hand, in vitro and in vivo experiments indicate that baicalin could have neurogenic promotion and neuroprotective effects (Li et al., 2010; Tu, Yang, Shi, Wang, & Chen, 2009). Baicalin has strong antioxidant ability through directly scavenging superoxide and peroxynitrite. Baicalin protected the neuronal cells from extraneous and endogenous peroxynitrite-induced neurotoxicity. In ischemia-reperfused brains, baicalin inhibited the formation of 3-nitrotyrosine, reduced infarct size, and attenuated apoptotic cell death, indicating that baicalin has neuroprotective effects against cerebral ischemia reperfusion injury (Xu et al., 2013). In our previous study, the effects of baicalin on promoting neurogenesis was explored in cultured NPCs and found that baicalin dose dependently promoted the neuronal differentiation via downregulating the expression of p-stat3 and Hes1 and promoting the expressions of NeuroD1 and Mash1 in the NPCs (Li et al., 2012). An in vivo animal experiment revealed that baicalin treatment for 3 weeks significantly increased the number of newly generated cells in hippocampus after transient cerebral ischemia (Zhuang et al., 2013). Furthermore, a recent study reported that baicalin treatment significantly ameliorated memory impairment in the amyloid-β (Aβ) (1–42) protein-induced AD mouse model (Chen et al., 2015). Those lines of evidence indicate that baicalin is a promoting herbal compound with the pharmacological effects of modulating cell metabolism and neurogenesis.

Berberine is an isoquinoline alkaloid presented in roots, rhizome, and stem bark of a number of medicinal plant species including *Berberis vulgaris* (刺檗 Ci Bo), *Hydrastis canadensis* (茅根 Mao Gen), and *Coptis chinensis* (黄连 Huang Lian) (Imanshahidi & Hosseinzadeh, 2008). Berberine treatment reduced liver weight and hepatic and plasma triglyceride, and cholesterol contents in obese *db/db* and *ob/ob* mice (Kim et al., 2009). Berberine also reduced body weight and plasma triglycerides and improved insulin action in HFD Wistar rats (Lee et al., 2006). The neuroprotective effects of berberine were found in different neurological disease models. In MPTP-induced PD animal model, berberine showed the neuroprotective effects on improving motor balance and preventing dopaminergic neuronal damage (Kim et al., 2014). Berberine also enhanced the survival rates of neurons against the MK801-induced excitatory neurotoxicity (Lee, Heo, & Kim Kwon, 2010). Treatment of berberine improved the hippocampal synaptic plasticity and electrophysiology impairments in STZ-induced AD model (Kalalian-Moghaddam et al., 2013), suggesting that the neuroprotective effects of berberine are closely related to its antidiabetic effects. However, the neurogenic promoting effects of berberine remain to be explored.

In summary, those active compounds from different medicinal plants have similar pharmacological functions of modulating lipid and glucose metabolisms and inducing neuroprotection and/or neurogenic promotion. Evidences are accumulated that many cellular metabolic factors have neurogenesis-promoting effects. The underlying mechanisms of neurogenesis promotion could be, at least in part, attributed to the regulation of cellular metabolic factors. Moreover, common pathologies between metabolic disorders and neurological diseases provide a new clue for drug development. Metabolic factors could function as neurogenesis regulators.

It is a great value to remark that TCM practitioners make prescriptions by using CHM formulas instead of using single compounds for disease treatments. We could not ignore the complicated interactions among different medicinal compounds in the CHM formulas during the preparation of herbal decoction and the complex metabolisms and distributions after oral administration of CHM herbal decoctions. Current knowledge built from the in vivo and in vitro studies is still far beyond the real pharmacological actions and their interactions. Therefore, more attention should be paid to CHM formula and use big data approaches and network pharmacology to further understanding complex metabolic process and therapeutic principles of TCM formula and related drug development.

Recent studies indicate the CHM formulas generally used for metabolism disorders can also provide neuroprotective effects. For example, Liuwei Dihuang Decoction (LWDHD) is a representative CHM formula for the treatment of "Kidney-Yin" deficiency syndrome for centuries. The formula is traditionally used for diabetics. In STZ diabetic animal model, LWDHD treatment improved cognitive impairments, upregulated the expression of IGF-1 and BDNF, downregulated the expression of caspase-3 and Aβ deposition, and attenuated neural apoptotic cell death in the hippocampus and cerebral cortex (Liu et al., 2013). The study indicates that LWDHD could have neuroprotective effects in a classic diabetic model with impaired neurological functions. LWDHD also revealed to promote hippocampal neurogenesis and improve cognitive function in rats (Zhang et al., 2014), suggests that LWDHD could mediate neurogenesis and have neuroprotective effects although the CHM formula is originally used for diabetes. Notably, CHM formulas contain multiple herbal items with complex chemical ingredients and have multiple targets involved in their pharmacological actions. It is important to integrate advanced technologies to elucidate the interactions of different active compounds and their metabolomics and explore the network pharmacological effects. Recent developments of advanced chemical analytic technologies, systematic biological approaches and bioinformatics, facilitate the studies of CHM formulas and bring novel insight into understanding the systematic regulations of CHM formulas on multiple targets for their therapeutic effects.

## 4.2 Network Pharmacology: New Application for TCM and Drug Discovery

The relationship between metabolism and adult neurogenesis regulation provides the hint that many effective drugs act via modulation of multiple proteins rather than single targets, which makes these drugs applicable to different diseases. So, it is necessary to conduct investigation on the drug targets involved in TCM therapeutic principles. Network Biology and polypharmacology hold the promise of expanding the current opportunity space for understanding multiple drug targets (Hopkins, 2008). The novel approach named CHM Network Pharmacology has been established by updating the research paradigm from the current "one target, one drug" mode to a new "network target, multicomponents" mode (Li & Zhang, 2013). This method provides a useful tool to understand the complicated therapeutic principles of TCM, and consequently enhances efficacy and precision for TCM drug development.

A schematic map of the integrative platform for CHM Network Pharmacology can be established by combination of the networks targets of certain CHM formulas and different diseases. For example, by analyzing the corresponding predicted oral bioavailability of *Radix Curcumae* (姜黄 Jiang Huang) formula as well as identifying their targets, it could establish a T-D network that *Radix Curcumae* formula with 147 diseases. Most of these diseases belong to neoplasms and neural system disorders, and additionally *Radix Curcumae* formula also targeted on nutritional and metabolic diseases (Tao et al., 2013). Through database analysis, the new discovery could be used for expanding the T-D network for linking the CHM formulas original used for metabolism disorders with neural system diseases. Fig. 2 elucidates the procedure for TCM T-D network. Furthermore, the T-D based CHM network pharmacology can facilitate to explore therapeutic mechanisms for specific neurological diseases and metabolic disorders. For example, in combination the molecular networks related to neurogenesis, CHM based network could be an important database to link certain CHM compounds with those neurodegenerative and metabolic diseases. Moreover, the network analysis could explore the mechanisms certain diseases by specifically pinpointing the certain pathway or targeting the particular pathways

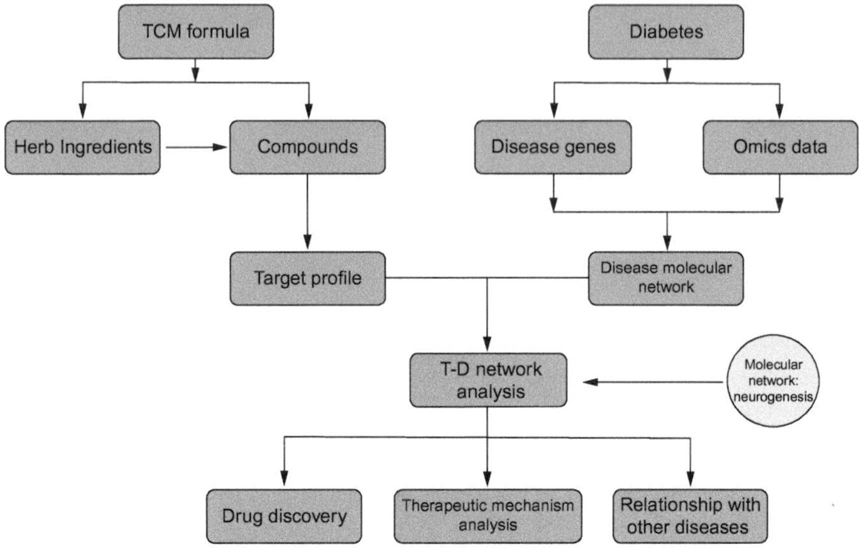


**Fig. 2** Process of TCM network pharmacological application on diabetic-related neurological disorders. Through systematic analysis of the herbal ingredients and compounds from the certain TCM formula, it can get the correlated target profile. On the other side, disease molecular network would be obtained by exploring the genes and omics data of the diabetes. Target-drug (T-D) map will be generated through pinpointing the certain disease factors to the TCM targets. Combining the neurogenesis network will help to transform the TCM original antidiabetic targets into neuroprotective effects. This network analysis will be applicable into drug discovery, therapeutic mechanic analysis of TCM formula. In addition, this will also provide the hints about the relationships among different diseases.

network. The network database provides a large amount of information for drug exploring and enhances the evidence to confirm the relationship among different diseases. With the help of network database, scientists can identify the drug related targets more efficiently and confirm the effects of the drug more precisely. Therefore, combination of advanced analytic approaches and network pharmacology methodology will create new opportunities to understand the complex therapeutic principles for TCM and related drug discovery in future.

## 5. CONCULSION AND PERSPECTIVE

Adult neurogenesis is an excellent therapeutic target for brain repair and recovering neurological functions in different neurological diseases, psychiatric disorders, and metabolic diseases. Accumulating evidences indicate that targeting cellular metabolic factors or neurogenesis/neuroprotective factors would impact on the treatments of neurological diseases, psychiatric disorders, and metabolic diseases. Although precise underlying mechanisms are not fully understood, recent progress indeed provide strong support to the hypothesis that crosstalk and interactions of different cellular signaling molecules contribute to the overall modulation for cellular metabolisms, neurogenesis, and neuroprotection. By specifically regulating the target molecules, scientists would be able to control the process of neurogenesis at the specific stage and position. To make this come true, identifying specific compounds targeting different modulating factors for neurogenesis regulation are essential. Exploration of the relationship between metabolism and neurogenesis provides the new thought to explore the new neurogenic factors, bring novel insight into developing therapeutic strategies for neurological diseases and metabolic disorders. Through T-D-based network pharmacology, a multiple-compound/multiple-targeted map can be established in order to further clarify the neurogenic functions of different compounds extracted from TCM formulas.

TCM practitioners commonly make prescriptions and therapeutic protocols according to integrative medicinal principle and standing a systematic angle, aiming to restore *Yin-Yang* balance in human body. Since multiple active compounds are included in the CHM formulas, scientists generally view this therapeutic strategy as a puzzle. Unfortunately, many chronic diseases have overall disruptions at multiple signaling pathways. The conventional therapeutic strategy based on "Single Drug and Single Target" approaches is commonly unsatisfied to treat those diseases such as neurological diseases and metabolic disorders. Recent years, attentions have been switched to use classic TCM approaches to treat those diseases since multiple targets could be overall modulated by the multiple components of CHM formulas. The developments of the advanced chemical analytic platforms and the high-throughput approaches of proteomics, metabolomics, and bioinformatics provide powerful tools for overcoming the obstacles to understand the therapeutic principles of CHM formulas. With the in-depth understanding of the molecular targets and active compounds, more and more scientists and physicians with conventional medicine training background will adopt those "old" but "novel" therapeutic strategies in their practice and research. We should note that many metabolic factors have neurogenesis-promoting effects. Those regeneration therapies could be due to the direct modulations of targeting metabolic factors to affect NPCs growth or come from the indirect actions of network regulations. Linkage between metabolic disorder and neurogenesis impairments helps to expanding the horizon of TCM application when choosing the appropriate formulas to neurological patients. In addition, through T-D network analysis, it could provide a theoretic basement for one TCM formula applicable for different types of diseases. Furthermore, the network map also could provide a solid guidance for adding or removing the certain ingredients when treating different diseases.

As an important physiology to maintain the body functions, metabolism can affect many aspects including immunology, circulation, and neurological functions. Many metabolic factors function as the important factors to regulate neurogenesis. This overlap functions enhance the intrinsic association between metabolic disorders and neurological diseases. Further investigations are warranted to understand the mechanisms of adult neurogenesis and bring new targets for drug discovery.

## ACKNOWLEDGMENTS

This work was supported by RGC GRF grants from Hong Kong Research Grants Council (GRF No. 777313M, 776512M, J.S.) and National Natural Science Foundation of China (NSFC)—General Program (No. 31270902, J.S.).

## REFERENCES

Abbas, T., Faivre, E., & Holscher, C. (2009). Impairment of synaptic plasticity and memory formation in GLP-1 receptor KO mice: Interaction between type 2 diabetes and Alzheimer's disease. *Behavioural Brain Research*, *205*, 265–271.

Aberg, M. A., Aberg, N. D., Hedbacker, H., Oscarsson, J., & Eriksson, P. S. (2000). Peripheral infusion of IGF-I selectively induces neurogenesis in the adult rat hippocampus. *The Journal of Neuroscience: The Official Journal of the Society for Neuroscience*, *20*, 2896–2903.

Aimone, J. B., Li, Y., Lee, S. W., Clemenson, G. D., Deng, W., & Gage, F. H. (2014). Regulation and function of adult neurogenesis: From genes to cognition. *Physiological Reviews*, *94*, 991–1026.

Anacker, C., Zunszain, P. A., Cattaneo, A., Carvalho, L. A., Garabedian, M. J., Thuret, S., et al. (2011). Antidepressants increase human hippocampal neurogenesis by activating the glucocorticoid receptor. *Molecular Psychiatry*, *16*, 738–750.

Anderson, M. F., Aberg, M. A., Nilsson, M., & Eriksson, P. S. (2002). Insulin-like growth factor-I and neurogenesis in the adult mammalian brain. *Brain Research. Developmental Brain Research*, *134*, 115–122.

Avraham, Y., Davidi, N., Lassri, V., Vorobiev, L., Kabesa, M., Dayan, M., et al. (2011). Leptin induces neuroprotection neurogenesis and angiogenesis after stroke. *Current Neurovascular Research*, *8*, 313–322.

Battista, D., Ferrari, C. C., Gage, F. H., & Pitossi, F. J. (2006). Neurogenic niche modulation by activated microglia: Transforming growth factor beta increases neurogenesis in the adult dentate gyrus. *The European Journal of Neuroscience*, *23*, 83–93.

Bernabeu, R., & Sharp, F. R. (2000). NMDA and AMPA/kainate glutamate receptors modulate dentate neurogenesis and CA3 synapsin-I in normal and ischemic hippocampus. *Journal of cerebral blood flow and metabolism: official journal of the International Society of Cerebral Blood Flow and Metabolism*, *20*, 1669–1680.

Bordet, R., Ouk, T., Petrault, O., Gele, P., Gautier, S., Laprais, M., et al. (2006). PPAR: A new pharmacological target for neuroprotection in stroke and neurodegenerative diseases. *Biochemical Society Transactions*, *34*, 1341–1346.

Bradford, P. G. (2013). Curcumin and obesity. *Biofactors*, *39*, 78–87.

Brezun, J. M., & Daszuta, A. (1999). Depletion in serotonin decreases neurogenesis in the dentate gyrus and the subventricular zone of adult rats. *Neuroscience*, *89*, 999–1002.

Cameron, H. A., McEwen, B. S., & Gould, E. (1995). Regulation of adult neurogenesis by excitatory input and NMDA receptor activation in the dentate gyrus. *The Journal of Neuroscience: The Official Journal of the Society for Neuroscience*, *15*, 4687–4692.

Cao, L., Jiao, X., Zuzga, D. S., Liu, Y., Fong, D. M., Young, D., et al. (2004). VEGF links hippocampal activity with neurogenesis, learning and memory. *Nature Genetics*, *36*, 827–835.

Chen, C., Li, X., Gao, P., Tu, Y., Zhao, M., Li, J., et al. (2015). Baicalin attenuates alzheimer-like pathological changes and memory deficits induced by amyloid beta1-42 protein. *Metabolic Brain Disease*, *30*, 537–544.

Cheng, B., Gong, H., Li, X., Sun, Y., Chen, H., Zhang, X., et al. (2013). Salvianolic acid B inhibits the amyloid formation of human islet amyloid polypeptide and protects pancreatic beta-cells against cytotoxicity. *Proteins*, *81*, 613–621.

Cho, K. O., Lybrand, Z. R., Ito, N., Brulet, R., Tafacory, F., Zhang, L., et al. (2015). Aberrant hippocampal neurogenesis contributes to epilepsy and associated cognitive decline. *Nature Communications*, *6*, 6606.

Clelland, C. D., Choi, M., Romberg, C., Clemenson, G. D., Jr., Fragniere, A., Tyers, P., et al. (2009). A functional role for adult hippocampal neurogenesis in spatial pattern separation. *Science*, *325*, 210–213.

Clemmons, D. R. (2004). The relative roles of growth hormone and IGF-1 in controlling insulin sensitivity. *The Journal of Clinical Investigation*, *113*, 25–27.

Collingridge, G. L., & Bliss, T. V. (1995). Memories of NMDA receptors and LTP. *Trends in Neurosciences*, *18*, 54–56.

Cottrell, J. R., Dube, G. R., Egles, C., & Liu, G. (2000). Distribution, density, and clustering of functional glutamate receptors before and after synaptogenesis in hippocampal neurons. *Journal of Neurophysiology*, *84*, 1573–1587.

Drucker, D. J., & Nauck, M. A. (2006). The incretin system: Glucagon-like peptide-1 receptor agonists and dipeptidyl peptidase-4 inhibitors in type 2 diabetes. *Lancet*, *368*, 1696–1705.

Durairajan, S. S., Yuan, Q., Xie, L., Chan, W. S., Kum, W. F., Koo, I., et al. (2008). Salvianolic acid B inhibits Abeta fibril formation and disaggregates preformed fibrils and protects against Abeta-induced cytotoxicty. *Neurochemistry International*, *52*, 741–750.

Emsley, J. G., & Hagg, T. (2003). Endogenous and exogenous ciliary neurotrophic factor enhances forebrain neurogenesis in adult mice. *Experimental Neurology*, *183*, 298–310.

Faivre, E., Gault, V. A., Thorens, B., & Holscher, C. (2011). Glucose-dependent insulinotropic polypeptide receptor knockout mice are impaired in learning, synaptic plasticity, and neurogenesis. *Journal of Neurophysiology*, *105*, 1574–1580.

Farmer, J., Zhao, X., van Praag, H., Wodtke, K., Gage, F. H., & Christie, B. R. (2004). Effects of voluntary exercise on synaptic plasticity and gene expression in the dentate gyrus of adult male Sprague-Dawley rats in vivo. *Neuroscience*, *124*, 71–79.

Garza, J. C., Guo, M., Zhang, W., & Lu, X. Y. (2008). Leptin increases adult hippocampal neurogenesis in vivo and in vitro. *The Journal of Biological Chemistry*, *283*, 18238–18247.

Garza, J. C., Guo, M., Zhang, W., & Lu, X. Y. (2012). Leptin restores adult hippocampal neurogenesis in a chronic unpredictable stress model of depression and reverses

glucocorticoid-induced inhibition of GSK-3beta/beta-catenin signaling. *Molecular Psychiatry, 17*, 790–808.
Hakansson, M. L., Brown, H., Ghilardi, N., Skoda, R. C., & Meister, B. (1998). Leptin receptor immunoreactivity in chemically defined target neurons of the hypothalamus. *The Journal of Neuroscience: The Official Journal of the Society for Neuroscience, 18*, 559–572.
Hamilton, A., Patterson, S., Porter, D., Gault, V. A., & Holscher, C. (2011). Novel GLP-1 mimetics developed to treat type 2 diabetes promote progenitor cell proliferation in the brain. *Journal of Neuroscience Research, 89*, 481–489.
Heins, N., Malatesta, P., Cecconi, F., Nakafuku, M., Tucker, K. L., Hack, M. A., et al. (2002). Glial cells generate neurons: The role of the transcription factor Pax6. *Nature Neuroscience, 5*, 308–315.
Hopkins, A. L. (2008). Network pharmacology: The next paradigm in drug discovery. *Nature Chemical Biology, 4*, 682–690.
Hunter, K., & Holscher, C. (2012). Drugs developed to treat diabetes, liraglutide and lixisenatide, cross the blood brain barrier and enhance neurogenesis. *BMC Neuroscience, 13*, 33.
Imanshahidi, M., & Hosseinzadeh, H. (2008). Pharmacological and therapeutic effects of Berberis vulgaris and its active constituent, berberine. *Phytotherapy Research: PTR, 22*, 999–1012.
Imayoshi, I., Sakamoto, M., Yamaguchi, M., Mori, K., & Kageyama, R. (2010). Essential roles of notch signaling in maintenance of neural stem cells in developing and adult brains. *The Journal of Neuroscience: The Official Journal of the Society for Neuroscience, 30*, 3489–3498.
Jang, M. H., Bonaguidi, M. A., Kitabatake, Y., Sun, J., Song, J., Kang, E., et al. (2013). Secreted frizzled-related protein 3 regulates activity-dependent adult hippocampal neurogenesis. *Cell Stem Cell, 12*, 215–223.
Jin, K., LaFevre-Bernt, M., Sun, Y., Chen, S., Gafni, J., Crippen, D., et al. (2005). FGF-2 promotes neurogenesis and neuroprotection and prolongs survival in a transgenic mouse model of Huntington's disease. *Proceedings of the National Academy of Sciences of the United States of America, 102*, 18189–18194.
Jin, K., Peel, A. L., Mao, X. O., Xie, L., Cottrell, B. A., Henshall, D. C., et al. (2004). Increased hippocampal neurogenesis in Alzheimer's disease. *Proceedings of the National Academy of Sciences of the United States of America, 101*, 343–347.
Jin, K., Wang, X., Xie, L., Mao, X. O., Zhu, W., Wang, Y., et al. (2006). Evidence for stroke-induced neurogenesis in the human brain. *Proceedings of the National Academy of Sciences of the United States of America, 103*, 13198–13202.
Jin, K., Zhu, Y., Sun, Y., Mao, X. O., Xie, L., & Greenberg, D. A. (2002). Vascular endothelial growth factor (VEGF) stimulates neurogenesis in vitro and in vivo. *Proceedings of the National Academy of Sciences of the United States of America, 99*, 11946–11950.
Jones, J. R., Barrick, C., Kim, K. A., Lindner, J., Blondeau, B., Fujimoto, Y., et al. (2005). Deletion of PPARgamma in adipose tissues of mice protects against high fat diet-induced obesity and insulin resistance. *Proceedings of the National Academy of Sciences of the United States of America, 102*, 6207–6212.
Kadowaki, T., Yamauchi, T., Kubota, N., Hara, K., Ueki, K., & Tobe, K. (2006). Adiponectin and adiponectin receptors in insulin resistance, diabetes, and the metabolic syndrome. *The Journal of Clinical Investigation, 116*, 1784–1792.
Kahn, S. E., Hull, R. L., & Utzschneider, K. M. (2006). Mechanisms linking obesity to insulin resistance and type 2 diabetes. *Nature, 444*, 840–846.
Kalalian-Moghaddam, H., Baluchnejadmojarad, T., Roghani, M., Goshadrou, F., & Ronaghi, A. (2013). Hippocampal synaptic plasticity restoration and anti-apoptotic effect underlie berberine improvement of learning and memory in streptozotocin-diabetic rats. *European Journal of Pharmacology, 698*, 259–266.

Kapadia, R., Yi, J. H., & Vemuganti, R. (2008). Mechanisms of anti-inflammatory and neuroprotective actions of PPAR-gamma agonists. *Frontiers in Bioscience*, *13*, 1813–1826.

Kaufman, J., DeLorenzo, C., Choudhury, S., & Parsey, R. V. (2016). The 5-HT1A receptor in major depressive disorder. *European Neuropsychopharmacology: The Journal of the European College of Neuropsychopharmacology*, *26*, 397–410.

Kempermann, G. (2011). The pessimist's and optimist's views of adult neurogenesis. *Cell*, *145*, 1009–1011.

Kim, M., Cho, K. H., Shin, M. S., Lee, J. M., Cho, H. S., Kim, C. J., et al. (2014). Berberine prevents nigrostriatal dopaminergic neuronal loss and suppresses hippocampal apoptosis in mice with Parkinson's disease. *International Journal of Molecular Medicine*, *33*, 870–878.

Kim, W. S., Lee, Y. S., Cha, S. H., Jeong, H. W., Choe, S. S., Lee, M. R., et al. (2009). Berberine improves lipid dysregulation in obesity by controlling central and peripheral AMPK activity. *American Journal of Physiology. Endocrinology and Metabolism*, *296*, E812–E819.

Kim, J. Y., Liu, C. Y., Zhang, F., Duan, X., Wen, Z., Song, J., et al. (2012). Interplay between DISC1 and GABA signaling regulates neurogenesis in mice and risk for schizophrenia. *Cell*, *148*, 1051–1064.

Kjems, L. L., Holst, J. J., Volund, A., & Madsbad, S. (2003). The influence of GLP-1 on glucose-stimulated insulin secretion: Effects on beta-cell sensitivity in type 2 and nondiabetic subjects. *Diabetes*, *52*, 380–386.

Kobayashi, T., Ahlenius, H., Thored, P., Kobayashi, R., Kokaia, Z., & Lindvall, O. (2006). Intracerebral infusion of glial cell line-derived neurotrophic factor promotes striatal neurogenesis after stroke in adult rats. *Stroke*, *37*, 2361–2367.

Komiya, Y., & Habas, R. (2008). Wnt signal transduction pathways. *Organogenesis*, *4*, 68–75.

Kotani, S., Yamauchi, T., Teramoto, T., & Ogura, H. (2008). Donepezil, an acetylcholinesterase inhibitor, enhances adult hippocampal neurogenesis. *Chemico-Biological Interactions*, *175*, 227–230.

Kuhn, H. G., Dickinson-Anson, H., & Gage, F. H. (1996). Neurogenesis in the dentate gyrus of the adult rat: Age-related decrease of neuronal progenitor proliferation. *The Journal of Neuroscience: The Official Journal of the Society for Neuroscience*, *16*, 2027–2033.

Kuhn, H. G., Winkler, J., Kempermann, G., Thal, L. J., & Gage, F. H. (1997). Epidermal growth factor and fibroblast growth factor-2 have different effects on neural progenitors in the adult rat brain. *The Journal of Neuroscience: The Official Journal of the Society for Neuroscience*, *17*, 5820–5829.

Lau, B. W., Lee, J. C., Li, Y., Fung, S. M., Sang, Y. H., Shen, J., et al. (2012). Polysaccharides from wolfberry prevents corticosterone-induced inhibition of sexual behavior and increases neurogenesis. *PloS One*, *7*, e33374.

Lavado, A., Lagutin, O. V., Chow, L. M., Baker, S. J., & Oliver, G. (2010). Prox1 is required for granule cell maturation and intermediate progenitor maintenance during brain neurogenesis. *PLoS Biol*, *8*(8), e1000460.

Lee, T., Heo, H., & Kim Kwon, Y. (2010). Effect of Berberine on cell survival in the developing rat brain damaged by MK-801. *Experimental Neurobiology*, *19*, 140–145.

Lee, Y. S., Kim, W. S., Kim, K. H., Yoon, M. J., Cho, H. J., Shen, Y., et al. (2006). Berberine, a natural plant product, activates AMP-activated protein kinase with beneficial metabolic effects in diabetic and insulin-resistant states. *Diabetes*, *55*, 2256–2264.

Li, H., Hu, J., Ma, L., Yuan, Z., Wang, Y., Wang, X., et al. (2010). Comprehensive study of baicalin down-regulating NOD2 receptor expression of neurons with oxygen-glucose deprivation in vitro and cerebral ischemia-reperfusion in vivo. *European Journal of Pharmacology*, *649*, 92–99.

Li, Y., Perry, T., Kindy, M. S., Harvey, B. K., Tweedie, D., Holloway, H. W., et al. (2009). GLP-1 receptor stimulation preserves primary cortical and dopaminergic neurons in cellular and rodent models of stroke and Parkinsonism. *Proceedings of the National Academy of Sciences of the United States of America, 106*, 1285–1290.

Li, X., Song, D., & Leng, S. X. (2015). Link between type 2 diabetes and Alzheimer's disease: From epidemiology to mechanism and treatment. *Clinical Interventions in Aging, 10*, 549–560.

Li, S., & Zhang, B. (2013). Traditional Chinese medicine network pharmacology: Theory, methodology and application. *Chinese Journal of Natural Medicines, 11*, 110–120.

Li, Y., Zhuang, P., Shen, B., Zhang, Y., & Shen, J. (2012). Baicalin promotes neuronal differentiation of neural stem/progenitor cells through modulating p-stat3 and bHLH family protein expression. *Brain Research, 1429*, 36–42.

Lie, D. C., Colamarino, S. A., Song, H. J., Desire, L., Mira, H., Consiglio, A., et al. (2005). Wnt signalling regulates adult hippocampal neurogenesis. *Nature, 437*, 1370–1375.

Lin, D., He, H., Ji, H., Willis, J., Willard, L., Jiang, Y., et al. (2014). Wolfberries potentiate mitophagy and enhance mitochondrial biogenesis leading to prevention of hepatic steatosis in obese mice: The role of AMP-activated protein kinase alpha2 subunit. *Molecular Nutrition & Food Research, 58*, 1005–1015.

Liu, J. P., Feng, L., Zhang, M. H., Ma, D. Y., Wang, S. Y., Gu, J., et al. (2013). Neuroprotective effect of Liuwei Dihuang decoction on cognition deficits of diabetic encephalopathy in streptozotocin-induced diabetic rat. *Journal of Ethnopharmacology, 150*, 371–381.

Liu, H., & Song, N. (2016). Molecular mechanism of adult neurogenesis and its association with human brain diseases. *Journal of Central Nervous System Disease, 8*, 5–11.

Lopez-Lopez, C., LeRoith, D., & Torres-Aleman, I. (2004). Insulin-like growth factor I is required for vessel remodeling in the adult brain. *Proceedings of the National Academy of Sciences of the United States of America, 101*, 9833–9838.

Luo, P., Tan, Z., Zhang, Z., Li, H., & Mo, Z. (2008). Inhibitory effects of salvianolic acid B on the high glucose-induced mesangial proliferation via NF-kappaB-dependent pathway. *Biological & Pharmaceutical Bulletin, 31*, 1381–1386.

Lupien, S. B., Bluhm, E. J., & Ishii, D. N. (2003). Systemic insulin-like growth factor-I administration prevents cognitive impairment in diabetic rats, and brain IGF regulates learning/memory in normal adult rats. *Journal of Neuroscience Research, 74*, 512–523.

MacDonald, B. T., Tamai, K., & He, X. (2009). Wnt/beta-catenin signaling: Components, mechanisms, and diseases. *Developmental Cell, 17*, 9–26.

Mahanthappa, N. K., & Schwarting, G. A. (1993). Peptide growth factor control of olfactory neurogenesis and neuron survival in vitro: Roles of EGF and TGF-beta s. *Neuron, 10*, 293–305.

Malberg, J. E., Eisch, A. J., Nestler, E. J., & Duman, R. S. (2000). Chronic antidepressant treatment increases neurogenesis in adult rat hippocampus. *The Journal of Neuroscience: The Official Journal of the Society for Neuroscience, 20*, 9104–9110.

Mitchell, A. B., Cole, J. W., McArdle, P. F., Cheng, Y. C., Ryan, K. A., Sparks, M. J., et al. (2015). Obesity increases risk of ischemic stroke in young adults. *Stroke, 46*, 1690–1692.

Morris, A. M., Curtis, B. J., Churchwell, J. C., Maasberg, D. W., & Kesner, R. P. (2013). Temporal associations for spatial events: The role of the dentate gyrus. *Behavioural Brain Research, 256*, 250–256.

Morrison, S. J., Perez, S. E., Qiao, Z., Verdi, J. M., Hicks, C., Weinmaster, G., et al. (2000). Transient notch activation initiates an irreversible switch from neurogenesis to gliogenesis by neural crest stem cells. *Cell, 101*, 499–510.

Nakagawa, S., Kim, J. E., Lee, R., Malberg, J. E., Chen, J., Steffen, C., et al. (2002). Regulation of neurogenesis in adult mouse hippocampus by cAMP and the cAMP response element-binding protein. *The Journal of Neuroscience: The Official Journal of the Society for Neuroscience, 22*, 3673–3682.

Nicolas, S., Veyssiere, J., Gandin, C., Zsurger, N., Pietri, M., Heurteaux, C., et al. (2015). Neurogenesis-independent antidepressant-like effects of enriched environment is dependent on adiponectin. *Psychoneuroendocrinology*, *57*, 72–83.

Nusse, R., & Varmus, H. E. (1992). Wnt genes. *Cell*, *69*, 1073–1087.

Olson, A. K., Eadie, B. D., Ernst, C., & Christie, B. R. (2006). Environmental enrichment and voluntary exercise massively increase neurogenesis in the adult hippocampus via dissociable pathways. *Hippocampus*, *16*, 250–260.

Oomen, C. A., Mayer, J. L., de Kloet, E. R., Joels, M., & Lucassen, P. J. (2007). Brief treatment with the glucocorticoid receptor antagonist mifepristone normalizes the reduction in neurogenesis after chronic stress. *The European Journal of Neuroscience*, *26*, 3395–3401.

Ormerod, B. K., Hanft, S. J., Asokan, A., Haditsch, U., Lee, S. W., & Palmer, T. D. (2013). PPARgamma activation prevents impairments in spatial memory and neurogenesis following transient illness. *Brain, Behavior, and Immunity*, *29*, 28–38.

Park, H. R., Park, M., Choi, J., Park, K. Y., Chung, H. Y., & Lee, J. (2010). A high-fat diet impairs neurogenesis: Involvement of lipid peroxidation and brain-derived neurotrophic factor. *Neuroscience Letters*, *482*, 235–239.

Pelleymounter, M. A., Cullen, M. J., Baker, M. B., Hecht, R., Winters, D., Boone, T., et al. (1995). Effects of the obese gene product on body weight regulation in ob/ob mice. *Science*, *269*, 540–543.

Ramanan, S., Kooshki, M., Zhao, W., Hsu, F. C., Riddle, D. R., & Robbins, M. E. (2009). The PPARalpha agonist fenofibrate preserves hippocampal neurogenesis and inhibits microglial activation after whole-brain irradiation. *International Journal of Radiation Oncology, Biology, Physics*, *75*, 870–877.

Redmond, L., Oh, S. R., Hicks, C., Weinmaster, G., & Ghosh, A. (2000). Nuclear Notch1 signaling and the regulation of dendritic development. *Nature Neuroscience*, *3*, 30–40.

Rosen, E. D., Sarraf, P., Troy, A. E., Bradwin, G., Moore, K., Milstone, D. S., et al. (1999). PPAR gamma is required for the differentiation of adipose tissue in vivo and in vitro. *Molecular Cell*, *4*, 611–617.

Saaksjarvi, K., Knekt, P., Mannisto, S., Lyytinen, J., & Heliovaara, M. (2015). Prospective study on the components of metabolic syndrome and the incidence of Parkinson's disease. *Parkinsonism & Related Disorders*, *21*, 1148–1155.

Sahay, A., Scobie, K. N., Hill, A. S., O'Carroll, C. M., Kheirbek, M. A., Burghardt, N. S., et al. (2011). Increasing adult hippocampal neurogenesis is sufficient to improve pattern separation. *Nature*, *472*, 466–470.

Santarelli, L., Saxe, M., Gross, C., Surget, A., Battaglia, F., Dulawa, S., et al. (2003). Requirement of hippocampal neurogenesis for the behavioral effects of antidepressants. *Science*, *301*, 805–809.

Scharfman, H., Goodman, J., Macleod, A., Phani, S., Antonelli, C., & Croll, S. (2005). Increased neurogenesis and the ectopic granule cells after intrahippocampal BDNF infusion in adult rats. *Experimental Neurology*, *192*, 348–356.

Schlett, K. (2006). Glutamate as a modulator of embryonic and adult neurogenesis. *Current Topics in Medicinal Chemistry*, *6*, 949–960.

Segura, S., Efthimiadi, L., Porcher, C., Courtes, S., Coronas, V., Krantic, S., et al. (2015). Leptin-dependent neurotoxicity via induction of apoptosis in adult rat neurogenic cells. *Frontiers in Cellular Neuroscience*, *9*, 350.

Sergent, T., Vanderstraeten, J., Winand, J., Beguin, P., & Schneider, Y.-J. (2012). Phenolic compounds and plant extracts as potential natural anti-obesity substances. *Food Chemistry*, *135*, 68–73.

Shehzad, A., Ha, T., Subhan, F., & Lee, Y. S. (2011). New mechanisms and the anti-inflammatory role of curcumin in obesity and obesity-related metabolic diseases. *European Journal of Nutrition*, *50*, 151–161.

Shimomura, I., Hammer, R. E., Ikemoto, S., Brown, M. S., & Goldstein, J. L. (1999). Leptin reverses insulin resistance and diabetes mellitus in mice with congenital lipodystrophy. *Nature*, *401*, 73–76.

Song, J., Zhong, C., Bonaguidi, M. A., Sun, G. J., Hsu, D., Gu, Y., et al. (2012). Neuronal circuitry mechanism regulating adult quiescent neural stem-cell fate decision. *Nature*, *489*, 150–154.

Sonksen, P., & Sonksen, J. (2000). Insulin: Understanding its action in health and disease. *British Journal of Anaesthesia*, *85*, 69–79.

Sugiura, T., Imai, E., Moriyama, T., Horio, M., & Hori, M. (2000). Calcium channel blockers inhibit proliferation and matrix production in rat mesangial cells: Possible mechanism of suppression of AP-1 and CREB activities. *Nephron*, *85*, 71–80.

Tang, L., Zhang, Y., Jiang, Y., Willard, L., Ortiz, E., Wark, L., et al. (2011). Dietary wolfberry ameliorates retinal structure abnormalities in db/db mice at the early stage of diabetes. *Experimental Biology and Medicine (Maywood, N.J.)*, *236*, 1051–1063.

Tao, W., Xu, X., Wang, X., Li, B., Wang, Y., Li, Y., et al. (2013). Network pharmacology-based prediction of the active ingredients and potential targets of Chinese herbal Radix Curcumae formula for application to cardiovascular disease. *Journal of Ethnopharmacology*, *145*, 1–10.

Thiyagarajan, M., & Sharma, S. S. (2004). Neuroprotective effect of curcumin in middle cerebral artery occlusion induced focal cerebral ischemia in rats. *Life Sciences*, *74*, 969–985.

Tiwari, S. K., Agarwal, S., Seth, B., Yadav, A., Nair, S., Bhatnagar, P., et al. (2014). Curcumin-loaded nanoparticles potently induce adult neurogenesis and reverse cognitive deficits in Alzheimer's disease model via canonical Wnt/beta-catenin pathway. *ACS Nano*, *8*, 76–103.

Trejo, J. L., Llorens-Martin, M. V., & Torres-Aleman, I. (2008). The effects of exercise on spatial learning and anxiety-like behavior are mediated by an IGF-I-dependent mechanism related to hippocampal neurogenesis. *Molecular and Cellular Neurosciences*, *37*, 402–411.

Tsuchida, A., Yamauchi, T., Takekawa, S., Hada, Y., Ito, Y., Maki, T., et al. (2005). Peroxisome proliferator-activated receptor (PPAR)alpha activation increases adiponectin receptors and reduces obesity-related inflammation in adipose tissue: Comparison of activation of PPARalpha, PPARgamma, and their combination. *Diabetes*, *54*, 3358–3370.

Tu, X. K., Yang, W. Z., Shi, S. S., Wang, C. H., & Chen, C. M. (2009). Neuroprotective effect of baicalin in a rat model of permanent focal cerebral ischemia. *Neurochemical Research*, *34*, 1626–1634.

Waisundara, V. Y., Hsu, A., Huang, D., & Tan, B. K. (2008). Scutellaria baicalensis enhances the anti-diabetic activity of metformin in streptozotocin-induced diabetic Wistar rats. *The American Journal of Chinese Medicine*, *36*, 517–540.

Waisundara, V. Y., Hsu, A., Tan, B. K., & Huang, D. (2009). Baicalin improves antioxidant status of streptozotocin-induced diabetic Wistar rats. *Journal of Agricultural and Food Chemistry*, *57*, 4096–4102.

Weisberg, S. P., Leibel, R., & Tortoriello, D. V. (2008). Dietary curcumin significantly improves obesity-associated inflammation and diabetes in mouse models of diabesity. *Endocrinology*, *149*, 3549–3558.

Whitman, M. C., & Greer, C. A. (2009). Adult neurogenesis and the olfactory system. *Progress in Neurobiology*, *89*, 162–175.

Winner, B., Kohl, Z., & Gage, F. H. (2011). Neurodegenerative disease and adult neurogenesis. *The European Journal of Neuroscience*, *33*, 1139–1151.

Xu, M., Chen, X., Gu, Y., Peng, T., Yang, D., Chang, R. C., et al. (2013). Baicalin can scavenge peroxynitrite and ameliorate endogenous peroxynitrite-mediated neurotoxicity in cerebral ischemia-reperfusion injury. *Journal of Ethnopharmacology*, *150*, 116–124.

Xu, Y., Ku, B., Cui, L., Li, X., Barish, P. A., Foster, T. C., et al. (2007). Curcumin reverses impaired hippocampal neurogenesis and increases serotonin receptor 1A mRNA and brain-derived neurotrophic factor expression in chronically stressed rats. *Brain Research, 1162*, 9–18.

Yamauchi, T., Kamon, J., Waki, H., Terauchi, Y., Kubota, N., Hara, K., et al. (2001). The fat-derived hormone adiponectin reverses insulin resistance associated with both lipoatrophy and obesity. *Nature Medicine*, 7, 941–946.

Yau, S. Y., Li, A., Hoo, R. L., Ching, Y. P., Christie, B. R., Lee, T. M., et al. (2014). Physical exercise-induced hippocampal neurogenesis and antidepressant effects are mediated by the adipocyte hormone adiponectin. *Proceedings of the National Academy of Sciences of the United States of America, 111*, 15810–15815.

Yoshimura, S., Takagi, Y., Harada, J., Teramoto, T., Thomas, S. S., Waeber, C., et al. (2001). FGF-2 regulation of neurogenesis in adult hippocampus after brain injury. *Proceedings of the National Academy of Sciences of the United States of America, 98*, 5874–5879.

Zhang, F., Basinski, M. B., Beals, J. M., Briggs, S. L., Churgay, L. M., Clawson, D. K., et al. (1997). Crystal structure of the obese protein leptin-E100. *Nature, 387*, 206–209.

Zhang, E., Shen, J., & So, K. F. (2014). Chinese traditional medicine and adult neurogenesis in the hippocampus. *Journal of Traditional Complementary Medicine, 4*, 77–81.

Zhang, R. L., Zhang, Z. G., Lu, M., Wang, Y., Yang, J. J., & Chopp, M. (2006). Reduction of the cell cycle length by decreasing G1 phase and cell cycle reentry expand neuronal progenitor cells in the subventricular zone of adult rat after stroke. *Journal of Cerebral Blood Flow and Metabolism: Official Journal of the International Society of Cerebral Blood Flow and Metabolism, 26*, 857–863.

Zhao, W. Q., & Alkon, D. L. (2001). Role of insulin and insulin receptor in learning and memory. *Molecular and Cellular Endocrinology, 177*, 125–134.

Zhao, C., Deng, W., & Gage, F. H. (2008). Mechanisms and functional implications of adult neurogenesis. *Cell, 132*, 645–660.

Zhu, D. Y., Lau, L., Liu, S. H., Wei, J. S., & Lu, Y. M. (2004). Activation of cAMP-response-element-binding protein (CREB) after focal cerebral ischemia stimulates neurogenesis in the adult dentate gyrus. *Proceedings of the National Academy of Sciences of the United States of America, 101*, 9453–9457.

Zhuang, P. W., Cui, G. Z., Zhang, Y. J., Zhang, M. X., Guo, H., Zhang, J. B., et al. (2013). Baicalin regulates neuronal fate decision in neural stem/progenitor cells and stimulates hippocampal neurogenesis in adult rats. *CNS Neuroscience & Therapeutics, 19*, 154–162.

Zhuang, P., Zhang, Y., Cui, G., Bian, Y., Zhang, M., Zhang, J., et al. (2012). Direct stimulation of adult neural stem/progenitor cells in vitro and neurogenesis in vivo by salvianolic acid B. *PloS One*, 7, e35636.

CHAPTER SEVEN

# Antiinflammatory and Hepatoprotective Medicinal Herbs as Potential Substitutes for Bear Bile

**Sandra Appiah*[,1], Mike Revitt*, Huw Jones*, Milan Vu*, Monique Simmonds[†], Celia Bell***

*Faculty of Science and Technology, Middlesex University, The Burroughs, Hendon, London, United Kingdom
[†]Royal Botanic Gardens, Kew, Surrey, United Kingdom
[1]Corresponding author: e-mail address: s.appiah@mdx.ac.uk

## Contents



## Abstract

Practitioners of traditional Chinese medicine (TCM) commonly prescribe medicinal formulations relying on the purported synergism of a combination of plant species, sometimes incorporating animal parts and minerals. Bear bile, obtained from either wild or farmed bears, is a commonly used constituent of traditional medicine formulations. With several bear species now listed under Convention on International Trade in Endangered Species of Wild Fauna and Flora as threatened with extinction and with bear farming being actively campaigned against on ethical grounds, it is important to seek and promote alternatives to the use of bear bile as medicine. This chapter describes and evaluates the scientific data relating to the efficacy of bear bile and potential alternatives to its use, including the use of bile from other animal species, the use of synthetic chemical alternatives, and the use of herbal substitutes.

*International Review of Neurobiology*, Volume 135
ISSN 0074-7742
http://dx.doi.org/10.1016/bs.irn.2017.02.008

Scientific studies have confirmed the efficacy of bear bile as an antiinflammatory and a hepatoprotective agent. Ursodeoxycholic acid (UDCA), the active component of bear bile is used in a synthetic form in Western medicine and can serve as an alternative to bear bile in the treatment and management of certain cholestatic liver conditions. In TCM practice, bile from domesticated animal species (such as cattle, chicken, and pig) has been used as a substitute for bear bile. Following evaluation of TCM literature and pharmacological/clinical data, the authors propose six plant species, either as single herbs or in combination, *Gardenia jasminoides* (zhī zi; 栀子), *Scutellaria baicalensis* (huáng qín; 黄芩), *Coptis chinensis* (huáng lián, 黄连), *Phellodendron amurense* (huáng bǎi; 黄柏), *Andrographis paniculata* (chuan xin lian; 穿芯连), and *Rheum palmatum* (dà huang; 大簧), two medicinal Kampo formulations, Orengedokuto, Dia-Orengedokuto (which originated from traditional Chinese herbal formula Huanglian Jiedu Tang, 黄蓮解毒湯), and two individual phytochemicals (berberine and andrographolide) as alternatives to bear bile. The proposed herbal alternatives are frequently found listed in traditional formulations also containing bear bile, usually with different therapeutic roles ascribed to them. The existing evidence base for the effectiveness of herbal alternatives is sufficiently strong for TCM practitioners and consumers to consider using these without the addition of bear bile. This consideration is driven by the imperative to protect populations of bears from overexploitation in the wild and when farmed. However, for the identified alternatives to be accepted by users, it is essential that researchers and TCM practitioners collaborate effectively to initiate consumer behavior change.

## 1. INTRODUCTION AND OVERVIEW

Traditional medicine (TM) has maintained its influence in all regions of the developing world and is gaining in popularity in industrialized countries (Leonti & Casu, 2013; Robinson & Zhang, 2011). The global market for TM rose from US$ 60 billion in 2000 (WHO, 2004) to an estimated US$ 83 billion in 2008 (Robinson & Zhang, 2011), and this trend is expected to continue. In China, TM forms an important part of the health system and according to the World Health Organization, there were 907 million visits to traditional practitioners in China in 2009 (WHO, 2013). Due to the continued medicinal use of several endangered species as medicine, signatories to the Convention on International Trade in Endangered Species of Wild Fauna and Flora (CITES) have expressed concern over the potential threat to the long-term survival of these species and the need to develop the use of TMs on a more sustainable basis (CITES, 1997). It has been recognized that problems of overexploitation must be addressed within the context of an improved understanding of the significance of TMs in the world's healthcare systems. A resolution was agreed at the

10th meeting of the Conference of Parties (CITES, 1997) calling for more research into potential substitutes for products derived from endangered species used in TM and this was strongly supported by the United Kingdom, represented by the Global Wildlife Division of the Department for Environment, Food and Rural Affairs. The International Fund for Animal Welfare (IFAW) first began to address the issue of endangered species use in TM products in 1983 and have also emphasized the importance of research into herbal alternatives.

One of the most widely used constituents of TMs from endangered species is bear bile (derived from dried gallbladder), the use of which is common in Asia and in Asian communities throughout the world. Bear bile (xióng dǎn; 熊胆) was originally obtained from two members of the Ursidae family, namely, *Ursus thibetanus* (*Selenarctos thibetanus*; Asiatic black bear; Himalayan black bear) and *Ursus arctos* (Brown bear, grizzly bear) (Bensky & Gamble, 1993; Chang & But, 1987). Species of bears such as *Ursus americanus* (American black bear) and *Helarctos malayanus* (sun bear) have also been exploited for medicinal use (Lin, Chang, Chen, & Chen, 1997). *U. thibetanus* and *U. arctos* (those found in Bhutan, China, Mexico, and Mongolia) and *H. malayanus* are among bears listed under CITES Appendix I and *U. americanus* is listed under The CITES Appendices (2016). Species listed under CITES Appendices are threatened with extinction and CITES member countries (currently 182 countries, also known as "Parties") restrict their trade unless under exceptional circumstances (CITES, 2016). To protect populations of bears in the wild from overexploitation for medicine, the Chinese government, in the 1980s, commissioned the farming of bears for the purpose of extracting bile and harvesting their gallbladder for medicine. However, bear farming is not considered to be an ethical solution to protect wild bears. To prevent the further depletion of endangered species in the wild and to abolish bear farming, it is essential to find alternatives to the use of bear bile products in TM. This chapter describes and evaluates the scientific data relating to efficacy of bear bile and potential alternatives as well as the use of bile from other animal species, the use of synthetic chemical alternatives to the active constituents of bear bile and the potential of herbal substitutes.

## 1.1 Bear Bile: Usage and Pharmacology

Dried bear gallbladder or bile (xióng dǎn; 熊胆; Fel Ursi) is used in traditional Chinese medicine (TCM) for the treatment of gallstones, cholestatic hepatic diseases, fever (with febrile convulsions), pharyngolaryngitis, conjunctivitis, traumatic injuries, swelling, pain, sprains, fractures, hemorrhoids, and cardiovascular diseases (Bensky & Gamble, 1993; Chang & But, 1987). Bear bile is prescribed either singly or in combination with medicinal herbs. The underlying pathology of several of the conditions treated with bear bile is inflammation and the antiinflammatory effect of the bile has been described in children suffering from recurrent parotitis (Ruan, Huang, He, Zhang, & Tao, 2013) and also in rat models (when combined with Radix Astragali; huáng qí; 黄芪) (Li, Zhu, But, & Yeung, 1995) (Table 1). Two of the main constituents of bear bile, taurochenodeoxycholic acid and taurocholic acid, have been shown to aid fat digestion through the stimulation of pancreatic lipase (Watanabe et al., 2009). Bear bile (200 mg/kg) potently inhibited ($P < 0.05$) serum aspartate amino transferase in rats with carbon tetrachloride ($CCl_4$)-induced fibrosis compared to nontreated rats (Wang et al., 2012). Although serum aspartate amino transferase levels are elevated during liver injury, the enzyme is not selectively expressed in the liver. Such antifibrotic properties of bear bile have been confirmed from histological observations of liver sections of fibrotic rats (Wang et al., 2012). The antiinflammatory and hepatoprotective effect of bear bile has been attributed to the unique bile acid profile leading to further investigation in other animal species to identify suitable alternatives (Table 2).

## 1.2 Bile Acid Composition of Bear Bile and Bile Derived From Other Animal Species

Bile acids/salts are the end products of enzymatic catabolism of cholesterol and serve as a means of maintaining cholesterol homeostasis. They facilitate the digestion and absorption of dietary fats, including cholesterol and the fat-soluble vitamins, as well as stimulating bile flow (Boatright, Nickerson, Moring, & Pardue, 2009). The liver is the central organ for detoxifying almost all compounds ingested including pharmaceutical drugs, and bile acids aid in the disposal of toxins from this organ. Primary bile acids are synthesized in the liver and modified by cholangiocytes before being stored in the gallbladder ready to be transported to the duodenum to be further metabolized by enteric bacteria into conjugated secondary bile salts. In general, bile contains water, ions (including $Fe^{2+}$, $Mg^{2+}$, $Ca^{2+}$, $Cu^{2+}$, $Zn^{2+}$, $Na^{+}$, $K^{+}$, $Cl^{-}$, and $HCO_3^{-}$), proteins, carbohydrates, lipids (including phospholipids and cholesterol), vitamins, and minerals (Jones, Alpini, & Francis, 2015; Jones & Zollman, 1997) with about 50%–70% of the total bile acid

**Table 1** Comparative Pharmacology of Bear Bile and Bile From Other Animal Species

| Author | Animal Species | Study Design | Main Results | | | |
|---|---|---|---|---|---|---|
| *Antiinflammatory effect of bear bile* | | | | | | |
| Ruan et al. (2013) | Bear bile in formulation with Radix Astragali | Prospective, controlled randomized 1-month study in children with recurrent parotitis | Group A: Massage and vitamin C supplementation | | | |
| | | | Group B: bear bile (0.05 g) and Radix Astragali | | | |
| | | | Group C: combined treatment of A and B. Difference in recurrent rate between: | | | |
| | | | Groups A and B = $P$ = 0.001 (Group B had higher efficacy) | | | |
| | | | Groups A and C = $P$ = 0.003 (Group C had higher efficacy) | | | |
| | | | Groups B and C = $P$ = 0.701 (nonsignificant difference) | | | |
| Li et al. (1995) | Pig bile from *Sus scrofa domestica* and powdered artificially drained bear bile from *Selenarctos thibetanun* | In vivo studies in rat models | Effect compared to control (untreated rats) | | | |
| | | | Antiinflammation induced by croton oil | | Anticonvulsion | |
| | | | Bear bile | Pig bile | Bear bile | Pig bile |
| | | | Inhibitory effect | | Delayed rate | |
| | | | 200 mg/kg (i.p.) 49.8%; $P<0.01$ | 46.9%; $P<0.01$ | 93% | 81.40% |
| | | | 200 mg/kg (s.c.) 25.5%; $P<0.01$ | 25.9%; $P<0.05$ | — | — |
| | | | 1000 mg/kg (p.o.) 26.8%; $P<0.05$ | 32.1%; $P<0.01$ | 90.30% | 137.20% |
| | | | Bear and pig bile demonstrated comparable antiinflammatory and anticonvulsant effects | | | |

*Continued*

**Table 1** Comparative Pharmacology of Bear Bile and Bile From Other Animal Species—cont'd

| Author | Animal Species | Study Design | Main Results |
|---|---|---|---|
| *Hepatoprotective effect of bear bile* | | | |
| Watanabe and Tsuneyama (2012) | Bear bile | In vivo studies in mice models—study on liver injury | After dietary supplementation of mice feed with 1% (w/w) of bile acid for 4 weeks: |
| | Pig bile | | Cattle bile and cholic acid (but not bear bile or pig bile) induced lipid abnormalities ($P<0.001$) and fatty liver injury in mice, caused by the hydrophobic nature of cholic acid |
| | Cattle bile | | |
| | Cholic acid | | |
| Wang et al. (2012) | Bear bile (comparative study with Rhizoma Coptidis) | In vivo studies in acute liver damage rat model | Decrease in serum aspartate amino transferase compared to nontreated rats with $CCl_4$-induced fibrosis: |
| | | | • Aqueous extract of Rhizoma Coptidis (600 mg/kg; $P<0.01$)<br>• Berberine (120 mg/kg; a constituent of the herb; $P<0.05$)<br>• Bear bile (200 mg/kg; $P<0.05$) |
| *Effect of bear bile on the gastrointestinal system* | | | |
| Watanabe et al. (2009) | Bear bile and cattle bile | Incubated bile with safflower oil emulsified with 5% gum Arabic and porcine pancreatic lipase | *Effect on facilitating fat digestion*: |
| | | | • 0.1–0.5 mg/mL of bear bile and cattle bile-stimulated pancreatic lipase activity by similar amounts (120%–200% of control)<br>• The bile acid constituents of bear bile, TCDCA, and TCA (but not TUDCA) also potently stimulated pancreatic lipase activity but with lower effect than bear bile.<br>• Constituent of cattle bile, GDCA, and TDCA exhibited greatest effect on lipase stimulation |

Route of administration: *GDCA*, glycodeoxycholic acid; *i.p.*, intraperitoneal; *p.o.*, oral; *s.c.*, subcutaneous; *TCA*, taurocholic acid; *TCDCA*, taurochenodeoxycholic acid; *TDCA*, taurodeoxycholic acid; *TUDCA*, tauroursodeoxycholic acid.

**Table 2** Composition of Predominant Secondary Bile Acids in the Gallbladder of Bear and Other Domesticated Animal Sources Used in Traditional Chinese Medicine

| Animal Species | Individual Bile Acid (as % of Total) and Their Hydroxyl Position | | | | | | | |
|---|---|---|---|---|---|---|---|---|
| | TUDCA 3α7β | TCDCA 3α7α | TCA 3α,7α,12α | TDCA 3α12α | GCA 3α7α12α | GCDCA 3α7α | GDCA 3α12α | GHDCA 3α6α |
| | Taurine Conjugates | | | | Glycine Conjugates | | | |
| *Black Bear | 38.6[a,b] | 33[a,b] | 3.5[a,b] | | | | | |
| | 30.0[c] | 15.9[c] | 34.1[c] | | | | | |
| | 39.0[d] | 29.6[d] | 31.2[d] | | | | | |
| Cattle | | | 19.5[a,b] | 4.9[a,b] | 14[a,b] | | 1.7[a,b] | |
| | | 9.9[c] | 15.1[c] | | 18.0[c] | 1.8[c] | | |
| Chicken | | 7.2[c] | | 11.4[c] | | | | |
| Pig | | 16.5[b] | | | | 7.6[b] | | 21.2[b] |
| | | | | | | 34.7[c] | | |
| Rabbit | | | | | | | 18.4[c] | |

[a]Watanabe et al. (2009).
[b]Watanabe and Tsuneyama (2012).
[c]Wang et al. (2011).
[d]Hagey et al. (1993).
*TUDCA decreases by about half in bears during periods of dormancy (from mid-November to 1 April) in black bears (Jones & Zollman, 1997). The bile acid profile of black bears (Asiatic bears) contain the highest levels of TUDCA; 39% ($n=7$) of the bile acid pool compared to 19% ($n=6$) in brown bears, 17% ($n=4$) in polar bears and 9% in Sun bears (Hagey et al., 1993). Levels of the other bioactive bile acid, TCDCA is highest in the Sun bear (64%), compared to 30%, 6%, and 20% in black, brown, and polar bears, respectively (Hagey et al., 1993).
*GCA*, glycocholic acid; *GCDCA*, glycochenodeoxycholic acid; *GDCA*, glycodeoxycholic acid; *GHDCA*, glycohyodeoxycholic acid; *TCA*, taurocholic acid; *TCDCA*, taurochenodeoxycholic acid; *TDCA*, taurodeoxycholic acid; *TUDCA*, tauroursodeoxycholic acid.
The predominant bile acids in humans are chenodeoxycholic acid (CDCA; 30%–50%) and cholic acid (CA; 27%–48%) with only about 0.4%–5% being ursodeoxycholic acid (UDCA) (Rossi, Converse, & Hofmann, 1987). Bears are unique in being able to produce TUDCA from bacterial modification of TCDCA in the gut (Gamboa, Tian, Massaad, Reshamwala, & Cai, 2011).

pool being conjugated bile salts (Yerigui, Wu, Wang, & Ma, 2016). Under physiological conditions, bile acids exist as ionized salts known as cholates.

The bile acid pool of Asiatic black bears in the gallbladder is composed mainly of conjugated bile salts, namely, tauroursodeoxycholic (TUDCA; 30%–39%), taurochenodeoxycholic (TCDCA; 16%–30%), and taurocholic acid (TCA; 3%–34%) (Hagey et al., 1993; Wang, Feng, et al., 2011; Watanabe & Tsuneyama, 2012; Watanabe et al., 2009) (Table 2; Fig. 1). The predominant bile acids in humans are chenodeoxycholic acid

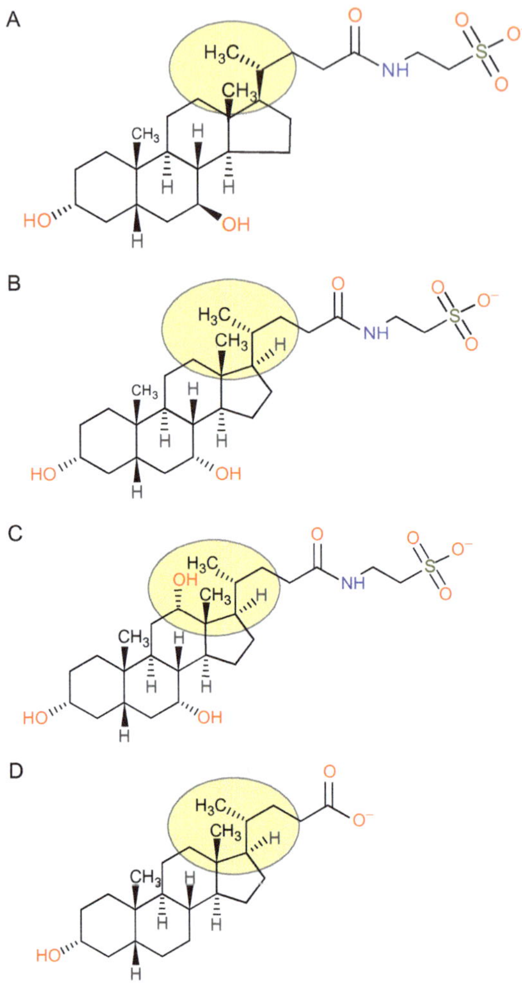


**Fig. 1** The chemical structures of bile acids in bear bile.

| | |
|---|---|
| A: Tauroursodeoxycholate | 2-[(3α,7β-dihydroxy-24-oxo-5β-cholan-24-yl)amino] ethanesulfonate |
| B: Taurochenodeoxycholate | 2-[(3α,7α-dihydroxy-24-oxo-5β-cholan-24-yl)amino] ethanesulfonate |
| C: Taurocholate | 2-[(3α,7α,12α-trihydroxy-24-oxo-5β-cholan-24-yl) amino]ethanesulfonate |
| D: Lithocholate | 3α-hydroxy-5β-cholan-24-oic acid |

The bile acids in the gallbladder of bears are tauroursodeoxycholate (30%–39%), taurochenodeoxycholate (16%–30%), and taurocholate (3%–34%). These can be further metabolized into lithocholate under certain physiological conditions.

Denotes location of structure differences.

(CDCA; 30%–50%) and cholic acid (CA; 27%–48%) with only about 0.4%–5% being ursodeoxycholic acid (UDCA) (Rossi et al., 1987). TUDCA is produced from bacterial modification of TCDCA in the gut of bears (Gamboa et al., 2011) and the bile acid profile of black bears is unique in that it contains about 10 times more TUDCA to that of humans, and about 30%–40% to that of cattle, chicken, pig, and rabbit (Hagey et al., 1993; Wang, Feng, et al., 2011; Watanabe et al., 2009; Watanabe & Tsuneyama, 2012) (Table 2). Another bioactive constituent of bear bile, TCDCA, is found in appreciable concentrations in cattle (10%), chicken (7%), and pig (17%), with TCA also present in the cattle gallbladder (15%–20%) (Table 2). All the aforementioned animal species have been studied as potential alternatives to bear bile (Wang, Feng, et al., 2011; Watanabe et al., 2009; Watanabe & Tsuneyama, 2012). Recently, when some medicines sold as bear bile were tested, only trace levels of TUDCA and TCDCA were found in the preparations (Yerigui et al., 2016) raising issues of quality control in bear bile products available on the market.

## 1.3 Ursodeoxycholic Acid

Kanazawa, Shimazaki, Sato, and Hoshino (1954) synthesized UDCA which was subsequently developed into pharmaceutical products marketed in the United Kingdom as Destolit®, Urdox®, Ursofalk®, and Ursogal, and used mainly for the dissolution of cholesterol gallstones and the treatment of chronic cholestatic liver conditions. UDCA is comparatively more hydrophilic than other bile acids and several pharmacological activities of the drug have been attributed to this unique property. UDCA reduces the retention of potentially toxic bile acids in cells by displacing hydrophobic bile acids (Yang et al., 2012), altering the hydrophobicity of bile and thereby preventing premature death of healthy cells (Bessone & Roma, 2016; Fickert et al., 2013; Rodrigues, Fan, Ma, Kren, & Steer, 1998). The cytoprotective nature of UDCA and TUDCA makes them potentially therapeutically relevant in diverse diseases including neurodegenerative conditions (such as Alzheimer's disease, Parkinson's disease, ocular disease, and stroke (Vang, Longley, Steer, & Low, 2014), as well as liver conditions and cancer. The anticancer effect of UDCA has been shown in a mouse model (Liu et al., 2015) and in cell lines (Lim, Choi, Kang, & Si, 2009). In the absence of liver disease, UDCA (8–10 mg/kg/day) was effective in reducing the occurrence of high risk dysplasia (precancerous stage which can develop into invasive colorectal carcinoma) in a phase III trial, but did not have a significant effect on the overall total colorectal carcinoma recurrence in the same study (Alberts et al., 2005). At high concentrations, UDCA (28–30 mg/kg) has been linked to increased side effects (Lindor et al., 2009) and is a risk factor in colorectal cancer patients suffering from primary sclerosing cholangitis (PSC) (Eaton et al., 2011) due to the production of a more hydrophobic metabolite of UDCA known as lithocholic acid (Fig. 1). This bile acid exerts the same negative effects as other hydrophobic bile acids, such as cytotoxicity and stimulation of proinflammatory proteins. It is apparent that UDCA is contraindicated at doses greater than 25 mg/day for certain chronic liver conditions (Goode & Rushbrook, 2016; Poupon, 2012).

Hepatocytes and cholangiocytes are two cell types involved in bile formation, with hepatocytes being responsible for the secretion of the main biliary constituents. Cholangiocytes, epithelial cells lining ducts of the biliary tree, contain several transporters and their role is to aid the transport of bile acids and ductal bicarbonate secretion (Jones et al., 2015; Lazaridis & LaRusso, 2015; O'Hara, Gradilone, Masyuk, Tabibian, & LaRusso, 2014), thereby regulating the fluidity and alkalinity of bile. Activation of cholangiocytes, mediated by injury, produce inflammatory response molecules (interleukin-6 (IL-6), IL-8, tumor necrosis factor-alpha (TNF-α), chemokines, and various growth factors), which influence apoptosis, proliferation, and senescence of the cells, resulting in inflammation, angiogenesis, and fibrosis (Lazaridis & LaRusso, 2015). Activated cholangiocytes can result in the development of several cholestatic disorders (cholangiopathies). UDCA has been reported to inhibit cell death of hepatocytes caused by the hydrophobic bile acid, deoxycholic acid (Rodrigues et al., 1998), as well as having a direct effect on adaptive immunity (Goode & Rushbrook, 2016).

Cholestasis develops when there is biliary obstruction, resulting in disruption of flow of bile from the liver to the duodenum. This can be due to the presence of gallstones, inflammation of the bile ducts (biliary inflammation), an infection of the bile duct (cholangitis), scarring of liver tissue (cirrhosis and fibrosis), and tumors (Wu et al., 2016). Chronic cholestatic liver diseases include primary biliary cholangitis (PBC; previously known as primary biliary cirrhosis) and PSC both of which are heterogeneous autoimmune diseases influenced by genetic and environmental factors (Beery, Vaziri, & Forouhar, 2014; Eaton, Talwalkar, Lazaridis, Gores, & Lindor, 2013; O'Hara et al., 2014; Sclair, Little, & Levy, 2015). Biliary inflammation can lead to end-stage liver failure requiring liver transplantation (Fickert et al., 2013) and/or development into the primary liver cancer, cholangiocarcinoma (Jones et al., 2015). Cholestasis often leads to pruritus (itchiness) due to the accumulation of hydrophobic bile acids in the systematic circulation (Hegade et al., 2016). Patients with intrahepatic cholestasis of pregnancy characterized by pruritus and elevation of alanine aminotransferase and total fasting bile acids had a good obstetric outcome with UDCA medication in a 5-year observational study (Joutsiniemi, Timonen, Linden, Suvitie, & Ekblad, 2015). Increased total bile acid in pregnancy can lead to poor fetal outcome in rat models, and UDCA administered for intrahepatic cholestasis of pregnancy improved survival rates by displacing hydrophobic bile acids and inhibiting the transcriptional factor nuclear factor-kappaB (NF-κB) (Zhang et al., 2016).

UDCA is, to date, the only approved pharmaceutical drug for the management of PBC in Europe (Goode & Rushbrook, 2016; O'Hara et al., 2014; Sclair et al., 2015; Zhu et al., 2015). A recent meta-analysis of 31 randomized trials indicated that, when used alone, UDCA caused very few adverse effects for PBC treatments compared to combinational therapies of UDCA and other drugs (Zhu et al., 2015). PBC displays autoimmune properties where small intrahepatic chalongiocytes and interlobular bile ducts are destroyed by the immune system (Lazaridis & LaRusso, 2015; O'Hara et al., 2014). Over 90% of cases display highly specific antimitochondrial antibodies in serological tests, making this a hallmark feature for PBC (Lazaridis & LaRusso, 2015; Sclair et al., 2015). Additionally, antinuclear antibodies and autoreactive T-cells are commonly present (Beery et al., 2014; O'Hara et al., 2014; Sclair et al., 2015). Other key features in PBC are reduced secretin-stimulated bicarbonate secretion and anion exchanger 2 biliary expressions (AE2/SLC4A2), which promote bile alkalization (O'Hara et al., 2014).

Similar to PBC, PSC is characterized by chronic inflammation and progressive fibrosis (Beery et al., 2014; Lazaridis & LaRusso, 2015; Sclair et al., 2015) and also results in the stricturing of intrahepatic and/or extrahepatic bile ducts. The bile ducts eventually narrow, diffuse, and obliterate, leading to bile duct obstruction and biliary cirrhosis (Beery et al., 2014; Sclair et al., 2015). Although UDCA is the most prescribed drug for cholestatic hepatopathies, it is contraindicated in obstructive cholestasis, including PSC and end-stage liver fibrosis/cirrhosis, particularly at high concentrations of the drug (>25 mg/kg/day) (Goode & Rushbrook, 2016; Poupon, 2012). The choleretic effect of UDCA results in increased bile flow (Bessone & Roma, 2016) which consequently causes increased biliary pressure with aggravation of bile infarcts in obstructed bile ducts (Fickert et al., 2013).

For obstructive cholestasis, an epimer of UDCA, nor UDCA has been reported to be more effective than UDCA (Poupon, 2012). Although there is evidence that UDCA is useful in the treatment of certain gastrointestinal and liver conditions, it can only be obtained via prescription in certain countries, including the United Kingdom.

# 2. MEDICINAL HERBS AS POTENTIAL SUBSTITUTES FOR BEAR BILE

## 2.1 Selection of Herbs as Alternatives to Bear Bile

An important aspect of research into herbs that could potentially replace medicinal products derived from endangered animal species is the requirement of effective communication between researchers and TM practitioners and an appreciation of the TCM philosophy underlying, historical and current usage. This should be combined with knowledge of biomedical modes of action to determine an effective herbal alternative. TCM practitioners commonly prescribe mixtures or formulations of plant species, with the formulation relying on purported synergism of a combination of herbs, sometimes with animal parts and minerals. Therefore, the characteristics of a constituent in TCM terms, as well as its role in the formulation should be considered when seeking alternatives.

Bear gallbladder or bile is described in TCM as having "heat-clearing," detoxifying, and "fire-purging" properties as well as having a "bitter" taste and being "cold" in nature. In seeking potential plant-based alternatives to bear bile, Appiah (2006) devised criteria to select herbs for further study based on both the properties of bear bile as described in TCM literature and the known pharmacological properties of bear bile and UDCA (Table 3). This enabled the list of over 100 "heat-clearing" herbs listed in TCM Materia Medicas, pharmacology books, and Chinese and Japanese Pharmacopoeias to be considerably reduced based on the reported pharmacological properties of the herbs and the existence of constituents possessing similar structures and functions to UDCA. In addition, known relevant biological activities of the crude extracts were considered, with hepatoprotective effects being given top priority. Consideration was also given to ensure that no endangered plant species were included (those restricted by CITES) and the continued use of the selected herbs in TCM practice. Following discussions with TCM practitioners, seven medicinal herbs were identified selected for further study (Table 3) and with the exception of *Rhizoma Anemarrhenae* (zhī mŭ, 知母), fulfil all the

**Table 3** Herbs Selected for Investigation as Potential Substitutes for Bear Bile Following Evaluation of TCM Literature and Pharmacological/Clinical Data

| Plant Species With Similar Properties to Bear Bile (Plant Family, Chinese Names; Plant Part) | Criteria Considered | | | | | |
|---|---|---|---|---|---|---|
| | TCM Properties | | | Pharmacological Properties | | |
| | Cold Nature | Bitter Taste | Heat Clearing | Antiinflammatory | Hepatoprotective | Anticancer |
| Fructus Gardenia (*Gardenia augusta* Merr. syn.: *G. jasminoides* Ellis; Rubiaceae; zhī zi; 栀 子); Cape jasmine dried fruit) | * | * | * | Ga, Gb, Gc | Gb, Gd | Ge |
| Radix Scutellariae (*Scutellaria baicalensis* Georgi (Lamiaceae; huáng qín; 黄芩, dried root) | * | * | * | Sa, Sb, Sc | Sd, Se, Sf, Sg | Sg, Sh |
| Rhizoma Coptidis (*Coptis chinensis* Franch. syn.: *C. deltoidea* C. Y. Cheng et Hsiao, *C. teetoides* C. Y. Cheng, *C. omeiensis* (Chen) C.Y. Cheng., Ranunculaceae; huáng lián, 黄连, goldthread rhizome) | * | * | * | Sc, Ca | Cb | Cc |
| Cortex Phellodendri (*Phellodendron amurense* Rupr Rutaceae; huáng bǎi; 黄柏; dried bark) | * | * | * | Sc, Pa, Pb | Pa, Pc | Pd, Pe |
| Herba Andrographitis (*Andrographis paniculata* (Burm. f) Nees (Acanthaceae; chuan xin lian; 穿芯连; green chiretta, dried aerial parts) | * | * | * | Aa | Ab, Ac, Ad | Ae |
| Radix et Rhizoma Rhei (*Rheum palmatum* L. (*R. tanguticum* Maim. Ex Balf., *R. officinale* Bail, *R. coreanum* Nakai and *R. undulatum* L., Polygonaceae; dà huang; 大簧, dried root and rhizome, rhubarb) | * | * | * | Sc, Ra | Ra, Rb | Rc, Rd |
| ¥Rhizoma Anemarrhenae (*Anemarrhena asphodeloides* Bge, Anthericaceae; Zhī, mǔ,知母) | * | * | * | ‡ | | |

*TCM literature: Hsu et al. (1986); Chang and But (1987); Bensky and Gamble (1993); Chang and But (2001); [Ga]Chen et al. (2016); [Gb]Lin et al. (2015); [Gc]Jung et al. (2008); [Gd]Chen et al. (2012); [Ge]Koo et al. (2004); [Sa]Jung et al. (2012); [Sb]Appiah et al. (2006); [Sc]Cuéllar, Giner, Recio, Máñez, and Ríos (2001); [Sd]Dong et al. (2016); [Se]Pan, Wang, Leu, Wu, and Wu (2012); [Sf]Nan, Park, Kim, Ko, and Sohn (2002); [sg]Park et al. (2014); [Sh]Peng et al. (2014), [Ca]Choi et al. (2013); [Cb]Wang et al. (2012); [Cc]Zhu et al. (2011); [Pa]Choi et al. (2014); [Pb]Xian, Mao, Ip, Lin, and Che (2011); [Pc]Yotsumoto et al. (1997); [Pd]Ghosh et al. (2007); [Pe]Garcia et al. (2006); [Aa]Kishore et al. (2016); [Ab]Singh, Awasthi, Luqman, Singh, and Mani (2015); [Ac]Bardi et al. (2014); [Ad]Koh, Mokhtar, and Iqbal (2011); [Ae]Yue et al. (2015); [Ra]Zhang, Qiu, Wang, Long, and Mao (2015); [Rb]Yang et al. (2012); [Rc]Ma et al. (2015); [Rd]Nho, Chun, Lee, and Kim (2015); and ‡Yeum et al. (2007). ¥There is insufficient evidence of the crude extract *Anemarrhena asphodeloides* as a hepatoprotective agent, an important property of bear bile.

established criteria. While individual herbs may be suitable as replacements for bear bile, in TCM a mixture of different plant species is commonly prescribed, relying on synergistic effects between herbs to optimize their action in the individual patient. Orengedokuto and Dia-Orengedokuto are Kampo formulations (which originated from traditional Chinese herbal formula Huanglian Jiedu Tang, 黄蓮解毒湯) containing four and five, respectively, of the seven herbs listed in Table 3, and may represent suitable replacements for bear bile. Orengedokuto is composed of Radix Scutellariae (huáng qín; 黄芩), Rhizoma Coptidis (huáng lián, 黄连), Fructus Gardeniae (zhī zi; 栀子), Cortex Phellodendri (huáng bǎi; 黄柏). Dia-orengedokuto contains an additional herb, Radix et Rhizoma Rhei (dà huang; 大簧).

The overproduction of proinflammatory proteins has been linked to several conditions treated by bear bile, including liver diseases (Tacke & Zimmermann, 2014), fever (Ivanov & Romanovsky, 2004), cancer (Atretkhany, Drutskaya, Nedospasov, Grivennikov, & Kuprash, 2016), and cardiovascular diseases (Brown & Jones, 2004; Liu, Liu, et al., 2016; Liu, Yang, et al., 2016). In order to evaluate the effectiveness of the plant species as antiinflammatory agents, in vitro evaluation of their effect of on NF-κB was conducted (see Section 2.2).

## 2.2 The Effect of Selected Herbs on NF-κB Induced IL-6 Activity

In recent years, NF-κB has become a therapeutic target for the discovery of novel antiinflammatory agents from natural products. NF-κB is a transcriptional factor involved in the expression of several proinflammatory proteins including the cytokine IL-6. The single herbs selected as potential alternatives to bear bile were studied for their effect on NF-κB, as assessed by in vitro IL-6-promoter activity. Crude water extracts (100 μg/mL) of *Coptis chinensis* Coptidis (huáng lián, 黄连), *Phellodendron amurense* (huáng bǎi; 黄柏), and *Rheum palmatum* (dà huang; 大簧) all reduced NF-κB-induced IL-6-promoter activity but statistical significance was detected only for *R. palmatum* ($P < 0.05$ compared to control cells stimulated to produce IL-6 but not treated with a drug) (Bell, Simmonds, Howes, & Appiah, 2006). Ethyl acetate but not water extracts of *Scutellaria baicalensis* (huáng qín; 黄芩) (potently inhibited IL-6-promoter activity ($P < 0.01$ compared to control cells)), and bioactivity-guided fractionations led to the isolation and characterization of the active compounds, chrysin, baicalein, wogonin, and scutellarein (Appiah et al., 2006; Bell et al., 2006). The bioactivity assays demonstrate the potential of *R. palmatum* and *S. baicalensis* as antiinflammatory agents. Crude water extracts (100 μg/mL) of *Anemarrhena asphodeloides* (zhī mǔ, 知母) and

*Andrographis paniculata* (chuan xin lian; 穿芯连) did not affect IL-6-promoter activity. There is insufficient scientific data on *A. asphodeloides* as a liver protective herb and in addition, a constituent of the herb, timosaponin, has been reported to trigger liver injury at 100 mg/kg/day (Wu et al., 2014). Therefore, *A. asphodeloides* was omitted from the list of alternatives to bear bile in the treatment of liver conditions until dose-dependent hepatotoxicity tests are conducted on the crude extract. It was important to establish marker proteins for hepatic injury as summarized in Section 2.3.

## 2.3 Hepatic Injury and Inflammation

Chronic hepatic injury may progress to liver fibrosis resulting in clinical complications and impairment of the organ (Moreira, 2007). The underlying cause of liver fibrosis is inflammation and excessive deposition of extracellular matrix (ECM) on liver parenchyma, characterized by an imbalance between collagen fiber synthesis, decomposition, and the presence of nonelastic fibrotic tissue (Gressner & Weiskirchen, 2006; Moreira, 2007; Richter & Kietzmann, 2016). Hepatic stellate cells (HSCs) are recognized as the main matrix-producing cells involved in the pathogenesis of liver fibrosis (Fig. 2). HSCs have a dual function as they play both protective and damaging roles in liver tissue repair (Schachtrup, Le Moan, Passino, & Akassoglou, 2011). Platelet-derived growth factors (PDGF) mediate the proliferation of HSCs and hence influence the level of ECMs produced (Friedman, 2008; Liu, Liu, et al., 2016; Liu, Yang, et al., 2016; Moreira, 2007; Schachtrup et al., 2011). Activated HSCs stimulate the production of IL-6, resulting in increased synthesis of collagen 1 (Friedman, 2008; Schachtrup et al., 2011; Tacke & Zimmermann, 2014). Transforming growth factor-β (TGF-β1) is another cytokine stimulated by reactive oxygen species in activated HSCs, whose production results in overexpressing collagen 1 and other ECM proteins thus playing a crucial part in the fibrotic pathway (Barnes & Gorin, 2011; Schachtrup et al., 2011). Inducing apoptosis in HSCs has been suggested to be a potential antifibrotic strategy (Moreira, 2007; Pan et al., 2012). Activated HSCs markedly upregulate expression of α-smooth muscle actin (α-SMA) cytoskeletal proteins (which are absent in other hepatic cells), making α-SMA a good marker for reflecting the degree of hepatic fibrosis and injury (Friedman, 2008; Liu, Liu, et al., 2016; Liu, Yang, et al., 2016). Other markers of hepatic injury include aminotransferases, alkaline phosphatase, bilirubin, γ-glutamyl transferase, autotaxin with specific fibrotic markers being hydroxyproline and α-smooth muscle actin (Table 4).

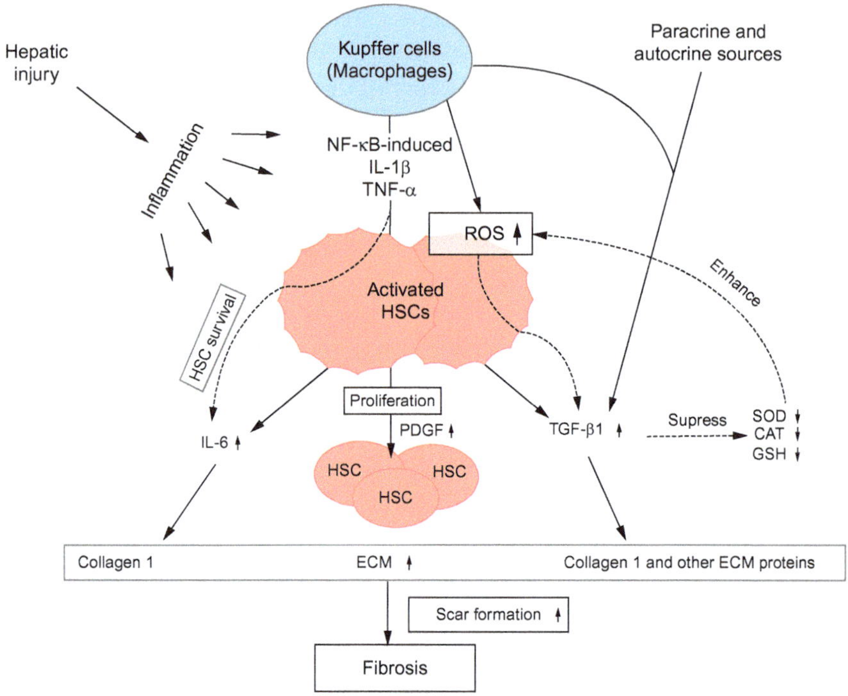


**Fig. 2** Fibrogenesis in liver induced by activated hepatic stellate cells. Liver fibrosis is characterized by the chronic stimulation of proinflammatory proteins. Hepatic stellate cells (HSCs) are the main matrix-producing cells involved in the pathogenesis of liver fibrosis and injury to the liver activates HSCs. Activated HSCs which cannot reach resolution phase (apoptosis or reversion) form proliferating myofibroblast-like cells creating a cycle of liver repair and damage, promoting fibrosis. Platelet-derived growth factors (PDGF, a main mediator of proliferation) and transforming growth factor-β (TGF-β1, a key profibrotic cytokine) secretion are elevated in activated HSCs during liver injury. TGF-β1 from paracrine and autocrine sources stimulates production of collagen I and other ECM proteins, increasing matrix production per cell and ECM deposition. HSCs secrete inflammatory cytokines such as IL-6, which promotes collagen production, further augmenting ECM progression. Macrophages (Kupffer cells), recruited by HSCs, enhance scaring by nuclear factor kappa-induced IL-1β and TNF-α production which potentially stimulate IL-6, promoting HSC survival. Injured hepatocytes and liver macrophages produce reactive oxygen species (ROS) which contributes to HSCs activation and inflammatory cells recruitment, leading to TGF-β1 secretion. TGF-β1 response to increased ROS may be linked to the suppression of several enzymes such as superoxide dismutase (SOD), catalase (CAT), and glutathione (GSH) which majorly participates in antioxidant defense system. This further enhances ROS activity, ECM progression and deposition, leading to accumulated scar formation and ultimately fibrosis.

## 2.4 Pharmacological Activities of Medicinal Herbs as Potential Alternatives to Bear Bile

### *2.4.1 Fructus Gardeniae (zhī zi; 栀子)*

The traditional use of *Gardenia jasminoides* as an antiinflammatory agent has been confirmed in mice models (Chen et al., 2016; Jung et al., 2008) and in vitro (Lin et al., 2015), through the attenuation of proinflammatory cytokines, IL-1β, IL-6, and TNF-α, the enzyme cyclooxygenase-2 and NF-κB. A constituent of the herb (geniposidic acid) has the ability to protect hepatocytes from cell injury and premature death by suppressing the apoptotic proteins, Bax, caspase 3, and caspase 8 protein expression, with a corresponding increase in the antiapoptotic protein, Bcl-XL (Kim et al., 2013). Crude water extracts of *G. jasminoides* (10–100 mg/kg in rats) reduced liver fibrosis and resulted in a decrease in the levels of the liver injury markers, ALT and AST (Chen et al., 2012; Lin et al., 2015), hydroxyproline and smooth muscle actin, a marker of active HSCs (Chen et al., 2012). In both in vivo studies and in human HSC lines, the crude extract reduced mRNA levels of the fibrotic genes, TGF-β1, and collagen 1, showing a possible mechanism of hepatoprotective action (Chen et al., 2012). The herbal extract further protected liver cells from lipid peroxidation as evidenced by a decrease in malondialdehyde and increase in the antioxidant, superoxide dismutase (SOD) in vivo (Chen et al., 2016).

### *2.4.2 Radix Scutellariae (huáng qín; 黄芩)*

The antifibrotic effect of *S. baicalensis* has been reported in vitro (Chen, Liang, Lee, Huang, & Lin, 2013) and in vivo (Nan et al., 2002; Pan et al., 2012). Liver tissues of rats with induced fibrosis but without any therapeutic intervention showed significant accumulations of collagen, α-SMA, ALT, and AST which were ameliorated in rats treated with ethanolic extracts of *S. baicalensis* (20 mg/kg for 4 weeks) (Pan et al., 2012). The mechanism of antifibrotic action was through the induction of HSC apoptosis; cells were arrested in $G_2M$ (genomic integrity check point ($G_2$) phase and the cell division (M phase)) with upregulation of the proapoptotic proteins (Bax, p21, and p53) and downregulation of cdc2, cyclin B1, and bcl2 (Pan et al., 2012). In hepatocellular carcinoma cell lines, *S. baicalensis* showed an antimetastatic effect by regulation of metalloproteinase 2 and FOXM1 activities (Park et al., 2014). Constituents from the herb have also been found to be effective in the treatment of ocular disorders (Xiao, Do, & To, 2014).

**Table 4** Summary of Biomarkers for Identifying Hepatic Injury and Fibrosis

| | Biomarkers | Normal Ranges in Humans | Presence and Function | Disease/Conditions | |
|---|---|---|---|---|---|
| Aminotransferases (transamination reactions) | Alanine aminotransferase (ALT) | 7–56 U/L* (serum) | • Cytoplasmic enzyme<br>• Predominantly in liver tissues<br>• Present in kidney, heart, muscle | ↑ ALT in most types of liver injury <300 U/L is nonspecific >500 U/L in primary hepatocyte diseases | AST/ALT ratio<br>↑ In liver function impairment |
| | Aspartate aminotransferase (AST) | 0–35 U/L* (serum) | • Enzyme of two isoforms (mitochondrial and cytoplasmic)<br>• Predominantly in heart tissues<br>• Present in liver, smooth muscle, kidney | ↑AST (mitochondrial) in chronic liver diseases | >1 U/L cirrhosis/fibrosis |
| Alkaline phosphatase (ALP) | | 41–133 U/L* (serum) | • Enzyme involved in lipid transportation in intestine and bone calcification<br>• Present in cells lining bile ducts and canaliculi, small intestine, kidney, bone, and placenta | ↑ ALP greatly in cholestatic diseases (released in respond to bile obstruction and salt accumulation) | |

| | | | |
|---|---|---|---|
| Autotaxin (ATX) | 2.8 ± 1.4 nmol/mL/min[a] (serum activity) | Enzyme which hydrolyses lysophosphatidylcholine to lysophosphatidic acid | ↑ ATX in pruritus associated with cholestasis (PBC and PSC)<br>↑ In liver dysfunction correlated with the severity of cholestatic liver diseases (novel prognostic marker) |
| Bilirubin | 2–21 μmol/L* (serum total bilirubin)<br><12 μmol/L* (serum indirect/unconjugated bilirubin)<br><8 μmol/L* (serum direct/conjugated bilirubin) | • Catabolic product of hemoglobin<br>• Released in unconjugated form, converts to conjugated forms in the liver by UDP-glucuronyltransferase | ↑ Conjugated bilirubin levels in hepatocellular liver injury<br>↑ Total serum bilirubin in parenchymal liver diseases, incomplete extrahepatic (bile duct) obstructions |
| γ-Glutamyl transferase (γ-GT) | 9–85 U/L* (serum) | • Cell membrane enzyme<br>• Transports peptides<br>• Involved in glutathione metabolism<br>• Present in hepatocytes, biliary epithelial cells, renal tubules, pancreas, intestine | ↑ γ-GT in:<br>Cholestatic disorders (usefulness in ruling out bone disease)<br>Acute viral hepatitis other unrelated liver conditions |

*Continued*

**Table 4** Summary of Biomarkers for Identifying Hepatic Injury and Fibrosis—cont'd

| | Biomarkers | Normal Ranges in Humans | Presence and Function | Disease/Conditions |
|---|---|---|---|---|
| Fibrotic markers | Hydroxyproline | 113–175 nmol/mg[b] (total, urine) | • Nonessential amino acid<br>• Major component in collagen | ↑ In fibrotic diseases<br>↓ In poor wound healing |
| | α-Smooth-muscle actin (α-SMA) | 2.9% ± 4.0%[c] (HSCs) | • Cytoskeletal proteins in smooth muscle cells<br>• Upregulated in HSCs but absent in other hepatic cells | ↑ In hepatic fibrosis<br>Reflects the degree of hepatic fibrosis and injury |

*HSCs*, hepatic stellate cells; *Gowda et al. (2009).
Other sources of information:
[a]Wunsch et al. (2016)—from clinical study.
[b]Inoue, Kohashi, and Tsuruta (1998)—from clinical study; Liu, Liu, et al. (2016); and Liu, Yang, et al. (2016).
[c]Carpino et al. (2005)—from clinical study and Friedman (2008).

### 2.4.3 Cortex Phellodendri (huáng bǎi; 黄柏)

Berberine (Fig. 3), an active component of both *C. chinensis* and *P. amurense*, has been shown to exhibit cholesterol lowering properties and hepatoprotective effects in animal models (Feng et al., 2010; Zhao, Zhang, Tong, Chen, & Luo, 2012) via inhibition of TNF-α and the proinflammatory enzymes cyclooxygenase-2 and inducible nitric oxide (Feng et al., 2010). Cortex Phellodendri has been reported to possess hepatoprotective (Yotsumoto et al., 1997) and antiinflammatory properties. However, the hepatoprotective effects of this herb appear to be less effective than Rhizoma Coptidis and Radix Scutellariae (Yotsumoto et al., 1997).

### 2.4.4 Rhizoma Coptidis (huáng lián; 黄连)

Wang et al. (2012) compared the effects of aqueous extracts of Rhizoma Coptidis (600 mg/kg), berberine (120 mg/kg; a constituent of the herb), and bear bile (200 mg/kg) on acute liver damage in the rat. Liver tissues of $CCl_4$-induced liver fibrotic rats showed a statistically significant decrease of serum AST in rats treated with either bear bile ($P<0.01$) or berberine ($P<0.05$), compared to nontreated rats. The test compounds did not have a statistically significant effect on ALT serum levels. ALT and AST both have limitations as indicators of liver fibrosis, and their levels are used only as part of the evaluation of liver damage. Histological examination of liver tissues confirmed the antifibrotic effect of Rhizoma Coptidis in rat models and mechanisms of action included the stimulation of the antioxidant, SOD (Feng et al., 2011; Wang et al., 2012) and inhibition of the antiapoptotic protein Erk1/2 (Feng et al., 2011).

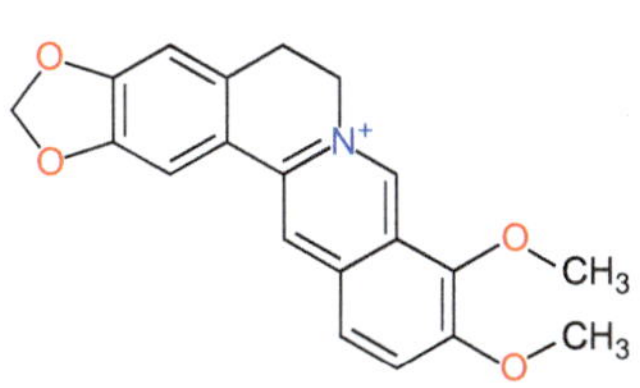


**Fig. 3** Molecular structure of berberine. Berberine is a hepatoprotective alkaloid present in *Coptis chinensis* (huáng lián, 黄连) and *Phellodendron amurense* (huáng bǎi; 黄柏).

### 2.4.5 Herba Andrographis (chuan xin lian; 穿芯连)

The hepatoprotective effect of water extracts of *A. paniculata* showed significant reduction in liver tissue injury when administered to rat models of liver cirrhosis compared to untreated controls (Bardi et al., 2014; Koh et al., 2011) through induction of the endogenous antioxidant, glutathione (Koh et al., 2011). *A. paniculata* also showed an antioxidant effect through the stimulation of SOD levels and an antiproliferative effect of a human cancer cell line, HepG2 (Bardi et al., 2014). The main liver protective constituent of the herb is a diterpene lactone known as andrographolide which attenuates HSC activation through inhibition of NF-κB expression in rats (Khamphaya et al., 2016). Another constituent, 14-deoxyandrographolide reduced the sensitivity of hepatocytes to TNF-α-induced apoptosis showing a potential as a hepatoprotective compound (Roy, Mandal, Sen, Mukhopadhyay, & Biswas, 2010). Hydrophobic bile acids can induce proinflammatory genes in trophoblasts through the upregulation of phosphoinositide 3-kinase (P13K)/NF-κB pathway, and this effect can be inhibited by UDCA (Zhang et al., 2016), a constituent of bear bile. A comparative study showed that andrographolide was more potent than UDCA in improving the survival rate of pregnant rats with obstructive cholestasis (Zhang et al., 2016) (Fig. 4).

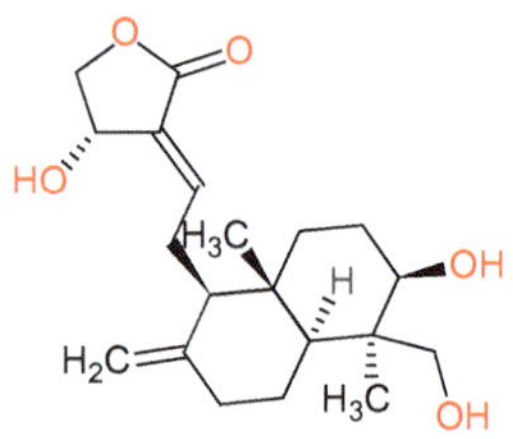


**Fig. 4** Molecular structure of andrographolide. Andrographolide, a hepatoprotective terpenoid found in *Andrographis paniculata* (chuan xin lian; 穿芯连).

### *2.4.6 Radix et Rhizoma Rhei (dà huang; 大黄)*

Radix et Rhizoma Rhei possesses cardioprotective properties with an ability to reduce cholesterol levels through the inhibition of HMG-CoA reductase activity (Kim et al., 2002). A crude ethanolic extract of *R. palmatum* (2–40 mg/kg) caused a statistically significant decrease in liver injury markers, including ALT and AST, with 5 and 15 mg/kg being the optimum doses in $CCl_4$-induced liver injury in rats (Wang, Zhao, et al., 2011). Non-significant reduction in hepatic serum TGF-β1 was reported for lower doses of the herbal extract but was significantly elevated at the higher concentration of 40 mg/kg ($P < 0.05$ compared to nontreated groups). Significant elevation of ALT and TGF-β1 was reported for nondiseased rats administered with 40 mg/kg *R. palmatum* indicating the potential of the herb to induce hepatic fibrosis at high concentrations (Wang, Zhao, et al., 2011). When administered to rats, *R. palmatum* (40 mg/kg/day) resulted in more accumulation of anthraquinone constituents (aloe-emodin, rhein, and emodin) in liver tissues of normal rats than rats induced with liver disease, indicating that the prophylactic use of *R. palmatum* should be avoided.

### 2.4.7 Orengedokuto and Dia-Orengedokuto

Orengedokuto (Huanglian Jiedu Tang, 黃蓮解毒湯; TJ-15) is an approved medicine by the Ministry of Health and Welfare of Japan and the formulation is comparable to bear bile and UDCA in exhibiting antiinflammatory (Fukutake et al., 2000; Oshima, Narukawa, Hada, & Kiuchi, 2013), hepatoprotective (Fujimoto et al., 2008; Maclean & Taylor, 2000; Ohta et al., 1998; Sekiya et al., 2002), and anticancer (Fukutake et al., 2000) effects. Orengedokuto has been reported to inhibit hepatic cholesterol ester formation by inhibiting the activity of acyl-coenzyme A: cholesterol acyltransferase (ACAT) in vitro (Yotsumoto et al., 1997). In addition, extracts of Radix Scutellariae (huáng qín; 黄芩), Rhizoma Coptidis (huáng lián, 黄连), and Cortex Phellodendri (huáng bǎi; 黄柏) were found to decrease ACAT activity, whereas Fructus Gardeniae (zhī zi; 梔子) had no significant effect (Yotsumoto et al., 1997).

Dia-Orengedokuto with an additional herb (Radix et Rhizoma Rhei, dà huang; 大黃) is used to treat cardiovascular diseases (Cho et al., 2004; Kim et al., 2002). Water and ethanol extracts of Dia-Orengedokuto inhibited the activity of HMG-CoA reductase more potently than extracts of Orengedokuto (Kim et al., 2002). Of the constituents of Dia-Orengedokuto, Rhizoma Coptidis was most potent at reducing HMG-CoA reductase activity, followed by Radix et Rhizoma Rhei (Kim et al., 2002). Constituents of bear bile (CDCA and cholic acid) are also inhibitors of HMG-CoA reductase, which leads to reduction in hepatic cholesterol levels (Björkhem, Andersson, Sudjama-Sugiaman, Eggertsen, & Hylemon, 1993).

## 2.5 Alternatives to Bear Bile: Summary and Concluding Comments

In this chapter, we have described and evaluated the current evidence to support potential alternatives to the use of bile sourced from wild bears in TM, with particular emphasis given to the potential of herbal substitutes. The main groups of potential alternatives include bile from farmed bears and domesticated animals, chemically synthesized UDCA as well as medicinal herbs and their constituents. The authors initially proposed seven herbs (Table 3) and two herbal formulations (Orengedokuto (Huanglian Jiedu Tang, 黄蓮解毒湯) and Dia-Orendokuto) as potential alternatives to bear bile. On the basis of the updated evaluation of TCM literature and pharmacological/clinical data presented in this chapter, we now propose six plant species, either as single herbs or in combination (*G. jasminoides* (zhī zi; 栀子)), *S. baicalensis* (huáng qín; 黄芩), *C. chinensis* (huáng lián, 黄连), *P. amurense* (huáng bǎi; 黄柏), *A. paniculata* (chuan xin lian; 穿芯连), and *R. palmatum* (dà huang; 大簧), two medicinal formulations, Orengedokuto, Dia-Orengedokuto, and two individual phytochemicals (berberine and andrographolide) as alternatives to bear bile. These six were included in Bell et al. (2006), but the updated list excludes *A. asphodeloides* (zhī mǔ, 知母) on the basis of the subsequent report (Wu et al., 2014) of it containing a possible hepatotoxic compound.

The method of obtaining bile from caged live bears is highly controversial for animal welfare reasons. There is also evidence that, in spite of this option being available, as well as the options of obtaining bile from domesticated animal species or using chemically synthesized UDCA, there continues to be an illegal trade in wild bile bear, further endangering the survival of the species. Research suggests that the introduction of farmed bear bile has had little impact on demand for wild bear bile and, at worst, suggests that it may have in fact increased demand (Dutton, Hepburn, & Macdonald, 2011).

It is recognized that the problems of overexploitation of endangered species in TM must be addressed in the context of an improved understanding of the significance of TMs in the world's health care systems. Research into the use of substitutes is seen as a vital step in the process of alternatives gaining acceptance through demonstration of similarities in actions to bear bile in terms of medicinal use. This research must place emphasis on possible alternatives based on the theory and philosophy of TM as well as investigations into similarities in chemical structure and/or activity in Western pharmacological terms.

Different strategies previously have been employed to identify herbal alternatives to the use of bear bile in TM. The Association of Chinese Medicine and Philosophy in collaboration with the EarthCare Society (Hong Kong) and IFAW suggested 54 herbs as alternatives to bear bile based on TCM philosophy (IFAW report, 1994). More recently Yuan et al. (2014) selected herbs as alternatives to bear bile based on similarities in their pharmacological properties using a drug property prediction model. With the exception of one of the plant species, *C. chinensis* (huáng lián, 黄连) all of the herbs suggested in this more recent research (Table 3) were also proposed in the IFAW report (1994) as alternatives to bear bile. The most convincing method for testing the effectiveness of a substitute to an animal product is through in vivo studies and clinical trials, where the biological activities of the medicines are compared. Wang et al. (2012) reported a comparative study of the action of bear bile and Rhizoma Coptidis (huáng lián, 黄连) on experimental liver fibrosis and concluded that a crude extract of the herb, as well as its active compound (berberine) could serve as substitutes for bear bile.

Most of the suggested herbal alternatives to bear bile were found to already form part of one or more traditional prescriptions described in herbal Material Medica and classical texts as containing the animal product (Zhu, 1989; Zou, 1989). This finding confirmed the practice in TCM of combining remedies with similar functions for their additive and synergistic effects. We propose that the existing evidence base of herb substitute effectiveness is sufficiently strong for TM practitioners and consumers to consider using these to replace bear bile. As Feng et al. (2009) have emphasized, bear bile has been used in TCM for thousands of years, and its use is rooted in history and culture. Its future use will depend as much on an appreciation of these elements as well as politics and economics. To provide alternatives which are acceptable to those consumers who currently use bear bile, many factors will need to be brought into play. More persuasive research will be one such factor, with researchers and TM practitioners needing to work in collaboration to communicate and implement changes that will ultimately lead to mitigation of ongoing sustainability, animal welfare, and ethical concerns.

## REFERENCES


Alberts, D. S., Martínez, M. E., Hess, L. M., Einspahr, J. G., Green, S. B., Bhattacharyya, A. K., et al. (2005). Phase III trial of ursodeoxycholic acid to prevent colorectal adenoma recurrence. *Journal of the National Cancer Institute*, *97*(11), 846–853.

Appiah, S. (2006). *Evaluation of herbs as potential alternatives for bear bile and rhino horn used in traditional Chinese medicines: Chemical and biological analysis*. Ph.D. thesis https://eprints.mdx.ac.uk/7897/1/Appiah-phd.pdf.

Appiah, S. S., Bremner, P., Kokubun, T., Russell, M., Bell, C. M., Heinrich, M., et al. (2006). Herbal alternatives to bear bile: Effects of *Scutellaria baicalensis* Georgi on IL-6 promoter and CYP3A4 activities. *Focus on Alternative and Complementary Therapies*, *11*(1), 1–3.

Atretkhany, K. N., Drutskaya, M. S., Nedospasov, S. A., Grivennikov, S. I., & Kuprash, D. V. (2016). Chemokines, cytokines and exosomes help tumors to shape inflammatory microenvironment. *Pharmacology & Therapeutics*, *168*, 98–112.

Bardi, D. A., Halabi, M. F., Hassandarvish, P., Rouhollahi, E., Paydar, M., Moghadamtousi, S. Z., et al. (2014). *Andrographis paniculata* leaf extract prevents thioacetamide-induced liver cirrhosis in rats. *PloS One*, *9*(10), e109424.

Barnes, J. L., & Gorin, Y. (2011). Myofibroblast differentiation during fibrosis: Role of NAD(P)H oxidase. *Kidney International*, *79*(9), 944–956.

Beery, R. M., Vaziri, H., & Forouhar, F. (2014). Primary biliary cirrhosis and primary sclerosing cholangitis: A review featuring a women's health perspective. *Journal of Clinical and Translational Hepatology*, *2*(4), 266–284.

Bell, C. M., Simmonds, M. S. J., Howes, M. R., & Appiah, S. S. (2006). *Plant substances as alternatives for animal products in traditional medicines*. Department for the Environment, Food and Rural Affairs.

Bensky, D., & Gamble, A. (1993). *Chinese herbal medicine materia medica*. Seattle, WA: Eastland Press Incorporated.

Bessone, F., & Roma, M. G. (2016). Is ursodeoxycholic acid detrimental in obstructive cholestasis? A propos of a case of malignant biliary obstruction. *Annals of Hepatology*, *15*(3), 442–447.

Björkhem, I., Andersson, U., Sudjama-Sugiaman, E., Eggertsen, G., & Hylemon, P. (1993). Studies on the link between HMG-CoA reductase and cholesterol 7 alpha-hydroxylase in lymph-fistula rats: Evidence for both transcriptional and post-transcriptional mechanisms for down-regulation of the two enzymes by bile acids. *Journal of Lipid Research*, *34*(9), 1497–1503.

Boatright, J. H., Nickerson, J. M., Moring, A. G., & Pardue, M. T. (2009). Bile acids in treatment of ocular disease. *Journal of Ocular Biology, Diseases, and Informatics*, *2*(3), 149–159.

Brown, M. A., & Jones, W. K. (2004). NF-kappaB action in sepsis: The innate immune system and the heart. *Frontiers in Bioscience*, *9*, 1201–1217.

Carpino, G., Morini, S., Ginanni Corradini, S., Franchitto, A., Merli, M., Siciliano, M., et al. (2005). Alpha-SMA expression in hepatic stellate cells and quantitative analysis of hepatic fibrosis in cirrhosis and in recurrent chronic hepatitis after liver transplantation. *Digestive and Liver Disease*, *37*(5), 349–356.

Chang, H. M., & But, P. P. H. (1987). In *Pharmacology and application of Chinese materia medica*, (Vol. 2). Hong Kong: World Scientific Publishing Co Pte. Ltd.

Chang, H. M., & But, P. P. H. (2001). Pharmacology and applications of Chinese materia medica. (Vols. 1 and 2). London: World Scientific.

Chen, Y., Lan, T., Li, J., Qiu, C., Wu, T., Gou, H., et al. (2012). *Gardenia jasminoides* attenuates hepatocellular injury and fibrosis in bile duct-ligated rats and human hepatic stellate cells. *World Journal of Gastroenterology*, *18*(48), 7158–7165.

Chen, H. J., Liang, T. M., Lee, I. J., Huang, Y. T., & Lin, Y. L. (2013). Scutellariae Radix suppresses LPS-induced liver endothelial cell activation and inhibits hepatic stellate cell migration. *Journal of Ethnopharmacology*, *150*(3), 835–842.

Chen, S., Zhao, X., Sun, P., Qian, J., Shi, Y., & Wang, R. (2016). Preventive effect of *Gardenia jasminoides* on HCl/ethanol induced gastric injury in mice. *Journal of Pharmacological Sciences*, *133*(1), 1–8.

Cho, K. H., Jung, W. S., Park, S. U., Moon, S. K., Ko, C. N., Ku, S., et al. (2004). Daio-Orengedokudo works as a cell-proliferating compound in endothelial cells. *Canadian Journal of Physiology and Pharmacology*, *82*(6), 380–386.

Choi, Y. Y., Kim, M. H., Cho, I., Kim, J. H., Hong, J., Lee, T. H., et al. (2013). Inhibitory effect of *Coptis chinensis* on inflammation in LPS-induced endotoxemia. *Journal of Ethnopharmacology*, *149*(2), 506–512.

Choi, Y., Kim, M., Han, J., Hong, J., Lee, T., Kim, S., et al. (2014). The anti-inflammatory potential of Cortex Phellodendron in vivo and in vitro: Down-regulation of NO and iNOS through suppression of NF-kappa B and MAPK activation. *International Immunopharmacology*, *19*(2), 214–220.

CITES. (1997). *The convention on international trade in endangered species of wild fauna and flora. In Resolutions of the conference of the parties. Conf. 10.19 (Rev. CoP12).* http://www.cites.org/eng/cop/10/E10-Res.pdf. Last accessed 19 February 2004.

Cuéllar, M. J., Giner, R. M., Recio, M. C., Máñez, S., & Ríos, J. L. (2001). Topical anti-inflammatory activity of some Asian medicinal plants used in dermatological disorders. *Fitoterapia*, *72*(3), 221–229.

Dong, Q., Chu, F., Wu, C., Huo, Q., Gan, H., Li, X., et al. (2016). *Scutellaria baicalensis* Georgi extract protects against alcohol-induced acute liver injury in mice and affects the mechanism of ER stress. *Molecular Medicine Reports*, *13*(4), 3052–3062.

Dutton, A. J., Hepburn, C., & Macdonald, D. W. (2011). A stated preference investigation into the Chinese demand for farmed vs. wild bear bile. *PloS One*, *6*(7), e21243, Hofreiter M.

Eaton, J. E., Silveira, M. G., Pardi, D. S., Sinakos, E., Kowdley, K. V., Luketic, V. A., et al. (2011). High-dose ursodeoxycholic acid is associated with the development of colorectal neoplasia in patients with ulcerative colitis and primary sclerosing cholangitis. *The American Journal of Gastroenterology*, *106*(9), 1638–1645.

Eaton, J. E., Talwalkar, J. A., Lazaridis, K. N., Gores, G. J., & Lindor, K. D. (2013). Pathogenesis of primary sclerosing cholangitis and advances in diagnosis and management. *Gastroenterology*, *14*(3), 521–536.

Feng, Y., Siu, K., Wang, N., Ng, K., Tsao, S., Nagamatsu, T., et al. (2009). Bear bile: Dilemma of traditional medicinal use and animal protection. *Journal of Ethnobiology and Ethnomedicine*, *5*(1), 2.

Feng, Y., Siu, K. Y., Ye, X., Wang, N., Yuen, M. F., Leung, C. H., et al. (2010). Hepatoprotective effects of berberine on carbon tetrachloride-induced acute hepatotoxicity in rats. *Chinese Medicine*, *5*, 33.

Feng, Y., Wang, N., Ye, X., Li, H., Feng, Y., Cheung, F., et al. (2011). Hepatoprotective effect and its possible mechanism of Coptidis rhizoma aqueous extract on carbon tetrachloride-induced chronic liver hepatotoxicity in rats. *Journal of Ethnopharmacology*, *138*(3), 683–690.

Fickert, P., Pollheimer, M. J., Silbert, D., Moustafa, T., Halilbasic, E., Krones, E., et al. (2013). Differential effects of norUDCA and UDCA in obstructive cholestasis in mice. *Journal of Hepatology*, *58*(6), 1201–1208.

Friedman, S. L. (2008). Hepatic stellate cells: Protean, multifunctional, and enigmatic cells of the liver. *Physiological Reviews*, *88*(1), 125–172.

Fujimoto, M., Tsuneyama, K., Kainuma, M., Sekiya, N., Goto, H., Takano, Y., et al. (2008). Evidence-based efficacy of Kampo formulas in a model of non-alcoholic fatty liver. *Experimental Biology and Medicine (Maywood, N.J.)*, *233*(3), 328–337.

Fukutake, M., Miura, N., Yamamoto, M., Fukuda, K., Iijima, O., Ishikawa, H., et al. (2000). Suppressive effect of the herbal medicine Orengedoku-to on cyclooxygenase-2 activity and azoxymethane-induced aberrant crypt foci development in rats. *Cancer Letters*, *157*(1), 9–14.

Gamboa, A., Tian, C., Massaad, J., Reshamwala, P., & Cai, Q. (2011). The therapeutic role of ursodeoxycholic acid in digestive diseases. *Annals of Gastroenterology & Hepatology*, *2*(1), 43–49.

Garcia, G. E., Nicole, A., Bhaskaran, S., Gupta, A., Kyprianouy, N., & Kumar, A. P. (2006). Akt- and CREB-mediated prostate cancer cell proliferation inhibition by nexrutine, a *Phellodendron amurense* extract. *Neoplasia*, *8*(6), 523–533.

Ghosh, R., Garcia, G. E., Crosbyy, K., Inouez, H., Thompson, I. M., Troyer, D. A., et al. (2007). Regulation of Cox-2 by cyclic AMP response element binding protein in prostate cancer: Potential role for nexrutine. *Neoplasia*, *9*(11), 893–899.

Goode, E. C., & Rushbrook, S. M. (2016). A review of the medical treatment of primary sclerosing cholangitis in the 21st century. *Therapeutic Advances in Chronic Disease*, 7(1), 68–85.
Gowda, S., Desai, P. B., Hull, V. V., Math, A. A. K., Vernekar, S. N., & Kulkarni, S. S. (2009). A review on laboratory liver function tests. *The Pan African Medical Journal*, *3*, 1–17.
Gressner, A. M., & Weiskirchen, R. (2006). Modern pathogenetic concepts of liver fibrosis suggest stellate cells and TGF-β as major players and therapeutic targets. *Journal of Cellular and Molecular Medicine*, *10*(1), 76–99.
Hagey, L. R., Crombie, D. L., Espinosa, E., Carey, M. C., Igimi, H., & Hofmann, A. F. (1993). Ursodeoxycholic acid in the Ursidae: Biliary bile acids of bears, pandas, and related carnivores. *Journal of Lipid Research*, *34*(11), 1911–1917.
Hegade, V., Kendrick, S., Dobbins, R., Miller, S., Richards, D., Storey, J., et al. (2016). BAT117213: Ileal bile acid transporter (IBAT) inhibition as a treatment for pruritus in primary biliary cirrhosis: Study protocol for a randomised controlled trial. *BMC Gastroenterology*, *16*(1), 71.
Hsu, H. Y., Shen, S. J., Chen, Y. P., Hsu, C. S., Chen, C. C., & Chang, H. C. (1986). *Oriental materia medica: A concise guide*. New Canaan, CT: Keats Publishing, Inc.
IFAW report. (1994). *Herbal alternatives to bear bile in Chinese medicine*. Hong Kong: The Association of Chinese Medicine and Philosophy and Earthcare Society.
Inoue, H., Kohashi, K., & Tsuruta, Y. (1998). Simultaneous determination of serum and urinary hydroxyproline and proline by liquid chromatography using two fluorescent labeling reagents. *Analytica Chimica Acta*, *365*(1), 219–226.
Ivanov, A. I., & Romanovsky, A. A. (2004). Prostaglandin E2 as a mediator of fever: Synthesis and catabolism. *Frontiers in Bioscience*, *9*, 1977–1993.
Jones, H., Alpini, G., & Francis, H. (2015). Bile acid signaling and biliary functions. *Acta Pharmaceutica Sinica B*, *5*(2), 123–128.
Jones, J., & Zollman, P. (1997). Black bear (Ursus americanus) bile composition: Seasonal changes. *Comparative Biochemistry and Physiology C: Toxicology & Pharmacology*, *118*(3), 387–390.
Joutsiniemi, T., Timonen, S., Linden, M., Suvitie, P., & Ekblad, U. (2015). Intrahepatic cholestasis of pregnancy: Observational study of the treatment with low-dose ursodeoxycholic acid. *BMC Gastroenterology*, *15*(1), 92.
Jung, W., Chae, Y., Kim, D., Seo, S., Park, H., Bae, G., et al. (2008). Gardenia jasminoides protects against cerulein-induced acute pancreatitis. *World Journal of Gastroenterology*, *14*(40), 6188–6194.
Jung, H., Kim, M. H., Gwak, N., Im, Y., Lee, K., Sohn, Y., et al. (2012). Antiallergic effects of *Scutellaria baicalensis* on inflammation in vivo and in vitro. *Journal of Ethnopharmacology*, *141*(1), 345–349.
Kanazawa, T., Shimazaki, A., Sato, T., & Hoshino, T. (1954). Syntheses of ursodeoxycholic acid and its conjugated bile acid. *Proceedings of the Japan Academy*, *30*(5), 391–392.
Khamphaya, T., Chansela, P., Piyachaturawat, P., Suksamrarn, A., Nathanson, M. H., & Weerachayaphorn, J. (2016). Effects of andrographolide on intrahepatic cholestasis induced by alpha-naphthylisothiocyanate in rats. *European Journal of Pharmacology*, *789*, 254–264.
Kim, Y. S., Jung, E. A., Shin, J. E., Chang, J. C., Yang, H. K., Kim, N. J., et al. (2002). Daio-Orengedokuto inhibits HMG-CoA reductase and pancreatic lipase. *Biological & Pharmaceutical Bulletin*, *25*(11), 1442–1445.
Kim, S., Kim, K., Park, J., Kwak, J., Kim, Y. S., & Lee, S. (2013). Geniposidic acid protects against D-galactosamine and lipopolysaccharide-induced hepatic failure in mice. *Journal of Ethnopharmacology*, *146*(1), 271–277.

Kishore, V., Yarla, N. S., Zameer, F., Nagendra Prasad, M. N., Santosh, M. S., More, S. S., et al. (2016). Inhibition of group IIA secretory phospholipase A2 and its inflammatory reactions in mice by ethanolic extract of *Andrographis paniculata*, a well-known medicinal food. *Pharmacognosy Research*, *8*(3), 213–216.
Koh, P. H., Mokhtar, R. A., & Iqbal, M. (2011). *Andrographis paniculata* ameliorates carbon tetrachloride (CCl(4))-dependent hepatic damage and toxicity: Diminution of oxidative stress. *Redox Report*, *16*(3), 134–143.
Koo, H., Lee, S., Shin, K., Kim, B., Lim, C., & Park, E. (2004). Geniposide, an anti-angiogenic compound from the fruits of *Gardenia jasminoides*. *Planta Medica*, *70*(5), 467–469.
Lazaridis, K., & LaRusso, N. (2015). The cholangiopathies. *Mayo Clinic Proceedings*, *90*(6), 791–800.
Leonti, M., & Casu, L. (2013). Traditional medicines and globalization: Current and future perspectives in ethnopharmacology. *Frontiers in Pharmacology*, *4*, 92.
Li, Y. W., Zhu, X. Y., But, P. P. H., & Yeung, H. W. (1995). Ethnopharmacology of bear gall bladder: I. *Journal of Ethnopharmacology*, *47*(1), 27–31.
Lim, S., Choi, J. E., Kang, H. S., & Si, H. (2009). Ursodeoxycholic acid switches oxaliplatin-induced necrosis to apoptosis by inhibiting reactive oxygen species production and activating p53-caspase 8 pathway in HepG2 hepatocellular carcinoma. *International Journal of Cancer*, *126*(7), 1582–1595.
Lin, D. L., Chang, H. C., Chen, C. P., & Chen, C. Y. (1997). Identification and differentiation of bear bile used in medicinal products in Taiwan. *Journal of Forensic Sciences*, *42*(5), 817–823.
Lin, W., Kuo, H., Ho, L., Tseng, M., Siao, A., Hung, C., et al. (2015). *Gardenia jasminoides* extracts and gallic acid inhibit lipopolysaccharide-induced inflammation by suppression of JNK2/1 signaling pathways in BV-2 cells. *Iranian Journal of Basic Medical Sciences*, *18*(6), 555–556.
Lindor, K. D., Kowdley, K. V., Luketic, V. A., Harrison, M. E., McCashland, T., Befeler, A. S., et al. (2009). High-dose ursodeoxycholic acid for the treatment of primary sclerosing cholangitis. *Hepatology*, *50*(3), 808–814.
Liu, X., Liu, R., Hou, F., Cui, L., Li, C., Chi, C., et al. (2016). Fibronectin expression is crucial for liver fibrogenesis in vivo and in vitro. *Molecular Medicine Reports*, *14*(4), 3669–3675.
Liu, H., Xu, H., Zhang, Y., Huang, Y., Han, G., Liang, T., et al. (2015). Ursodeoxycholic acid induces apoptosis in hepatocellular carcinoma xenografts in mice. *World Journal of Gastroenterology*, *21*(36), 10367–10374.
Liu, B., Yang, L., Zhang, B., Kuang, C., Huang, S., & Guo, R. (2016). NF-κB-dependent upregulation of NCX1 induced by angiotensin II contributes to calcium influx in rat aortic smooth muscle cells. *The Canadian Journal of Cardiology*, *32*(11), 1356. e11–1356.e20.
Ma, Y. S., Hsiao, Y. P., Lin, J. H., Hsu, S. C., Chueh, F. S., Weng, S. W., et al. (2015). Crude extract of *Rheum palmatum* L inhibits migration and invasion of LS1034 human colon cancer cells acts through the inhibition of matrix metalloproteinase-2/-9 by MAPK signaling. *Environmental Toxicology*, *30*(7), 852–863.
Maclean, W., & Taylor, K. (2000). *Chinese herbal patent medicines*. Sydney: Pangolin Press.
Moreira, R. K. (2007). Hepatic stellate cells and liver fibrosis. *Archives of Pathology & Laboratory Medicine*, *131*(11), 1728–1734.
Nan, J. X., Park, E. J., Kim, Y. C., Ko, G., & Sohn, D. H. (2002). *Scutellaria baicalensis* inhibits liver fibrosis induced by bile duct ligation or carbon tetrachloride in rats. *The Journal of Pharmacy and Pharmacology*, *54*(4), 555–563.
Nho, K., Chun, J., Lee, A., & Kim, H. (2015). Anti-metastatic effects of *Rheum Palmatum* L. extract in human MDA-MB-231 breast cancer cells. *Environmental Toxicology and Pharmacology*, *40*(1), 30–38.

O'Hara, S. P., Gradilone, S. A., Masyuk, T. V., Tabibian, J. H., & LaRusso, N. F. (2014). MicroRNAs in cholangiopathies. *Current Pathobiology Reports*, *2*(3), 133–142.

Ohta, Y., Sasaki, E., Nishida, K., Kongo, M., Hayashi, T., Nagata, M., et al. (1998). Inhibitory effect of Oren-gedoku-to (Huanglian-Jie-Du-Tang) extract on hepatic triglyceride accumulation with the progression of carbon tetrachloride-induced acute liver injury in rats. *Journal of Ethnopharmacology*, *61*(1), 75–80.

Oshima, N., Narukawa, Y., Hada, N., & Kiuchi, F. (2013). Quantitative analysis of anti-inflammatory activity of orengedokuto: Importance of combination of flavonoids in inhibition of PGE2 production in mouse macrophage-like cell line J774.1. *Journal of Natural Medicines*, *67*(2), 281–288.

Pan, T., Wang, P., Leu, Y., Wu, T., & Wu, T. (2012). Inhibitory effects of *Scutellaria baicalensis* extract on hepatic stellate cells through inducing G2/M cell cycle arrest and activating ERK-dependent apoptosis via Bax and caspase pathway. *Journal of Ethnopharmacology*, *139*, 829–837.

Park, H. S., Park, K. I., Hong, G. E., Nagappan, A., Lee, H. J., Kim, E. H., et al. (2014). Korean *Scutellaria baicalensis* Georgi methanol extracts inhibits metastasis via the forkhead Box M1 activity in hepatocellular carcinoma cells. *Journal of Ethnopharmacology*, *155*(1), 847–851.

Peng, Y., Guo, C., Li, P., Fu, Z., Gao, L., Di, Y., et al. (2014). Immune and anti-oxidant functions of ethanol extracts of *Scutellaria baicalensis* Georgi in mice bearing U14 cervical cancers. *Asian Pacific Journal of Cancer Prevention*, *15*(10), 4129–4133.

Poupon, R. (2012). Ursodeoxycholic acid and bile-acid mimetics as therapeutic agents for cholestatic liver diseases: An overview of their mechanisms of action. *Clinics and Research in Hepatology and Gastroenterology*, *36*(1), S3–S12.

Richter, K., & Kietzmann, T. (2016). Reactive oxygen species and fibrosis: Further evidence of a significant liaison. *Cell and Tissue Research*, *365*(3), 591–605.

Robinson, M. M., & Zhang, X. (2011). *The world medicines situation 2011: Traditional medicines: Global situation*. Geneva, Switzerland: WHO Press. http://www.who.int/medicines/areas/policy/world_medicines_situation/WMS_ch6_wPricing_v6.pdf.

Rodrigues, C. M. P., Fan, G., Ma, X., Kren, B. T., & Steer, C. J. (1998). A novel role for ursodeoxycholic acid in inhibiting apoptosis by modulating mitochondrial membrane perturbation. *The Journal of Clinical Investigation*, *101*(12), 2790–2799.

Rossi, S. S., Converse, J. L., & Hofmann, A. F. (1987). High pressure liquid chromatographic analysis of conjugated bile acids in human bile: Simultaneous resolution of sulfated and unsulfated lithocholyl amidates and the common conjugated bile acids. *Journal of Lipid Research*, *28*(5), 589–595.

Roy, D. N., Mandal, S., Sen, G., Mukhopadhyay, S., & Biswas, T. (2010). 14-Deoxyandrographolide desensitizes hepatocytes to tumour necrosis factor-alpha-induced apoptosis through calcium-dependent tumour necrosis factor receptor superfamily member 1A release via the NO/cGMP pathway. *British Journal of Pharmacology*, *160*(7), 1823–1843.

Ruan, W., Huang, M., He, X., Zhang, F., & Tao, H. (2013). Effects of Huangqi and bear bile on recurrent parotitis in children: A new clinical approach. *Journal of Zhejiang University. Science. B*, *14*(3), 253–258.

Schachtrup, C., Le Moan, N., Passino, M. A., & Akassoglou, K. (2011). Hepatic stellate cells and astrocytes: Stars of scar formation and tissue repair. *Cell Cycle*, *10*(11), 1764–1771.

Sclair, S., Little, E., & Levy, C. (2015). Current concepts in primary biliary cirrhosis and primary sclerosing cholangitis. *Clinical and Translational Gastroentrology*, *6*(8), e109.

Sekiya, N., Kogure, T., Kita, T., Kasahara, Y., Sakakibara, I., Goto, H., et al. (2002). Reduction of plasma triglyceride level and enhancement of plasma albumin concentration by Oren-gedoku-to administration. *Phytomedicine*, *9*(5), 455–460.

Singh, D., Awasthi, H., Luqman, S., Singh, S., & Mani, D. (2015). Hepatoprotective effect of a polyherbal extract containing *Andrographis paniculata*, *Tinospora cordifolia* and *Solanum nigrum* against paracetamol induced hepatotoxicity. *Pharmacognosy Magazine*, *11*(44), S375–S379.

Tacke, F., & Zimmermann, H. W. (2014). Macrophage heterogeneity in liver injury and fibrosis. *Journal of Hepatology*, *60*(5), 1090–1096.

The CITES Appendices. (2016). *The convention on international trade in endangered species of wild fauna and flora.* In *Resolutions of the conference of the parties.* Last accessed 23 September 2016 https://www.cites.org/eng/app/index.php.

Vang, S., Longley, K., Steer, C. J., & Low, W. C. (2014). The unexpected uses of urso- and tauroursodeoxycholic acid in the treatment of non-liver diseases. *Global Advances in Health and Medicine: Improving Healthcare Outcomes Worldwide*, *3*(3), 58–69.

Wang, N., Feng, Y., Cheung, F., Chow, O., Wang, X., Su, W., et al. (2012). A comparative study on the hepatoprotective action of bear bile and coptidis rhizoma aqueous extract on experimental liver fibrosis in rats. *BMC Complementary and Alternative Medicine*, *12*(1), 239.

Wang, N., Feng, Y., Xie, T. N., Su, W., Zhu, M., Chow, O., et al. (2011). Chemical and biological analysis of active free and conjugated bile acids in animal bile using HPLC-ELSD and MTT methods. *Experimental and Therapeutic Medicine*, *2*(1), 125–130.

Wang, J., Zhao, H., Zhao, Y., Jin, C., Liu, D., Kong, W., et al. (2011). Hepatotoxicity or hepatoprotection? Pattern recognition for the paradoxical effect of the Chinese herb Rheum palmatum L. in treating rat liver injury. *PloS One*, *6*(9), e24498.

Watanabe, S., Kamei, T., Tanaka, K., Kawasuji, K., Yoshioka, T., & Ohno, M. (2009). Roles of bile acid conjugates and phospholipids in in vitro activation of pancreatic lipase by bear bile and cattle bile. *Journal of Ethnopharmacology*, *125*(2), 203–206.

Watanabe, S., & Tsuneyama, K. (2012). Cattle bile but not bear bile or pig bile induces lipid profile changes and fatty liver injury in mice: Mediation by cholic acid. *The Journal of Toxicological Sciences*, *37*(1), 105–121.

World Health Organisation (WHO). (2004). *Medicinal plants—Guidelines to promote patient safety and plant conservation for a US$ 60 billion industry.* http://www.who.int/mediacentre/news/notes/2004/np3/en/. Last accessed September 2016.

World Health Organisation (WHO). (2013). *WHO traditional medicine strategy: 2014–2023.* http://apps.who.int/iris/bitstream/10665/92455/1/9789241506090_eng.pdf. Last accessed September 2016.

Wu, X., Liang, X., Du, Y., Zhang, Y., Yang, M., Gong, W., et al. (2016). Prevention of gallstones by Lidan Granule: Insight intounderlying mechanisms using a guinea pig model. *Biomedical Reports*, *5*, 50–56.

Wu, Z. T., Qi, X. M., Sheng, J. J., Ma, L. L., Ni, X., Ren, J., et al. (2014). Timosaponin A3 induces hepatotoxicity in rats through inducing oxidative stress and down-regulating bile acid transporters. *Acta Pharmacologica Sinica*, *35*(9), 1188–1198.

Wunsch, E., Krawczyk, M., Milkiewicz, M., Trottier, J., Barbier, O., Neurath, M. F., et al. (2016). Serum autotaxin is a marker of the severity of liver injury and overall survival in patients with cholestatic liver diseases. *Scientific Reports*, *6*, 1–12.

Xian, Y., Mao, Q., Ip, S., Lin, Z., & Che, C. (2011). Comparison on the anti-inflammatory effect of Cortex phellodendri chinensis and Cortex phellodendri amurensis in 12-O-tetradecanoyl-phorbol-13-acetate-induced ear edema in mice. *Journal of Ethnopharmacology*, *137*(3), 1425–1430.

Xiao, J. R., Do, C. W., & To, C. H. (2014). Potential therapeutic effects of baicalein, baicalin, and wogonin in ocular disorders. *Journal of Ocular Pharmacology and Therapeutics*, *30*(8), 605–614.

Yang, F., Xu, Y., Xiong, A., He, Y., Yang, L., Wan, Y. Y., et al. (2012). Evaluation of the protective effect of Rhei Radix et Rhizoma against α-naphthylisothiocyanate induced liver injury based on metabolic profile of bile acids. *Journal of Ethnopharmacology*, *144*(3), 599–604.

Yerigui, Wu, X., Wang, X., & Ma, C. (2016). Quantification of bile acids in traditional animal medicines and their preparations using ultra high-performance liquid chromatography-mass spectrometry in the multiple reaction monitoring mode. *Analytical Sciences*, *32*(5), 499–503.

Yeum, H., Lee, Y., Kim, S., Roh, S., Lee, J., & Seo, Y. (2007). *Fritillaria cirrhosa*, *Anemarrhena asphodeloides*, Lee-Mo-Tang and cyclosporine a inhibit ovalbumin-induced eosinophil accumulation and Th2-Mediated bronchial hyperresponsiveness in a murine model of asthma. *Basic & Clinical Pharmacology & Toxicology*, *100*(3), 205–213.

Yotsumoto, H., Yanagita, T., Yamamoto, K., Ogawa, Y., Cha, J. Y., & Mori, Y. (1997). Inhibitory effects of oren-gedoku-to and its components on cholesteryl ester synthesis in cultured human hepatocyte HepG2 cells: Evidence from the cultured HepG2 cells and in vitro assay of ACAT. *Planta Medica*, *63*(2), 141–145.

Yuan, B., Ren, Y., Ma, L., Gu, H., Wang, Y., & Qiao, Y. (2014). Analysis on replacement of traditional Chinese medicine bear bile with bile acids based on drug properties. *Zhongguo Zhongyao Zazhi*, *39*(4), 738–743 [Article in Chinese].

Yue, G., Lee, J., Li, L., Chan, K., Wong, E., Chan, J., et al. (2015). *Andrographis paniculata* elicits anti-invasion activities by suppressing TM4SF3 gene expression and by anoikis-sensitization in esophageal cancer cells. *American Journal of Cancer Research*, *5*(12), 3570–3587.

Zhang, Y., Pan, Y., Lin, C., Zheng, Y., Sun, H., Zhang, H., et al. (2016). Bile acids evoke placental inflammation by activating Gpbar1/NF-κB pathway in intrahepatic cholestasis of pregnancy. *Journal of Molecular Cell Biology*, *8*(6), 530–541.

Zhang, R., Qiu, H., Wang, N., Long, F., & Mao, D. (2015). Effect of Rheum palmatum L. on NF-kappa B signaling pathway of mice with acute liver failure. *Asian Pacific Journal of Tropical Medicine*, *8*(10), 823–829.

Zhao, X., Zhang, J., Tong, N., Chen, Y., & Luo, Y. (2012). Protective effects of berberine on doxorubicin-induced hepatotoxicity in mice. *Biological & Pharmaceutical Bulletin*, *35*(5), 796–800.

Zhu, C. H. (1989). *Clinical handbook of Chinese prepared medicines*. Brookline, MA: Paradigm Publications.

Zhu, G., Shi, K., Huang, S., Huang, G., Lin, Y., Zhou, Z., et al. (2015). Network meta-analysis of randomized controlled trials: Efficacy and safety of UDCA-based therapies in primary biliary cirrhosis. *Medicine*, *94*(11), e609.

Zhu, M., Wang, N., Tsao, S., Yuen, M., Feng, Y., Wan, T. S. K., et al. (2011). Up-regulation of microRNAs, miR21 and miR23a in human liver cancer cells treated with Coptidis rhizoma aqueous extract. *Experimental and Therapeutic Medicine*, *2*(1), 27–32.

Zou, J. (Ed.), (1989). *The Chinese materia medica*. China: Beijing University of Traditional Chinese Medicine

CHAPTER EIGHT

# Effect of Chinese Herbal Medicine on Molecular Imaging of Neurological Disorders

**Yao Yao**[*,†,‡,§,1], **Ting Chen**[*,†,‡,§,1], **Jing Huang**[*,†,‡,§,1], **Hong Zhang**[*,†,‡,§], **Mei Tian**[*,†,‡,§,2]

[*]The Second Affiliated Hospital of Zhejiang University School of Medicine, Hangzhou, Zhejiang, China
[†]Zhejiang University Medical PET Center, Hangzhou, Zhejiang, China
[‡]Institute of Nuclear Medicine and Molecular Imaging of Zhejiang University, Hangzhou, Zhejiang, China
[§]Key Laboratory of Medical Molecular Imaging of Zhejiang Province, Hangzhou, Zhejiang, China
[2]Corresponding author: e-mail address: meitian@zju.edu.cn

## Contents



## Abstract

Chinese herbal medicine has been used to treat a wide variety of neurological disorders including stroke, Alzheimer's disease, and Parkinson's disease. However, its mechanism behind the effectiveness remains unclear. Recently, molecular imaging technology has been applied for this purpose, since it can assess the cellular or molecular function in a living subject by using specific imaging probes and/or radioactive tracers, which enable efficient analysis and monitoring the therapeutic response repetitively. This chapter reviews the in vivo functional and metabolic changes after administration of Chinese herbal medicine in various neurological disorders and provides perspectives on the future evaluations of therapeutic response of Chinese herbal medicine.

[1] These authors contributed equally to this work.

*International Review of Neurobiology*, Volume 135
ISSN 0074-7742
http://dx.doi.org/10.1016/bs.irn.2017.02.009

# 1. INTRODUCTION

Herbal medicine and its derivatives were tested for many of pharmacological activities over the past decades (Vickers & Zollman, 1999). The existing literatures in China (i.e., Chinese Pharmacopoeia, Compendium of Materia Medica) have revealed thousands of herbal medicine or formulas, and presented the evidence of therapeutic value of Chinese herbal medicine. Although researchers have tried to understand the mechanism behind the effectiveness using modern research approaches (Gong & Sucher, 1999; Liu, Kong, Yuan, Liu, & Wang, 2014; Tang, Ye, Feng, & Quinn, 2016), findings are still inconclusive due to insufficient number of studies and the complexity of interactions of various active components.

Molecular imaging is an emerging technology, which allows visualization of molecular functions in a living subject by using specific molecular probe and/or radioactive tracer, and enables monitoring the therapeutic response repetitively (de Vries, de Jong, & Gietema, 2015). Table 1 is the summary of major imaging modalities including positron emission tomography (PET), single-photon emission computed tomography (SPECT), optical imaging, magnetic resonance imaging (MRI), X-ray, computed tomography (CT), and ultrasound. Although molecular imaging has gained increased attention in neurological research (Agdeppa et al., 2001; Alonso et al., 2007; Huang et al., 2001; Lee, Zhang, Zhu, Duff, & Pardridge,

**Table 1** Summary of Molecular Imaging Modalities

| Imaging Modality | Spatial Resolution | Depth of Penetration | Temporal Resolution | Sensitivity (mol/L) | Molecular Probe |
|---|---|---|---|---|---|
| PET | 1–2 mm | No limit | s–min | $10^{-11}$–$10^{-12}$ | ng |
| SPECT | 0.3–1 mm | No limit | min | $10^{-10}$–$10^{-11}$ | ng |
| Optical | | | | | |
| Bioluminescence | 3–5 mm | 1–2 mm | s–min | $10^{-15}$–$10^{-17}$ | ng–μg |
| Fluorescence | 2–3 mm | <1 mm | s–min | $10^{-9}$–$10^{-12}$ | ng–μg |
| FMT | 1–3 mm | <5 mm | min | $10^{-8}$–$10^{-12}$ | μg |
| MRI | 50–250 μm | No limit | min–h | $10^{-3}$–$10^{-5}$ | μg–mg |
| X-ray CT | 25–150 μm | No limit | min | ? | mg |
| Ultrasound | 30–500 μm | mm–cm | s–min | $10^{-8}$–$10^{-9}$ | μg–mg |

*CT*, computed tomography; *FMT*, fluorescence molecular tomography; *MRI*, magnetic resonance imaging; *PET*, positron emission tomography; *SPECT*, single-photon emission computed tomography.

2002; Marek et al., 2001; Reshef et al., 2008), only a few imaging studies on Chinese herbal medicine. Recently, we and other groups have investigated the effectiveness of Chinese herbal medicine by using molecular imaging modalities (Bi et al., 2011; Cheng et al., 2014; Li, Lu, & Zhang, 2013; Ren et al., 2016; Wan et al., 2008; Wang et al., 2014; Yuan et al., 2011; Zhang et al., 2013, 2015), which might shed new lights on neurological disorders. This chapter reviews the in vivo functional and metabolic changes after administration of Chinese herbal medicine in various neurological disorders and provides perspectives on the future evaluations of therapeutic response of Chinese herbal medicine.

## 2. MOLECULAR IMAGING IN CHINESE HERBAL MEDICINE

Neurological disorders including Alzheimer's disease (AD), epilepsy, multiple sclerosis, Parkinson's disease (PD), stroke, and traumatic brain injuries can lead to severe cognitive impairment and sensory function loss (Patel et al., 2015). Therefore, it is essential to seeking new treatment through innovative approach to improve the quality of life and minimize the impact of these diseases (Rocha, Carvalho, & Pego, 2016; Rojas & Boxer, 2016). In this section, we focus on the role of molecular imaging in evaluation of therapeutic potential of Chinese herbal medicine for stroke, AD, and PD (Table 2).

**Table 2** Molecular Imaging in Chinese Herbal Medicine for Neurological Disorders

| Disease Models | Drug or Other Therapy | Tracers | Modality | References |
|---|---|---|---|---|
| Stroke (rats) | Scu | $^{18}$F-FDG | PET | Li et al. (2013) |
| Stroke (rats) | QKL | $^{18}$F-FDG | PET | Zhang et al. (2015) |
| Stroke (rats) | TXL | $^{18}$F-FDG | PET | Cheng et al. (2014) |
| Stroke (rats) | Chuangxiongzine–puerarin | $^{18}$F-FDG | PET | Wan et al. (2008) |
| Stroke (rats) | DHI | $^{18}$F-FDG | PET | Wang et al. (2014) |
| AD (rats) | CBJC | $^{18}$F-FDG | PET | Zhang et al. (2013) |
| AD (human) | FZS | $^{18}$F-FDG | PET | Bi et al. (2011) |
| PD (rats) | SAFE | Gd-DTPA | MRI | Ren et al. (2016) |

*AD*, Alzheimer's disease; *CBJC*, baicalin, jasminoidin, and cholic acid; *DHI*, danhong injection; $^{18}$*F-FDG*, $^{18}$F-fluorodeoxyglucose; *FZS*, fuzhisan; *MRI*, magnetic resonance imaging; *PET*, positron emission tomography; *PD*, Parkinson's disease; *QKL*, qingkailing; *SAFE*, standardized flavonoid extract of safflower; *Scu*, scutellarin; *TXL*, tongxinluo.

## 2.1 Stroke

Scutellarin (or scutellarein-7-*O*-glucuronide) (黄芩素) is a natural drug found among the total flavonoids of barbed skullcap (*Scutellaria barbata*) and blue skullcap (*Scutellaria lateriflora*). It has been widely used for the treatment of cerebral infraction. Our group examined the neuroprotective effects of scutellarin (Scu) for cerebral ischemia with use of $^{18}$F-fluorodeoxyglucose ($^{18}$F-FDG) micro-positron emission tomography (micro-PET) (Li et al., 2013). Different Scu dosages (25 and 50 mg/kg) were orally administered for 2 weeks to rats with experimental stroke caused by middle cerebral artery occlusion (MCAO). In addition, the consistent evaluation of Scu treatment in rats was conducted by behavior test and micro-PET at 1, 7, 14, 21, and 28 days after MCAO. $^{18}$F-FDG accumulation in the ipsilateral cerebral infarction was found increased steadily over time in Scu (50 mg/kg) group compared with untreated and Scu (25 mg/kg) groups (Fig. 1). Enhanced neuronal maturation also observed in Scu (50 mg/kg) group was confirmed by immunohistochemical staining. In conclusion, Scu may have the potential to the prevention and treatment of ischemic stroke as it has shown antiischemia–reperfusion activity in a dose-dependent manner.

Tongxinluo (通心络, TXL) is a mixed compound of traditional Chinese herbs and insects, including Radix Ginseng (人参), Scorpio (全蝎), Hirudo (水蛭), Eupolyphaga seu Steleophage (土鳖虫粉), Scolopendra (蜈蚣), Periostracum Cicadae (蝉蜕), Radix Paeoniae Rubra (赤芍), and Borneolum Syntheticum (冰片), which has been used for the treatment of acute stroke and for neuroprotection in China. Cheng and colleagues have studied the neuroprotective effect of TXL in a rat MCAO model (Cheng et al., 2014). TXL suspension was administered intragastrically daily for 5 days before MCAO model establishment in the treatment group. $^{18}$F-FDG micro-PET imaging demonstrated increased glucose metabolism around the ischemic area in the treatment group compared with the sham-operated control group. TXL-treated rat had better performance in the neurological function test compared with the control. Their results suggest that preadministration of TXL could protect the neurological system and improve the outcome after ischemic stroke onset. By employing molecular imaging in their study, the metabolic situation in research animals can be clearly monitored and quantified over time.

Chuanxiongzine (川芎嗪), an active compound isolated from traditional Chinese medicine chuanxiong (川芎, *Ligusticum striatum*), is used to improve cerebral blood flow. Puerarin (葛根素) is an isoflavonoid found from the root of *Pueraria* (葛根, *Radix puerariae*) and has been used to improve

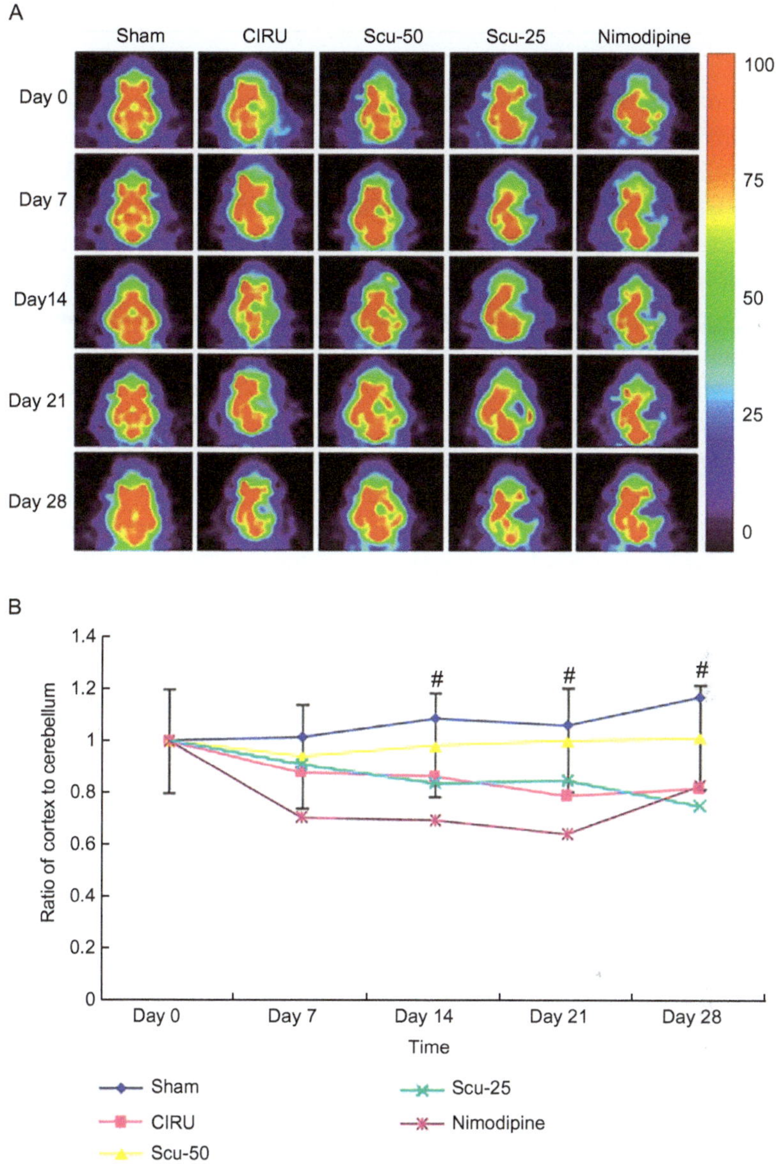


**Fig. 1** (A) $^{18}$FDG-PET images of activity of cerebral ischemic regions of the glucose metabolism. Sham group (sham), cerebral ischemia–reperfusion-untreated group (CIRU), scutellarin 50 mg/kg group (Scu-50), scutellarin 25 mg/kg group (Scu-25), and nimodipine 10 mg/kg group (nimodipine) in days 0, 7, 14, 21, and 28, respectively. (B) Ratios of cortex to cerebellum in different regions in the brain of the glucose metabolism sham group (sham), cerebral ischemia–reperfusion-untreated group (CIRU), scutellarin 50 mg/kg group (Scu-50), scutellarin 25 mg/kg group (Scu-25), and nimodipine 10 mg/kg group (nimodipine) in days 0, 7, 14, 21, and 28, respectively. *Reprinted from Li, J. H., Lu, J., & Zhang, H. (2013). Functional recovery after scutellarin treatment in transient cerebral ischemic rats: A pilot study with $^{18}$F-fluorodeoxyglucose microPET.* Evidence-Based Complementary and Alternative Medicine, 2013, *507091, Copyright (2013), with permission from Hindawi Publishing Corporation.*

blood circulation in cardiovascular and cerebrovascular diseases. Our group investigated the antioxidant and neuroprotective potential of chuangxiongzine–puerarin combination in a rat model of MCAO (Wan et al., 2008). $^{18}$F-FDG micro-PET imaging showed that glucose metabolism was significantly increased in the temporal and frontal lobes from the chuangxiongzine–puerarin combination group compared to the control group. We conclude that chuangxiongzine–puerarin played an important role in neuroprotection by reduced oxidant stress in the MCAO rat model. Glucose response measured by molecular imaging was strongly correlated with neurological assessment over time and immunocytochemistry.

Danhong injection (丹红注射液, DHI), a Chinese Materia Medica standardized product extracted from Danshen (丹参, *Radix Salviae Miltiorrhizae*) and Honghua (红花, *Flos Carthami*), has been reported to be effective for cardiovascular diseases. Ten components were identified in DHI through high-performance liquid chromatography coupled with diode array detector (HPLC-DAD) as protocatechuic aldehyde, caffeic acid, danshensu, 5-hydroxymethyl-2-furfural, salvianolic acid D, salvianolic acid B, salvianolic acid A, lithospermic acid, ferulic acid, and rosmarinic acid (Liu, Wu, Yang, Ding, & Wu, 2013). Our group has done a study on evaluating the effect of DHI in a rat model of MCAO using micro-PET scans with $^{18}$F-FDG (Wang et al., 2014). Different DHI dosages (1, 2, and 4 mL/kg) were intraperitoneally administered for 14 days to rats with experimental stroke caused by MCAO. In addition, the consistent evaluation of DHI treatment in rats was conducted by behavior test and micro-PET at 1, 7, and 14 days. Their results have shown that glucose metabolism in the area of cerebral infarction increased during treatment period, and the infraction volume was decreased. Improved neurological function was also observed in the treatment group. This micro-PET imaging study demonstrated the neuroprotective effects of DHI and metabolic recovery in the MCAO rat model.

Taken together, the above-mentioned studies verified that PET with $^{18}$F-FDG is feasible to measure the metabolic changes following herbal administration in ischemic stroke models.

## 2.2 Alzheimer's Disease

Qingkailing (清开灵, QKL) is one of the most well-known Chinese herb preparations with reported therapeutic effect on a broad spectrum of diseases, including high fever, coma, and acute inflammation. The active compounds in QKL are baicalin, jasminoidin, and cholic acid, and they are derived from Huangqin (黄芩, the root of *Scutellaria baicalensis*), Zhizi (栀子, the fruit of *Gardenia jasminoides* Ellis), and Niuhuang (牛黄, *Calculus bovis*), respectively. Zhang and colleagues have found that intragastric administration of a combination of baicalin, jasminoidin, and cholic acid (CBJC) can improve the ibotenic acid (IBO)-induced dementia in rat (Zhang et al., 2013). The solution of CBJC (1.25 mg/mL baicalin, 6.25 mg/mL jasminoidin, and 1.75 mg/mL cholic acid) was administered intragastrically at a dose of 3 mL/kg once a day from the IBO administration until the rats were sacrificed. The CBJC group showed a significantly decreased $^{18}$F-FDG uptake in the left olfactory bulb and the left anterior nucleus compared with the control group, and a significant increase in the left hippocampus. In addition to $^{18}$F-FDG signal changes, the escape latency time and the swim distance during the hidden platform and reverse platform tests in the CBJC group were also improved. CBJC may potentially act as an alternative treatment strategy for AD through the modulation of neuroprotective pathways.

Fuzhisan (复智散, FZS) is composed of ginseng root, baical skullcap root, rhizome of *Acorus calamus* L., and Radix Glycyrrhizae and has been used in the treatment of AD for more than 17 years in China. Bi and colleagues further investigate the effects of FZS (10 mg/day) on cerebral glucose metabolism and neuropsychological metrics (Bi et al., 2011). This was a 12-week, randomized, double-blind, placebo-controlled study conducted in patients with mild-to-moderate AD. 10 g FZS was given orally to FZS-treated patients daily for 12 weeks. FZS significantly increased the regional cerebral metabolic rate of glucose consumption (rCMRglc) in the bilateral temporal and parietal cortices, hippocampus, and posterior cingulate gyrus compared with placebo-treated subjects at week 12 (Fig. 2). This suggested that FZS may improve cognition and behavioral function in patients with mild-to-moderate AD.

Overall, imaging study to understand the effectiveness of Chinese herbal medicine in AD research is still limited. Thus, further study to investigate the potential of Chinese herbal medicine for the treatment of AD is needed.

## 2.3 Parkinson's Disease

Safflower (红花, *Carthamus tinctorius*) is an annual crop belonging to the Asteraceae family, cultivated mainly for its seed, which is used as vegetable oil. According to traditional Chinese medicine, safflower is generally

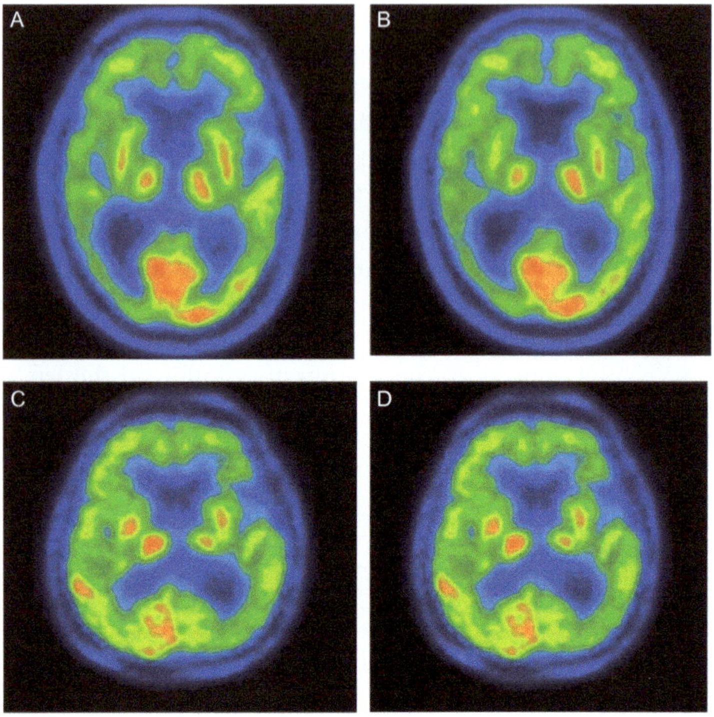


**Fig. 2** Representative $^{18}$F-FDG-PET scans from two subjects. Notes: The *upper figures* are baseline (A) and week 12 (B) scans of one representative FZS-treated patient. The scan was taken at the level of the corpus striatum and indicates slightly increased rCMRglc in the left temporal cortex. The *lower figures* are baseline (C) and week 12 (D) scans of one representative placebo-treated patient. The scan was taken at the level of the corpus striatum and indicates significantly decreased rCMRglc in the bilateral frontal cortices, temporal cortices, and posterior cingulated gyrus. *Reprinted from Bi, M., Tong, S., Zhang, Z., Ma, Q., Zhang, S., Luo, Z., … Wang, D. (2011). Changes in cerebral glucose metabolism in patients with mild-to-moderate Alzheimer's disease: A pilot study with the Chinese herbal medicine fuzhisan.* Neuroscience Letters, 501*(1), 35–40, Copyright (2011), with permission from Elsevier.*

considered to improve blood circulation by removing blood stasis, regulate menstruation, and alleviate pain. It is mainly used as alternative therapies for dysmenorrhea, amenorrhea, and mild joint injury. More than 200 different pharmaceutical compounds were extracted from safflower, including flavonoids, chalcones, phenylethanoid glycosides, coumarins, steroids, and safflower polysaccharides (Yao, Wang, Miao, & Wang, 2016). Ren and colleagues have studied the neuroprotective effect of a standardized flavonoid extract of safflower (SAFE) using PD cell and rat model

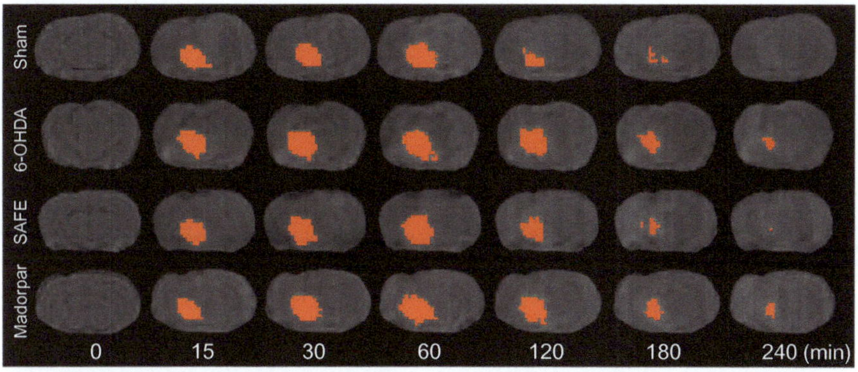


**Fig. 3** Coronal views of the MRI appearances of Gd-DTPA diffusion in the brain of 6-OHDA-induced rat model of PD after SAFE treatment. The animals were given an intraparenchymal injection of a 2 μL of 10 mM Gd-DTPA in the SN. The anesthetized rat was placed in a prone position and scanned in the 3.0-T MRI systems machine. For each subject, repeated scans with T1 3D MP-RAGE sequence were performed at 15 and 30 min postinjection and every hour thereafter. *Reprinted from Ren, R., Shi, C., Cao, J., Sun, Y., Zhao, X., Guo, Y., ⋯ Han, H. (2016). Neuroprotective effects of a standardized flavonoid extract of safflower against neurotoxin-induced cellular and animal models of Parkinson's disease.* Scientific Reports, *6, 22135, Copyright (2016), with permission from Nature Publishing Group.*

(Ren et al., 2016). The behavior performance was improved by oral administration of SAFE dripping pill daily for 3 weeks in the experimental rat group. The evaluation of extracellular diffusion following SAFE treatment in the PD rat model was conducted by MRI examination using Gd-DTPA as a tracer (Fig. 3). Their results show that SAFE is a potential therapeutic herbal for the treatment of PD. This study systematically evaluated in vitro and in vivo pharmacodynamics of SAFE with an MRI tracer-based technique.

## 3. SYNERGISTIC EFFECT BETWEEN CHINESE HERBAL MEDICINE AND STEM CELL

### 3.1 Stem Cell Therapy for Neurological Disorders

Many adult tissues containing population of stem cells that have the capacity for recovery from natural attrition or injury, since stem cells could proliferate, differentiate, and finally replace injured cells. However, the long-standing believe in neurobiology was that mammalian central nervous system does not share the same capacity, and therefore the loss of functional neural cells due to stroke, traumatic brain injury, or degenerative process will not be renewed. Until recent years, strong evidence from developmental neurobiology has accumulated that suggests otherwise (Eriksson et al., 1998; Gould, Reeves, Graziano, & Gross, 1999; van Praag et al., 2002). These encouraging findings demonstrate that an adult neural stem cell has a very broad developmental capacity and may potentially be used to treat various neurological disorders. Through this property, increasing attention has focused on developing new therapeutic strategies using neuronal stem cells (Chakradhar, 2016; Lepousez, Nissant, & Lledo, 2015).

Neural stem cell transplantation as a promising treatment for neurological disorders was initially studied in animal model for the past 30 years, and more recently in clinical trials (Rehman, 1996). As a result of neural stem cell transplantation, the neurological impairment result from ischemic stroke was improved, dynamic functional balance in GABAergic neurons was restored in epilepsy, and disease progression in AD was delayed (Cunningham et al., 2014; Lam, Lowry, Carmichael, & Segura, 2014; Lindvall & Kokaia, 2010; Upadhya et al., 2016). According to previous studies, the restoration of neurological function and the regeneration of neural tissue were obtained from the administration of neural stem cells (Corti et al., 2008; Engel, Do-Ha, Muñoz, & Ooi, 2016; Zweckberger, Ahuja, Yang, Jian, & Fehlings, 2016). There are largely two mechanisms of action following the administration of stem cells. First, the engrafted cells will migrate toward the injured area and differentiate into physiologically functional neural cells within the lesion; second, engrafted cells could improve local nervous system microenvironment and benefit injured neural cells by secreting neurotrophic factors (for example, BDNF, GDNF, EGF, bFGF, etc.) (Chen, Mallon, McKay, & Robey, 2014; Li, Christophersen, Hall, Soulet, & Brundin, 2008; Yu, Marchetto, & Gage, 2013).

Substantial advances have been made in the field of neural stem cells research for neurological disorders. Unfortunately, few of these results have been translated into clinical practice due to transplantation efficiency and safety concerns (Dimmeler, Ding, Rando, & Trounson, 2014; Upadhyay, Shankar, Srivastava, & Srivastava, 2014). The key challenges for a successful stem cell transplantation are survival rates in the targeted lesion. Although various studies were attempted to improve the transplantation efficiency, the long-term survival rate for engrafted stem cells is still quite low.

Apart from cell survival, the remaining stem cells may not have the mature phenotypes of existing neurons, which may result in subsequent negative outcomes. In order to achieve the maximum transplantation efficiency, combination therapy of stem cells and other adjuvants has been discussed.

As discussed earlier in this chapter, herbal medicine may potentially serve as adjuvant treatment in the combination therapy for neurological disorders. Ferulic acid is a hydroxycinnamic acid and is shown to promote the proliferation of neural stem/progenitor cells (NSC/NPCs) in the isolated telencephalon of embryonic day 14 rats, and increased number and size of secondary formed neurospheres were also observed (Yabe et al., 2010). Zhang and colleagues also reported that Salvianolic acid B isolated from Danshen could mediate the proliferation of NSC/NPCs (Zhuang et al., 2012). However, these studies suffer from a lack of information about the metabolic changes in the living animal model. Molecular imaging has already been applied to many standalone therapies (Cunha, Szigeti, Mathe, & Metello, 2014; Duncan, Winston, Koepp, & Ourselin, 2016; Willmann, van Bruggen, Dinkelborg, & Gambhir, 2008). Here, we present the first study of herbal medicine and stem cell combination therapy in cerebral ischemia.

## 3.2 Synergistic Combination Therapy Between Chinese Herbal Medicine and Stem Cell: A Pilot Study

As mentioned earlier, QKL injection has been widely used in China for treating stroke. Our group have conducted research on QKL and its combination with stem cell transplantation (iPSCs and NSCs) in a rat MCAO model using $^{18}$F-FDG micro-PET imaging (Zhang et al., 2015). Neurological outcomes in combination therapy were improved compared with the administration of stem cells alone. $^{18}$F-FDG imaging was employed to assess the metabolic recovery following ischemic stroke (Fig. 4). Glucose uptake in ischemic injured tissues was also increased, which was in good agreement with the neurological behavior test results. To further confirm the therapeutic effects in combination therapy, we performed immunochemical analysis in the injured cerebral hemisphere. The level of glucose transport-1 expression corresponds to the result of $^{18}$F-FDG micro-PET and neurological behavior test. The findings are encouraging and provide support for the development of molecular imaging methods to examine the potential therapeutic effect of combination therapy in neurological disorders.

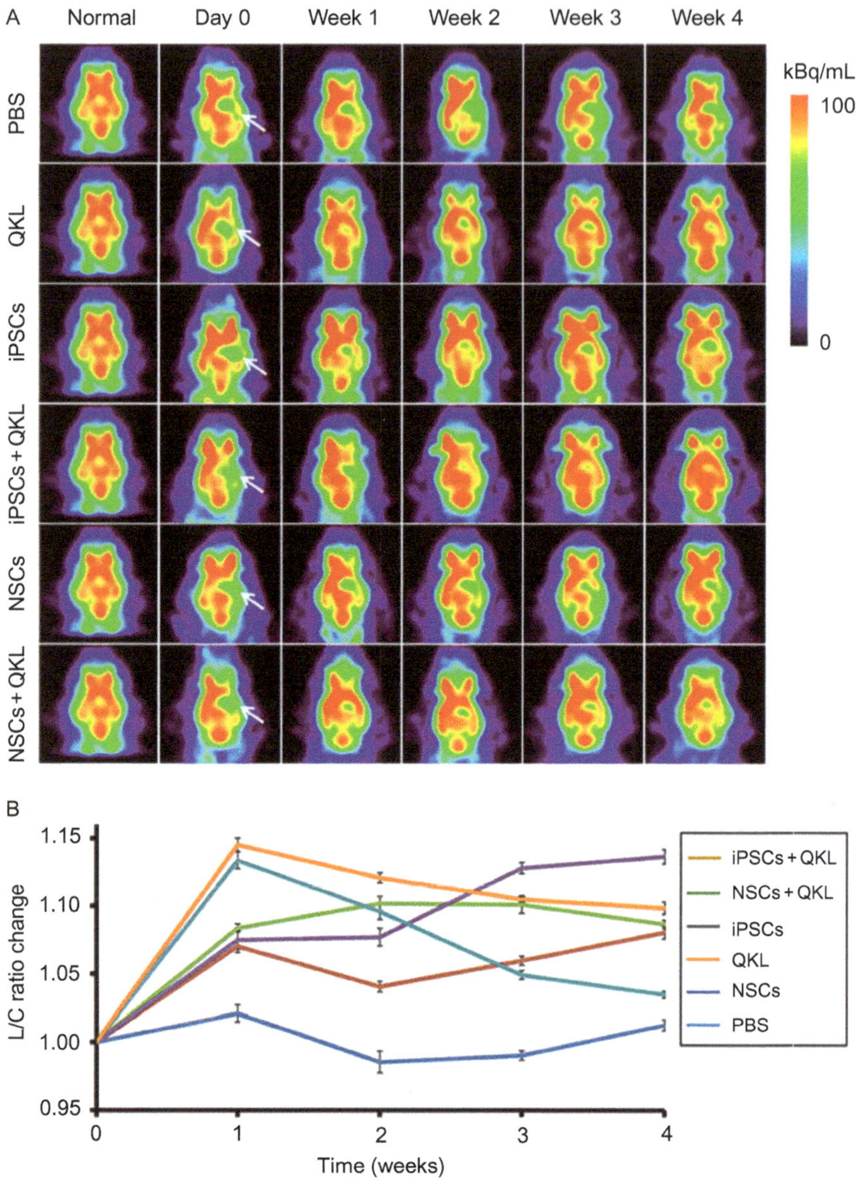


**Fig. 4** (A) Serial representative $^{18}$F-FDG-PET images demonstrate metabolic recovery after each treatment in rat model of MCAO. Rat brain images are shown in *axial view*. Ischemic areas are indicated by *white arrows*. Scale was set according to signal intensity. (B) Semiquantitative analysis of glucose metabolic change in each group. Data in various groups are shown as change of L/C ratio. *This research was originally published in Journal of Nuclear Medicine. Reprinted from Zhang, H., Song, F., Xu, C., Liu, H., Wang, Z., Li, J., ··· Tian, M. (2015). Spatiotemporal PET imaging of dynamic metabolic changes after therapeutic approaches of induced pluripotent stem cells, neuronal stem cells, and a Chinese patent medicine in stroke.* The Journal of Nuclear Medicine, 56*(11), 1774–1779, Copyright (2015), with permission from the Society of Nuclear Medicine and Molecular Imaging, Inc.*

## 4. PROSPECTS OF MOLECULAR IMAGING IN CHINESE HERBAL MEDICINE

The use of herbal medicine in neurological disorders is an emerging field that draws researchers across many different areas of expertise. Potential therapeutic effects of those herbal medicines have been proven. Unfortunately, despite the growing interest in this topic and the efforts of many laboratories, conventional research methods have not yet been shown to extract enough useful information from the tested subject or the animal disease model. In order to assess more information in living subject and to study the pharmaceutical response after administration of herbal medicine, molecular imaging approaches should be performed to better understand the molecular changes underlying those different responses (Heiss & Zimmermann-Meinzingen, 2012; Willmann et al., 2008). Furthermore, additional training for neurological researchers is needed to bridge the gap between molecular imaging and Chinese herbal medicine.

Molecular imaging differs from the structural imaging modalities in probes (also known as biomarkers), and is used to quantifying molecular interactions, biological processes, and protein concentration and distribution (Zurcher, Bhanot, McDougle, & Hooker, 2015). However, progress toward developing imaging probes has been slow (de Vries et al., 2015; Willmann et al., 2008). More studies need to be conducted to develop novel imaging probes, which would provide further information to correlate the imaging biomarker with the active components of Chinese herbal medicine. In summary, herbal medicine will not be adopted as a widely used clinical therapy for neurological disorders unless it can be scientifically proven to be safe and effective. Molecular imaging will play an important role in visualizing the underlying therapeutic mechanism of herbal medicine.

## REFERENCES

Agdeppa, E. D., Kepe, V., Liu, J., Flores-Torres, S., Satyamurthy, N., Petric, A., ... Barrio, J. R. (2001). Binding characteristics of radiofluorinated 6-dialkylamino-2-naphthylethylidene derivatives as positron emission tomography imaging probes for beta-amyloid plaques in Alzheimer's disease. *The Journal of Neuroscience*, *21*(24), RC189.

Alonso, A., Della Martina, A., Stroick, M., Fatar, M., Griebe, M., Pochon, S., ... Meairs, S. (2007). Molecular imaging of human thrombus with novel abciximab immunobubbles and ultrasound. *Stroke*, *38*(5), 1508–1514.

Bi, M., Tong, S., Zhang, Z., Ma, Q., Zhang, S., Luo, Z., ... Wang, D. (2011). Changes in cerebral glucose metabolism in patients with mild-to-moderate Alzheimer's disease: A pilot study with the Chinese herbal medicine fuzhisan. *Neuroscience Letters*, *501*(1), 35–40.

Chakradhar, S. (2016). An eye to the future: Researchers debate best path for stem cell-derived therapies. *Nature Medicine*, *22*(2), 116–119.
Chen, K., Mallon, B., McKay, R. G., & Robey, P. (2014). Human pluripotent stem cell culture: Considerations for maintenance, expansion, and therapeutics. *Cell Stem Cell*, *14*(1), 13–26.
Cheng, X., Luo, H., Zhou, L., Wang, L., Sun, J., Huang, Y., ... Cai, Y. (2014). Neuroprotective effect of the traditional Chinese herbal formula Tongxinluo: A PET imaging study in rats. *Neural Regeneration Research*, *9*(13), 1267–1274.
Corti, S., Nizzardo, M., Nardini, M., Donadoni, C., Salani, S., Ronchi, D., ... Del, B. R. (2008). Neural stem cell transplantation can ameliorate the phenotype of a mouse model of spinal muscular atrophy. *The Journal of Clinical Investigation*, *118*(10), 3316–3330.
Cunha, L., Szigeti, K., Mathe, D., & Metello, L. F. (2014). The role of molecular imaging in modern drug development. *Drug Discovery Today*, *19*(7), 936–948.
Cunningham, M., Cho, J. H., Leung, A., Savvidis, G., Ahn, S., Moon, M., ... Kim, K. S. (2014). HPSC-derived maturing GABAergic interneurons ameliorate seizures and abnormal behavior in epileptic mice. *Cell Stem Cell*, *15*(5), 559–573.
de Vries, E. G. E., de Jong, S., & Gietema, J. A. (2015). Molecular imaging as a tool for drug development and trial design. *The Journal of Clinical Oncology: official journal of the American Society of Clinical Oncology*, *33*(24), 2585–2587.
Dimmeler, S., Ding, S., Rando, T. A., & Trounson, A. (2014). Translational strategies and challenges in regenerative medicine. *Nature Medicine*, *20*(8), 814–821.
Duncan, J. S., Winston, G. P., Koepp, M. J., & Ourselin, S. (2016). Brain imaging in the assessment for epilepsy surgery. *The Lancet Neurology*, *15*(4), 420–433.
Engel, M., Do-Ha, D., Muñoz, S. S., & Ooi, L. (2016). Common pitfalls of stem cell differentiation: A guide to improving protocols for neurodegenerative disease models and research. *Cellular and Molecular Life Sciences*, *73*(19), 1–17.
Eriksson, P. S., Perfilieva, E., Bjork-Eriksson, T., Alborn, A. M., Nordborg, C., Peterson, D. A., & Gage, F. H. (1998). Neurogenesis in the adult human hippocampus. *Nature Medicine*, *4*(11), 1313–1317.
Gong, X., & Sucher, N. J. (1999). Stroke therapy in traditional Chinese medicine (TCM): Prospects for drug discovery and development. *Trends in Pharmacological Sciences*, *20*(5), 191–196.
Gould, E., Reeves, A. J., Graziano, M. S., & Gross, C. G. (1999). Neurogenesis in the neocortex of adult primates. *Science*, *286*(5439), 548–552.
Heiss, W. D., & Zimmermann-Meinzingen, S. (2012). PET imaging in the differential diagnosis of vascular dementia. *The Journal of the Neurological Sciences*, *322*(1–2), 268–273.
Huang, W. S., Lin, S. Z., Lin, J. C., Wey, S. P., Ting, G., & Liu, R. S. (2001). Evaluation of early-stage Parkinson's disease with 99mTc-TRODAT-1 imaging. *The Journal of Nuclear Medicine*, *42*(9), 1303–1308.
Lam, J., Lowry, W. E., Carmichael, S. T., & Segura, T. (2014). Delivery of iPS-NPCs to the stroke cavity within a hyaluronic acid matrix promotes the differentiation of transplanted cells. *Advanced Functional Materials*, *24*(44), 7053–7062.
Lee, H. J., Zhang, Y., Zhu, C., Duff, K., & Pardridge, W. M. (2002). Imaging brain amyloid of Alzheimer disease in vivo in transgenic mice with an Abeta peptide radiopharmaceutical. *The Journal of Cerebral Blood Flow & Metabolism*, *22*(2), 223–231.
Lepousez, G., Nissant, A., & Lledo, P.-M. (2015). Adult neurogenesis and the future of the rejuvenating brain circuits. *Neuron*, *86*(2), 387–401.
Li, J. Y., Christophersen, N. S., Hall, V., Soulet, D., & Brundin, P. (2008). Critical issues of clinical human embryonic stem cell therapy for brain repair. *Trends in Neurosciences*, *31*(3), 146–153.
Li, J. H., Lu, J., & Zhang, H. (2013). Functional recovery after scutellarin treatment in transient cerebral ischemic rats: A pilot study with $^{18}$F-fluorodeoxyglucose microPET. *Evidence-Based Complementary and Alternative Medicine*, *2013*, 507091.

Lindvall, O., & Kokaia, Z. (2010). Stem cells in human neurodegenerative disorders—Time for clinical translation? *The Journal of Clinical Investigation*, *120*(1), 29–40.

Liu, P., Kong, M., Yuan, S., Liu, J., & Wang, P. (2014). History and experience: A survey of traditional Chinese medicine treatment for Alzheimer's disease. *Evidence-Based Complementary and Alternative Medicine*, *2014*, 642128.

Liu, X., Wu, Z., Yang, K., Ding, H., & Wu, Y. (2013). Quantitative analysis combined with chromatographic fingerprint for comprehensive evaluation of Danhong injection using HPLC-DAD. *Journal of Pharmaceutical and Biomedical Analysis*, *76*, 70–74.

Marek, K., Innis, R., van Dyck, C., Fussell, B., Early, M., Eberly, S., … Seibyl, J. (2001). [123I]beta-CIT SPECT imaging assessment of the rate of Parkinson's disease progression. *Neurology*, *57*(11), 2089–2094.

Patel, V., Chisholm, D., Parikh, R., Charlson, F. J., Degenhardt, L., Dua, T., … Levin, C. (2015). Addressing the burden of mental, neurological, and substance use disorders: Key messages from disease control priorities, 3rd edition. *The Lancet*, *387*(10028), 1672–1685.

Rehman, K. L. (1996). Bone marrow transplantation for sickle cell disease. *New England Journal of Medicine*, *335*(24), 1845–1846.

Ren, R., Shi, C., Cao, J., Sun, Y., Zhao, X., Guo, Y., … Han, H. (2016). Neuroprotective effects of a standardized flavonoid extract of safflower against neurotoxin-induced cellular and animal models of Parkinson's disease. *Scientific Reports*, *6*, 22135.

Reshef, A., Shirvan, A., Waterhouse, R. N., Grimberg, H., Levin, G., Cohen, A., … Ziv, I. (2008). Molecular imaging of neurovascular cell death in experimental cerebral stroke by PET. *The Journal of Nuclear Medicine*, *49*(9), 1520–1528.

Rocha, D. N., Carvalho, E. D., & Pego, A. P. (2016). High-throughput platforms for the screening of new therapeutic targets for neurodegenerative diseases. *Drug Discovery Today*, *21*(9), 1355–1366.

Rojas, J. C., & Boxer, A. L. (2016). Neurodegenerative disease in 2015: Targeting tauopathies for therapeutic translation. *Nature Reviews Neurology*, *12*(2), 74–76.

Tang, C., Ye, Y., Feng, Y., & Quinn, R. J. (2016). TCM, brain function and drug space. *Natural Product Reports*, *33*(1), 6–25.

Upadhya, D., Hattiangady, B., Shetty, G. A., Zanirati, G., Kodali, M., & Shetty, A. K. (2016). Neural stem cell or human induced pluripotent stem cell-derived GABA-ergic progenitor cell grafting in an animal model of chronic temporal lobe epilepsy. *Current Protocols in Stem Cell Biology*, *38*, 2d.7.1–2d.7.47.

Upadhyay, G., Shankar, S., Srivastava, R. K., & Srivastava, R. K. (2014). Stem cells in neurological disorders: Emerging therapy with stunning hopes. *Molecular Neurobiology*, *52*(1), 1–16.

van Praag, H., Schinder, A. F., Christie, B. R., Toni, N., Palmer, T. D., & Gage, F. H. (2002). Functional neurogenesis in the adult hippocampus. *Nature*, *415*(6875), 1030–1034.

Vickers, A., & Zollman, C. (1999). ABC of complementary medicine: Herbal medicine. *British Medical Journal*, *319*(7216), 1050.

Wan, H., Zhu, H., Tian, M., Hu, X., Yang, J., Zhao, C., & Zhang, H. (2008). Protective effect of chuanxiongzine-puerarin in a rat model of transient middle cerebral artery occlusion-induced focal cerebral ischemia. *Nuclear Medicine Communications*, *29*(12), 1113–1122.

Wang, Z., Song, F., Li, J., Zhang, Y., He, Y., Yang, J., … Zhang, H. (2014). PET demonstrates functional recovery after treatment by Danhong injection in a Rat model of cerebral ischemic-reperfusion injury. *Evidence-Based Complementary and Alternative Medicine*, *2014*, 430757.

Willmann, J. K., van Bruggen, N., Dinkelborg, L. M., & Gambhir, S. S. (2008). Molecular imaging in drug development. *Nature Reviews Drug Discovery*, 7(7), 591–607.

Yabe, T., Hirahara, H., Harada, N., Ito, N., Nagai, T., Sanagi, T., & Yamada, H. (2010). Ferulic acid induces neural progenitor cell proliferation in vitro and in vivo. *Neuroscience*, *165*(2), 515–524.

Yao, D., Wang, Z., Miao, L., & Wang, L. (2016). Effects of extracts and isolated compounds from safflower on some index of promoting blood circulation and regulating menstruation. *The Journal of Ethnopharmacology*, *191*, 264–272.

Yu, D. X., Marchetto, M., & Gage, F. (2013). Therapeutic translation of iPSCs for treating neurological disease. *Cell Stem Cell*, *12*(6), 678–688.

Yuan, S. M., Gao, K., Wang, D. M., Quan, X. Z., Liu, J. N., Ma, C. M., ... Zhang, L. F. (2011). Evodiamine improves cognitive abilities in SAMP8 and APP(swe)/PS1(DeltaE9) transgenic mouse models of Alzheimer's disease. *Acta Pharmacologica Sinica*, *32*(3), 295–302.

Zhang, J., Li, P., Wang, Y., Liu, J., Zhang, Z., Cheng, W., & Wang, Y. (2013). Ameliorative effects of a combination of baicalin, jasminoidin and cholic acid on ibotenic acid-induced dementia model in rats. *PLoS One*, *8*(2), e56658.

Zhang, H., Song, F., Xu, C., Liu, H., Wang, Z., Li, J., ... Tian, M. (2015). Spatiotemporal PET imaging of dynamic metabolic changes after therapeutic approaches of induced pluripotent stem cells, neuronal stem cells, and a Chinese patent medicine in stroke. *The Journal of Nuclear Medicine*, *56*(11), 1774–1779.

Zhuang, P., Zhang, Y., Cui, G., Bian, Y., Zhang, M., Zhang, J., ... Lin, Y. (2012). Direct stimulation of adult neural stem/progenitor cells in vitro and neurogenesis in vivo by salvianolic acid B. *PLoS One*, 7(4), 2365–2373.

Zurcher, N. R., Bhanot, A., McDougle, C. J., & Hooker, J. M. (2015). A systematic review of molecular imaging (PET and SPECT) in autism spectrum disorder: Current state and future research opportunities. *Neuroscience & Biobehavioral Reviews*, *52*, 56–73.

Zweckberger, K., Ahuja, C. S., Yang, L., Jian, W., & Fehlings, M. G. (2016). Self-assembling peptides optimize the post-traumatic milieu and synergistically enhance the effects of neural stem cell therapy after cervical spinal cord injury. *Acta Biomaterialia*, *42*, 77–89.

CHAPTER NINE

# Herb–Drug Interactions of Commonly Used Chinese Medicinal Herbs

**Amrinder Singh, Kaicun Zhao**[1]
Traditional Chinese Herbal Medicine Programme, Middlesex University, The Borough, Hendon, London, United Kingdom
[1]Corresponding author: e-mail address: k.zhao@mdx.ac.uk

## Contents



## Abstract

With more and more popular use of traditional herbal medicines, in particular Chinese herbal medicines, herb–drug interactions have become a more and more important safety issue in the clinical applications of the conventional drugs. Researches in this area are increasing very rapidly.

Herb–drug interactions are complicated due to the fact that multiple chemical components are involved, and these compounds may possess diverse pharmacological activities. Interactions can be in both pharmacokinetics and pharmacodynamics. Abundant studies focused on pharmacokinetic interactions of herbs and drugs. Herbs may affect the behavior of the concomitantly used drugs by changing their absorption, distribution, metabolism, and excretion. Studies on pharmacodynamics interactions of herbs and drugs are still very limited.

Herb–drug interactions are potentially causing changes in drug levels and drug activities and leading to either therapeutic failure or toxicities. Sometime it can be fatal. The exposures to drugs, lacking of knowledge in the potential adverse herb–drug interactions, will put big risk to patients' safety in medical services. On the contrary, some interactions may be therapeutically beneficial. It may be used to help develop new therapeutic strategies in the future.

*International Review of Neurobiology*, Volume 135
ISSN 0074-7742
http://dx.doi.org/10.1016/bs.irn.2017.02.010

This chapter is trying to review the development in the area of herb–drug interactions based on the recently published research findings. Information on the potential interactions among the commonly used Chinese medicinal herbs and conventional drugs is summarized in this chapter.

## 1. INTRODUCTION

With long history of mankind fighting against human diseases, almost every culture in the world has developed its own tradition of herbal medicine. Investigations in 17 countries have shown that the prevalence in the use of traditional remedies is in a range of 6%–82% of the general populations (Harris, Cooper, Relton, & Thomas, 2012; Lotfi, Adib-Hajbaghery, Shahsavarloo, & Gandomani, 2016; Moses & McGuire, 2010; Ramos-Remus & Raut, 2008). It is reported that more than 80% of people in Africa and Asia using herbal medicines while as is estimated that 60%–70% of the American population is taking botanical products (Eisenberg et al., 1998). It is estimated around 80,000 plant species are in use as a complementary medicine all over the world (Obodozie, 2012). Among them, traditional Chinese herbal medicine (CHM) is the most popularly used herbal medicine system with about 11,145 plant species used by about 20% of total population in the world (Wang et al., 2009). In 2014 the global market of herbal medicines was USD 100 billion (Zhang, Onakpoya, Posadzki, & Eddouks, 2015). WHO estimated about the global market of herbal products would be around 5 trillion dollars by the year 2050 (Malagi, Kamath, & Adiga, 2014). It is clear that the global use of herbal medicines is exponentially increasing.

With wider use of herbal medicines, concomitant use of conventional drugs and herbal medicines poses a significant clinical concern as it potentiates the risk of drug–herb interactions (Fasinu, Bouic, & Rosenkranz, 2012). In particular, the use of Chinese herbal formulas further escalates the chances of clinical drug interactions due to administration of the complex mixture of multiple phytochemicals from different herbs composed in the formula (Fugh-Berman & Ernst, 2001). It has shown that the risk of drug interactions increases with the numbers of medical products consumed at the same time (Fasinu et al., 2012; Yaheya & Isamail, 2009).

Drug interactions can happen either pharmacodynamically and/or pharmacokinetically (Staines, 2011). Herbal medicines may increase or decrease the activities of concomitant drugs either through synergism/antagonism or through changing the disposition of drugs leading to changes in drug levels in the body (Hussain, 2011). While some combined use of drug and herbs may be beneficial and desirable, what more concerned are those unexpected drug–herb interactions that cause adverse drug reactions (ADRs) and toxicities. According to the China National Adverse Drug Reaction Monitoring Centre, about 13.8% and 17.3% of the reported ADR cases are caused by CHM and among them drug–herb interactions are believed to be one of the main causes (Wu, Ma, Ye, & Lin, 2016).

In this chapter, studies on drug–herb interactions are reviewed and their potential impacts to clinical practice are discussed. It is expected that the research discoveries will make awareness and provide some guidance for both the medical professionals and the public to consider when drugs may possibly be used with herbs concomitantly.

## 2. DRUG–HERB INTERACTIONS IN PARMACOKINETICS

Pharmacokinetic interactions between drugs and herbs are seen more frequently and more extensively studied. The drug–herb interactions may change the pharmacokinetic fate of a drug by altering its absorption and distribution processes, and/or interfering metabolic and excretory pathways. These pharmacokinetic changes caused by coadministered herbal medicines may potentially result in pronounced reduction or increase in the blood levels of the affected drugs and are possibly linked with clinical failure or toxicity (Brown, Heyneke, Brown, Van Wyk, & Hamman, 2008; Fasinu et al., 2012).

### 2.1 Gastrointestinal Absorption

The environmental conditions of gastrointestinal tract influence the absorption of drugs. Herbal medicines may affect the absorption of drugs through alteration of the gastrointestinal conditions, leading to changes in bioavailability of a drug. Some cases, this interaction may reduce therapeutic effects by decreasing drugs bioavailability. On the other hand, this interaction can be utilized to improve certain drugs bioavailability (Kesarwani, Gupta, & Mukerjee, 2013; Tian et al., 2016).

A number of factors play important roles in gastrointestinal absorption of drugs such as pH, gastric emptying, and intestinal peristalsis. For example, Banxia Xiexin Tang (BXT, 半夏泻心汤) is one of the most commonly used traditional Chinese herbal formula for treatment of gastrointestinal disorders. It is reported that oral coadministration of BXT significantly reduced the gastrointestinal absorption of the antiepileptic drug midazolam by 76% in mice. Evidence has shown that this was attributed to a decrease of pH in the stomach and intestine caused by BXT (Jo, Kim, Nam, Seung, & Lee, 2016). Fennel (*Foeniculum vulgare*, 小茴香, Xiaohuixiang) is a herb commonly used in traditional Chinese medicine for treatment of stomach and abdominal pain. Research showed that concomitant oral administration of the aqueous extract of fennel significantly lowered the absorption and reduced the bioavailability of the antibiotic drug ciprofloxacin. The underlying mechanism for this herb–drug interaction was suggested being due to the large amounts of metal cations from the herbal extract that may form a more lipophilic chelate with ciprofloxacin, leading to the changes in the absorption and dispositions of the drug (Zhu, Wong, & Li, 1999b). Another commonly used CHM Pugongying (*Taraxacum mongolicum*, 蒲公英) was found also to reduce bioavailability of ciprofloxacin. Pugongying is frequently used in treatment of infectious disease in CHM. Concurrently administration of the herb to rat reduced the maximum plasma concentration ($C_{max}$) of ciprofloxacin by 73%. However, the area under the plasma concentration–time curve (AUC) did not change significantly, indicating the herb was not only affected the absorption of the drug but also altered other pharmacokinetic behaviors of the drug, including distribution and elimination processes (Zhu, Wong, & Li, 1999a).

Most recently, Huang et al. (2016) reported interaction of Danggui Buxue Tang (DBT, 当归补血汤) with iron supplement ferrous sulfate. The study demonstrated the blood-toning properties of this traditional Chinese formula as it increased uptake of ferrous iron and promoted the utilization of irons through biosynthesis of hemoglobin. However, the authors found that coadministration of DBT with ferrous sulfate in fact decreased the uptake and utilization of ferrous iron as the DBT extract may form a complexation compound to reduce the level of free type of ferrous iron. This case shows the importance of awareness and understanding the potential drug–herb interactions in clinical practice.

While drug absorption is inhibited by some herbs or formulas, drug absorption may be promoted as well by other herbs or formulas such as *Zingiber officinale* (Shengjiang, 生姜) which has been shown to affect the gastrointestinal mucous membrane, facilitating the absorption of a number of drugs including some anticancer drugs like methotrexate and 5-fluorouracil, and some antibiotics like azithromycin, erythromycin, cephalexin, cefadroxil, amoxycillin, and cloxacillin (Ajazuddin et al., 2014; Dudhatra et al., 2012). The most popular herb Gancao (甘草), including species *Glycyrrhiza inflate*, *Glycyrrhiza glabra*, and *Glycyrrhiza uralensis*, is widely used as a harmonizing assistant herb in formulas of CHM. It is demonstrated that the major chemical component of this herb, Glycyrrhizin, enhances gastrointestinal absorption of various antimicrobial agents such as rifampicin, tetracycline, and nalidixic acids (Chen et al., 2009; Imai, Sakai, Ohtake, Azuma, & Otagiri, 2005). It is suggested that Glycyrrhizin enhanced drug absorption by increasing the permeability of intestinal epithelium (Imai, Sakai, Ohtake, Azuma, & Otagiri, 1999) and may also through inhibition of efflux transporter P-glycoprotein (P-gp) present in intestine (Chen et al., 2009).

## 2.2 Distribution

Following absorption, drugs are quickly distributed overall the body through blood circulation. Herb–drug interactions may also happen at this stage. Paclitaxel is an anticancer drug and frequently used in the treatment of breast cancer and ovarian cancer. In Chinese medicine, the traditional formula Jiawei Xiaoyao Wan (JXW, 加味逍遥丸) is a common remedy used for treatment of breast diseases including cancer. It has been demonstrated that pretreatment with JXW significantly attenuated the distribution of paclitaxel in rat lymphatic systems. By comparison of the AUCs of paclitaxel concentration in the lymph and plasma, the relative distribution of the drug in the lymphatic systems was found being increased from 39% to 45% (Hou, Lu, & Tsai, 2016).

Plasma protein binding is an important factor influencing drug in the distribution. As the most abundant protein component in the blood, human serum albumin (HSA) plays a key role in plasma protein binding of drugs. Other plasma proteins may be also involved in drug protein bindings such as α-1 acid glycoprotein and lipoproteins. Acidic drugs bind more extensively to albumin, and basic drugs bind to α-1 acid glycoprotein or lipoprotein (Caterina et al., 2013). Drugs in the bound form are normally not active. Only free drugs are able to reached the target tissue to exert pharmacological activities. Herb–drug interaction may happen on plasma protein binding when coadministered herb and drug are competing for the same plasma protein-binding site. This interaction will lead to an increase of the level of free drug which is directly related to its biological responses. Significant clinical implications may occur in case the drug has a higher degree of plasma protein binding (>90%) with a narrow therapeutic index such as warfarin and carbamazepine (Caterina et al., 2013; Milic et al., 2014; Staines, 2011; Yamasaki, Maruyama, Kragh-Hansen, & Otagiri, 1996). It was reported that a commonly used herb for improving blood circulation in

CHM *Salvia miltiorrhiza* (Danshen, 丹参) could intensify the warfarin effect putting potential risk of hemorrhage (Chan, 2001; Yu, Chan, & Sanderson, 1997). Recently, Shao, Ai, Xu, and Fan (2016) confirmed that several active components in *S. miltiorrhiza* (Danshen, 丹参) including the phenolic and quinone compounds were capable of decreasing the binding constant of warfarin to HSA by competing the warfarin-binding site of albumin. This may lead to higher level of free warfarin in the blood as the bound warfarin being replaced off the protein.

## 2.3 Metabolism

Metabolism is an important process in elimination of drugs. It is also the area where drug interactions including herb–drug interactions are most likely to occur. Drug metabolism generally includes two phases. Phase I metabolism normally introduces polar functional groups to drug molecule and is mainly mediated by cytochrome P450 (CYP). Phase II metabolism is a process where drugs and/or their phase I metabolites are conjugated with a polar compound by appropriate transferases such as UDP-glucuronosyltransferase, sulfotransferase, and glutathione-*S*-transferase (Fasinu et al., 2012; Saengtienchai et al., 2016). These metabolic enzymes are present mainly in the intestine and liver and variations in the expression level and activities of these enzymes will directly affect the pharmacokinetic fate of drugs (Sewradj et al., 2016).

Herb–drug interactions in metabolism occur mainly due to the overlapping of substrate specificity in the biotransformation pathways (Fasinu et al., 2012). The concomitantly administered herbs may induce or inhibit the intestinal and hepatic metabolic enzymes particularly the CYP enzyme family which are responsible for the metabolism the particular drugs, leading to increase or reduction of the drug levels in the body and changes in the drug pharmacokinetic disposition (Mazzari & Prieto, 2014; Meijerman, Beijnen, & Schellens, 2006; Nowack, 2008; Saengtienchai et al., 2016).

### *2.3.1 CYP P450 in Drug Metabolism and Herb–Drug Interactions*

CYP is superfamily of monooxygenase enzymes and present in the smooth endoplasmic reticulum of the liver and other extrahepatic tissues. Within the CYP superfamily, only about a dozen isozymes belonging to the subfamily of CYP1, 2, and 3 are responsible for the biotransformation of most foreign substances including 70%–80% of all drugs in clinical practice. The enzymes mediate various chemical reactions such as oxidation, reduction, hydrolysis, and hydration and play a key role in elimination of xenobiotics and drugs

(Mazzari & Prieto, 2014; Wanwomolruk & Prachayasittikul, 2014). It has been reported that most of the prescribed drugs are metabolized by the CYP1A2, CYP2C9, CYP2D6, CYP2E1, and CYP3A4, and about 50% of drug metabolism are solely carried out by CYP3A4 (Pandit et al., 2016). Herbs normally contain complicated constituents with various chemical structures. It is very common that one herb may influence multiple CYP isozymes. Herb–drug interactions would happen once the acitvities of CYP isozymes are altered by herbs. Table 1 lists some commonly used traditional Chinese medicinal herbs and the reported effects of these herbs on different CYP isozymes.

CYP1A2 is involved in metabolism of a number of commonly used medicines including theophylline, acetaminophen, propranolol, tamoxifen, verapamil, and warfarin. A number of antidepressants are also metabolized by CYP1A isozyme. Potential risks of toxicity or therapeutic failure may be put on if herbs affecting the activity of CYP1A2 were consumed while taking these medications. It has reported commonly used Chinese medicinal herb Yinchenhao (茵陈蒿, *Artemisia capillaries*) enhanced the metabolism of paracetamol via induction of CYP1A2 and CYP2E1 (Chan, Liu, Lin, Leung, & Wang, 2007). The overproduced metabolite of paracetamol *N*-acetyl-*p*-benzoquinone is a toxic compound which can cause cell death leading to liver failure (Kaur et al., 2015). However, on the contrary, some herbs including some food showed inhibitory effects to these metabolic enzymes such as *Rubia cordifolia* (Qiancao, 茜草) and its active component mollugin (Kim et al., 2013), *S. miltiorrhiza* (Danshen, 丹参) and its active components tanshinones (Qiu et al., 2008; Wang, Cheung, Lee, Or, & Yeung, 2010; Wang & Yeung, 2012; Zhou, Wang, et al., 2013), and *Schisandra chinensis* (Wuweizi, 五味子) and its active components schisandrins (Fan et al., 2014; Jiang, Fan, Wang, Tan, et al., 2015) exhibit significant inhibitory effect on CYP1A2 and/or CYP2E1. It was reported that Schisandrol B, an active constituent from *S. chinensis* (Wuweizi, 五味子), protects against acetaminophen-induced hepatotoxicity by inhibition of the CYPs-mediated bioactivation (Jiang, Fan, Wang, Chen, et al., 2015). Some food plants, such as *Momordica charantia* (bitter melon, 苦瓜), *Glycine max* (soybean, 大豆), *Mentha piperita* (peppermint, 薄荷), also showed inhibitory effects on CYP1A2 (Mazzari & Prieto, 2014).

In the CYP2C subfamily, CYP2C9 plays important role in the metabolism of nonsteroidal antiinflammatory drugs (ibuprofen), antihypersensitive (losartan), antidepressant (fluoxetine), antiepileptic (phenytoin), and antihypercholesterolemic drugs (fluvastatin) (Hirota & Ieiri, 2015; Mazzari &

**Table 1** Inducing (↑) or Inhibitory (↓) Effects of Commonly Used Chinese Medicinal Herbs on Cytochrome P450s (CYPs)

| Pinyin Name | Latin Name | Chemical Constituents | CYPs | Species | References |
|---|---|---|---|---|---|
| Baizhi 白芷 | *Angelicae dahuricae* | | CYP1A2↓ | in vivo, human | Yi et al. (2009) |
| Baizhi 白芷 | *Angelica dahurica* | | CYP3A↓ | In vitro, human liver microsomes | Guo, Taniguchi, Chen, Baba, and Yamazoe (2001) |
| Bu Zhong Yi Qi Tang 补中益气汤 | | | CYP2C11↓ | In vivo, rat | Yao et al. (2012) |
| Chaihu 柴胡 | *Bupleurum chinensis* | Saikosaponins | CYP2C9↓ | In vitro | Yu, Chen, Wang, Zhao, and Mao (2014) |
| Chuanxinlian 穿心莲 | *Andrographis paniculata* | | CYP2C19↓ | In vitro, human liver cytochrome | Pan et al. (2011) |
| Chuanxiong 川芎 | *Ligusticum chuanxiong* | | CYP2D6↓, CYP3A↓ | In vivo, rat | Tang, Zhang, Wu, and Li (2006) |
| Dahuang 大黄 | *Rheum palmatum* | | CYP1A2↑, CYP2D6↑, CYP2E1↑, CYP3A1/2↑ | In vivo, rat | Gao et al. (2013) and Tang et al. (2006) |
| Danggui 当归 | *Angelica sinensis* | | CYP2C19↓, CYP1A2↓ | In vitro, human liver microsomes | Sevior et al. (2010) |
| Danggui 当归 | *Angelica sinensis* | | CYP2E1↓, CYP3A1/2↑, CYP2D6↑ | In vivo and in vivo, rat | Li, Zhang, Li, Wang, and Wang (2008) and Tang et al. (2006) |
| Danggui 当归 | *Angelica sinensis* | | CYP3A↓ | In vitro, human liver microsomes | Guo et al. (2001) |

| | | | | | |
|---|---|---|---|---|---|
| Danshen 丹参 | *Salvia miltiorrhiza* | Tanshinones | CYP1A2↓, CYP2C9↓, CYP2E1↓, CYP3A4/5↓ | In vitro, human and rat liver microsomes | Wang, Cheung, et al. (2010) and Qiu et al. (2008) |
| Danshen 丹参 | *Salvia miltiorrhiza* | Miltirone, an abietane-type diterpene quinone | CYP1A2↓, CYP2C9↓, CYP2D6↓, CYP3A4/5↓ | In vitro, human cytochrome P450 isozymes | Zhou, Wang, et al. (2013) |
| Danshen 丹参 | *Salvia miltiorrhiza Bunge* | Tincture | CYP1A2↓, CYP3A1/2↓, CYP3A4/5↓ | In vitro, human and rat liver microsomes | Wang and Yeung (2012) |
| Dazao 大枣 | *Ziziphus jujuba* | | CYP1A2↑ | In vivo, rat | Jing, Peng, Wang, and Duane (2015) |
| Duhuo 独活 | *Angelica biserrata* | | CYP3A↓ | In vitro, human liver microsomes | Guo et al. (2001) |
| Fangfeng 防风 | *Saposhnikovia divaricata* | | CYP3A↓ | In vitro, human liver microsomes | Guo et al. (2001) |
| Gancao 甘草 | *Glycyrrhiza uralensis* | Glycyrrhetinic acid (GA) | CYP2C9↓, CYP2C19↓, CYP3A4/5↓ | In vitro, human liver microsomes; in vivo, mice | Lv et al. (2015) |
| Gancao 甘草 | *Glycyrrhiza uralensis* | Licochalcone A | CYP1A2↓, CYP2C9↓, CYP2C19↓, CYP2C8↓, CYP3A4↓ | In vitro, human liver microsomes | He et al. (2015) |
| Gegen 葛根 | *Puerariae lobatae* | | CYP1A2↑, CYP3A4/5↑ | In vivo, rat | Ge, Zhang, and Zuo (2016) |

*Continued*

**Table 1** Inducing (↑) or Inhibitory (↓) Effects of Commonly Used Chinese Medicinal Herbs on Cytochrome P450s (CYPs)—cont'd

| Pinyin Name | Latin Name | Chemical Constituents | CYPs | Species | References |
|---|---|---|---|---|---|
| Gouteng 钩藤 | *Uncaria tomentosa* | Decoction, tincture | CYP3A4/5↓ | In vitro, cells | Budzinski, Foster, Vandenhoek, and Arnason (2000) |
| Heye 荷叶 | *Nelumbo nucifera* | Alkaloid fraction, flavonoid fraction, nuciferine, *N*-nornuciferine, and 2-hydroxy-1-methoxyaporphine | CYP2D6↓ | In vitro, human liver microsomes | Ye et al. (2014) |
| Honghua 红花 | *Carthamus tinctorius* | | CYP1A2↑, CYP2E1↓, CYP3A1/2↑ | In vivo, rat | Li et al. (2008) |
| Hongjingtian 红景天 | *Rhodiola rosea* | | CYP3A4/5↓, CYP2D6↓, CYP1A2↓ | In vitro, human liver microsomes | Thu, Nilsen, and Hellum (2016) |
| Houpo 厚朴 | *Magnolia officinalis*; *Magnolia grandiflora* | Honokiol | CYP1A2↓, CYP2B6↓, CYP2C6↓, CYP2C8↓, CYP2C9↓, CYP2C19↓ | In vitro, human liver microsomes | Jeong et al. (2013) |
| Huanglian 黄连 | *Coptis chinensis* | Berberine | CYP1A2↑ | In vivo, mice | Guo, Pope, Cheng, Zhou, and Klaassen (2011) |
| Huanglian 黄连 | *Coptis chinensis* | | CYP1A2↓, CYP2D6↓, CYP3A4/5↓ | In vitro, human liver microsomes | Han et al. (2011) |
| Huanglian 黄连 | *Coptis chinensis* | Coptisine and epiberberine | CYP1A2↓, CYP2D6↓ | In vitro, human liver microsomes | Han et al. (2011) |
| Huanglian 黄连 | *Coptis chinensis* | Berberine, jateorrhizine, and palmatine | CYP2D6↓ | In vitro, human liver microsomes | Han et al. (2011) |

| | | | | | |
|---|---|---|---|---|---|
| Huanglian 黄连 | *Coptis chinensis* | Jateorrhizine | CYP3A4/5↓ | In vitro, human liver microsomes | Han et al. (2011) |
| Huanglian 黄连 | *Coptis chinensis* | Berberine, neferine, and tetrahydropalmatine | CYP1A2↓, CYP2D6↓, CYP3A4/5↓ | In vitro, human liver microsomes | Zhao, Hellum, Liang, and Nilsen (2015) |
| Huanglian Jiedu Tang 黄连解毒汤 | | Decoction, total iridoids, total flavonoids, and total alkaloids | CYP3A1↓, CYP2C6↓, CYP2E1↓, CYP1A2↓, CYP2D2↓ | In vitro, rat liver microsomes | Wang, Jiang, et al. (2014) |
| Huangqi 黄芪 | *Astragalus membranaceus* | Astragaloside IV | CYP1A2↓ | In vitro and in vivo, rat | Zhang, Zhang, Guo, Li, and Yu (2013) |
| Huangqin 黄芩 | *Scutellariae radix* | | CYP2C6↓, CYP2E1↑ | In vivo, human | Yi et al. (2009) |
| Hwang-Ryun-Hae-Dok-Tang (Huanglian Jiedu Tang) 黄连解毒汤 | | | CYP1A2↓, CYP2C19↓, CYP2D6↓, CYP3A4↓ | In vitro, human liver microsomes | Lee et al. (2015) |
| Jianghuang 姜黄 | *Curcuma longa* | Curcumin | CYP1A2↓, CYP2A6↑ | In vivo, human | Chen et al. (2010) |
| Jianghuang 姜黄 | *Curcuma longa* | | CYP3A4/5↓ | In vivo, rat; in vitro, cells | Kim, Cho, Cho, and Yoon (2015) and Graber-Maier et al. (2010) |
| Jingjie 荆芥 | *Schizonepeta tenuifolia* | Schizonepetin | CYP1A2↑↓, CYP2D6↓, CYP2E1↑↓, CYP3A1/2↑ | In vivo, rat | Bao et al. (2012) |
| Jinyinhua 金银花 | *Lonicera japonica* | Chlorogenic acid | CYP1A2↓, CYP2E1↓ | In vivo, mice | Pang, Sheng, Jiang, Wei, and Ji (2015) |

*Continued*

**Table 1** Inducing (↑) or Inhibitory (↓) Effects of Commonly Used Chinese Medicinal Herbs on Cytochrome P450s (CYPs)—cont'd

| Pinyin Name | Latin Name | Chemical Constituents | CYPs | Species | References |
|---|---|---|---|---|---|
| Kushen 苦参 | *Sophora flavescens* | | CYP1A2↑, CYP2B1/2↑, CYP2C11↑, CYP3A1/2↑, CYP3A4/5↑ | In vivo, rat | Ueng, Tsai, Lo, and Yun (2010) |
| Leigongteng 雷公藤 | *Tripterygium wilfordii* | Celastrol | CYP1A2↓, CYP2C11↓, CYP2D6↓, CYP2E1↓, CYP3A1/2↓ | In vitro, rat liver cytochrome | Sun, Tang, Ding, Liu, and Wang (2014) |
| Lingzhi 灵芝 | *Ganoderma lucidum* | Polysaccharide | CYP1A2↓, CYP2E1↓, CYP3A1/2↓ | In vivo, rat | Wang et al. (2007) |
| Mujing 牡荆 | *Vitex negundo* | | CYP3A1↓, CYP2C11↑ | In vivo, rat | Wang, Hu, et al. (2015) |
| Qiancao 茜草 | *Rubia cordifolia* | Mollugin | CYP1A2↓ | In vitro, human liver microsomes | Kim et al. (2013) |
| Qianghuo 羌活 | *Notopterygium incisum or N. forbesii* | | CYP3A↓ | In vitro, human liver microsomes | Guo et al. (2001) |
| Renshen 人参 | *Panax ginseng* | Ginsenosides, deglycosylated ginsenosides | CYP1A2↑, CYP3A4/5↑ | In vitro, HepG2 cells | Hao et al. (2011) |
| Renshen 人参 | *Panax ginseng* | | CYP3A1/2↑, CYP3A4/5↑ | In vivo, human | Malati et al. (2012) |
| Renshen 人参 | *Panax ginseng* | Total ginsenosides | CYP1A2↓ | In vivo, mice | Sun, Yan, Xu, Wan, and Liu (2016) |
| Sanqi 三七 | *Panax notoginseng* | | CYP2B1/2↓, CYP4A1↓ | In vivo, rat | Yang, Liao, Yang, Guo, and Gao (2009) |

| | | | | | |
|---|---|---|---|---|---|
| Shaoyao Gancao Tang 芍药甘草汤 | | Decoction | CYP1A2↓, CYP3A4/5↓ | In vitro, human liver microsomes | Wang, Tian, et al. (2015) |
| Shengdihuang 生地黄 | *Rehmannia glutinosa* | | CYP3A4/5↓ | In vitro, human liver microsomes | Or et al. (2012) |
| Shengjiang 生姜 | *Zingiber officinale* | | CYP3A4/5↓ | In vitro, cells | Kimura, Ito, and Hatano (2010) |
| Wuweizi 五味子 | *Schisandra chinensis* | Schisandrin A, Schisandrin B, Schisandrin C, Schisandrol A, Schisandrol B, and Schisantherin A | CYP1A2↓, CYP2E1↓ | In vivo, mice | Fan et al. (2014) and Jiang, Fan, Wang, Tan, et al. (2015) |
| Wuweizi 五味子 | *Schisandra chinensis* | | CYP1A2↓, CYP2C6↓, CYP2C11↓, CYP2D2↓, CYP2E1↓, CYP3A1/2↓ | In vitro and in vivo, rat | Wang, Yang, Hu, and Li (2014) |
| Wuzhuyu 吴茱萸 | *Evodia rutaecarpa* | Rutaecarpine | CYP1A2↑ | In vivo, rat | Ueng, Tsai, Don, Chen, and Chen (2005) |
| Wuzhuyu 吴茱萸 | *Evodia rutaecarpa* | Rutaecarpine | CYP1A2↓ | In vitro, cells | Zhang, He, et al. (2015) |
| Wuzhuyu Tang 吴茱萸汤 | | Rutaecarpine | CYP1A2↑ | In vivo, mice | Ueng et al. (2002) |
| Yangjinhua 洋金花 | *Datura metel* | | CYP2D6↑ | In vivo, rat | Geng et al. (2015) |

*Continued*

**Table 1** Inducing (↑) or Inhibitory (↓) Effects of Commonly Used Chinese Medicinal Herbs on Cytochrome P450s (CYPs)—cont'd

| Pinyin Name | Latin Name | Chemical Constituents | CYPs | Species | References |
|---|---|---|---|---|---|
| Yanhusuo 延胡索 | *Corydalis yanhusuo* | Total alkaloid | CYP2E1↑, CYP3A1/2↑, CYP1A2↑, CYP2C11↑ | In vivo, rat | Yan et al. (2014) |
| Yanhusuo 延胡索 | *Corydalis yanhusuo* | Tetrahydropalmatine | CYP3A4/5↓ | In vivo, dog | Zhao et al. (2016) |
| Yin Zhi Huang 茵栀黄 | | | CYP2C19↑, CYP3A4/5↑ | In vivo, human | Fan et al. (2007) |
| Yinxingye 银杏叶 | *Ginkgo biloba* | | CYP2C19↑ | In vivo, human | Yin, Tomlinson, Waye, Chow, and Chow (2004) |
| Yinxingye 银杏叶 | *Ginkgo biloba* | Bilobalide, ginkgolide A, B, quercetin, and kaempferol | CYP3A1↑, CYP1A2↑, CYP2E1↑, CYP2B1/2↑ | In vivo, rat | Deng et al. (2008) |
| Zexie 泽泻 | *Alisma plantago-aquatica* | | CYP1A2↑, CYP3A4/5↓ | In vivo, rat | Huang, Zheng, Xu, and Hou (2014) |
| Zhishi or Zhiqiao 枳实,枳壳 | *Citrus aurantium* | | CYP3A↓ | In vitro, human liver microsomes | Guo et al. (2001) |

Prieto, 2014). *Ginkgo biloba* (Yingxingye, 银杏叶) with terpene lactones, ginkgolides A, B, C and bilobalide as the main active components, is used to manage neurodegenerative disorders, tinnitus, some types of glaucoma and peripheral vascular disease (Brondino et al., 2013; Colalto, 2010). It was reported concomitant administration of warfarin and *G. biloba* caused bleeding episodes in clinical patients (Mohutsky, Anderson, Miller, & Elmer, 2006). In vitro studies with human liver microsome have shown that *Ginkgo* flavonoids (amentoflavone) have strong inhibitory effect on CYP2C9 (Colalto, 2010; Gaudineau, Beckerman, Welbourn, & Auclair, 2004). However, conflict finding was reported by Jiang et al. (2005). In their studies, no significant effect on CYP2C9 activity and warfarin metabolism was caused by *G. biloba*.

Hu et al. (2015) explored the effects of Danshen (*S. miltiorrhiza*, 丹参) and its active constituent tanshinones on CYP2C19 activity with pooled human liver microsomes by using omeprazole as a probe substrate. Danshen is commonly used in CHM in the treatment of cardiovascular and cerebrovascular disorders. It has demonstrated significant inhibition of CYP2C19 activity by the herb and its active components, implicating the potential herb–drug interactions between Danshen and therapeutic agents metabolized by CYP2C19 in clinical practice such as warfarin. It is reported that activities of CYP2C19 are closely related to the maintenance doses in patients (Zhang et al., 2016). Adverse reactions of concomitant use of Danshen with warfarin may be contributed to inhibition of the metabolic enzyme by the herb. In contrast, *G. biloba* (Yinxingye, 银杏叶) was found to exert a significant inductive effect on CYP2C19 activity as indicated by enhanced omeprazole hydroxylation (Yin et al., 2004). Coadministration of *G. biloba* with CYP2C19 substrates may significantly reduce their therapeutic effect (Colalto, 2010). It has known that CYP2C19 is involved in the metabolism of proton pump inhibitors (e.g., omeprazole, lansoprazole, pantoprazole, and rabeprazole), tricycle antidepressants (e.g., amitriptyline, imipramine, and clomipramine), antiepileptics (e.g., diazepam, phenytoin, nordazepam, and phenobarbital), warfarin, clopidogrel, propranolol, and cyclophosphamide (Mazzari & Prieto, 2014).

CYP3A is expressed about 30% in the liver and metabolizes about 50% of the drugs (Zanger & Schwab, 2013). Some commonly prescribed therapeutic drugs including benzodiazepines, antihistamines, calcium channel blockers, and statins are metabolized by CYP3A (Mazzari & Prieto, 2014; Mizuno et al., 2009; Neuvonen, Niemi, & Backman, 2006). Many chemical compounds from plants are active on the isozymes of this CYP subfamily. Lau, Mooiman, Maas-Bakker, and Beijnen (2013) conducted an in vitro study on the effects of four Chinese herbs Baihuasheshecao (*Oldenlandia diffusa*, 白花蛇舌草), Dangshen (*Codonopsis tangshen*, 党参), Dihuang (*Rehmannia glutinosa*, 地黄), and Huangqi (*Astragalus propinquus*, 黄芪) on human CYP3A4 isozyme. It was found that the water–alcohol extract of the herbs inhibited the activity of the isozyme. However, at the same time, Baihuasheshecao and Dihuang significantly induced the gene expression of CYP3A4. These results suggest that concomitant use of these herbs with those drugs that are CYP3A4 substrates may cause herb–drug interactions leading to either a reduced efficacy of drugs or an enhanced toxicity.

In addition, CYP2D6 and CYP2E1 are another two CYP isozymes actively involved in metabolism of medicines. It is known that CYP2D6 is involved in the metabolism of some key prescribed medicines including timolol, propafenone, amitriptyline, haloperidol, risperidone, and chlorphenamine (Mazzari & Prieto, 2014). While CYP2E1 is normally involved in the metabolism of small molecules such as ethanol, it also plays an important role in the metabolism of certain medicines including paracetamol, theophylline, chlorzoxazone, and zopiclone.

## 2.4 Excretion

The drugs are excreted through kidneys, liver, lungs, feces, sweat, saliva, and the major route of excretion is renal tubular excretion involves glomerular filtration, tubular reabsorption, and active tubular secretion and biliary excretion (Caterina et al., 2013). Herb–drug interactions can affect the excretion and liquorice can interfere with antihypertensive and antiarrhythmic agents and result in hypokalemia and water retention (Staines, 2011).

It has been found that the active components of Danshen (*S. miltiorrhiza*, 丹参) including tanshinol, polyphenols, rosmarinic acid, lithospermic acid, and salvianolic acid A are excreted from urine through an active transportation process. The acids exhibited strong inhibitory activities against the human organic anion transporters OAT1 or OAT3 (Jia et al., 2015; Wang & Sweet, 2013). Concomitant administration of the herb with medicines that are also excreted through the same process, such as the antibiotics of benzylpenicillin and antivirus medicine adefovir (Maeda et al., 2014), herb–drug interaction may happen causing reduced excretion and elevated drug concentration in the body. Similarly, the commonly used Chinese medicinal herbs Renshen (*Panax ginseng*, 人参) and Sanqi (*Panax notoginseng*, 三七) also demonstrated that the active chemicals ginsenosides in the herbs experienced both active renal and hepatobiliary excretion in rat (Jiang, Dong, et al., 2015). Cautions should be taken in clinical practice when use these herbs with certain conventional medicines as the herb–drug may competitively bind to the active transporters. This interaction may cause significant alterations to the excretions of the drugs, leading to therapeutic failure or toxicity. In the modern days, it has become more and more popular to use Sanqi and Renshen as healthy food supplements. People may use these herbs with a high dose and for a longtime. This increases the risk of herb–drug interactions happening. Changes in hepatobiliary and in renal excretory clearances may lead to markedly different systemic exposure to the drugs as changes in excretion will alter the drugs' elimination kinetics.

## 3. ROLES OF TRANSPORTERS IN HERB–DRUG INTERACTIONS

Drug transporters play an important role in modulating drug absorption, distribution, and excretion. Drug–drug or herb–drug interactions on these transporters will markedly influence the pharmacokinetic behavior of drugs. Hundreds of membrane-bound transporters have been found as they are playing important roles in the movement of drugs and endogenous molecules as well across the cell membrane. The transporters are classified into two major superfamilies. One is the ATP-binding cassette (ABC) family, and the other one is the solute carrier (SLC) family. ABC transporters are mainly efflux, while the SLC transporters are more likely influx (Li & Shu, 2014; Liang, Li, & Chen, 2015; Wu et al., 2016). The most extensively studied ABC transporters including P-gp, multidrug resistance-associated proteins (MRPs), and breast cancer resistance protein (BCRP) mediate the efflux of drugs, and SLC transporters including organic cation transporters, organic anion transporters (OATs), and organic anion-transporting polypeptides are accountable for uptake of substrates into cells (Wu et al., 2016). As these transporters act as a control of drugs out of or in to cells, modulation of the transporters by herbs may influence not only absorption but also distribution and excretion of the concomitantly used drugs and could cause adverse effects consequently.

*P. notoginseng* (Sanqi, 三七) is an important herb used for promoting blood circulation but also help to stop bleeding in CHM. It contains a group of ginsenosides. Study has found that both the ABC and SLC transporters are involved the movement of these ginsenosides. But only 20(*S*)-protopanaxatriol-type ginsenosides can be transported through cells. This type of ginsenosides may bring potential risk of herb–drug interactions when taking together with some drugs which are also transported by these transporters (Jiang, Dong, et al., 2015).

P-gp is widely expressed in the body such as in the liver, kidney, and intestine. It is involved in transport of many prescribed drugs for instance anticancer drugs and cardiovascular pharmaceuticals (Al-Khazaali & Arora, 2014; Kobori, Harada, Nakamoto, & Tokuyama, 2014). Studies also found that chemical compounds from a number of medicinal herbs are substrates of the efflux transporter and potentially cause interactions with concurrently administered conventional medicines (Li et al., 2014, 2016; Oga, Sekine, Shitara, & Horie, 2012).

Fan, Tao, et al. (2009) investigated the effects of *S. chinensis* (Wuweizi, 五味子) and *G. biloba* (Yinxingye, 银杏叶) on P-gp transporter by using a specific nonmetabolized P-gp substrate talinolol as a probe in 12 healthy male individuals. The results showed that the $AUC_{0-24}$ of talinolol was increased 47% and 21%, the $C_{max}$ of talinolol was up 51% and 33% by *S. chinensis* (Wuweizi, 五味子) and *G. biloba* (Yinxing, 银杏), respectively. The half-life ($t_{1/2}$) of talinolol was also increased by 7% and 11% by *S. chinensis* and *G. biloba*. This suggested that both *S. chinensis* and *G. biloba* inhibited the efflux function of P-gp transporter.

In contrast, Qiu et al. (2014) found that the AUC and $C_{max}$ of the antihistamine drug fexofenadine were reduced by 37.2% and 27.4%, respectively, in healthy volunteers following consumed Danshen (*S. miltiorrhiza*, 丹参) extract for 10 days. The mean clearance of fexofenadine was also increased by 104.9%. The underlying mechanism for this has been linked to the induction of intestinal P-gp transporter (Wu et al., 2016).

Baicalin is the major active components of a commonly used Chinese medicinal herb Huangqin (*Scutellariae baicalensis*, 黄芩). It has been shown that baicalin is a substrate of MRP2 (Wu, Li, Xu, Du, & Liu, 2013). The interaction of baicalin with specific ABC transporters was studied and the results showed strong inhibitory effects on BCRP, MRP3, and MRP4 and less potent inhibition to MDR1 and MRP2 (Kalapos-Kovacs et al., 2015). This study suggested that by inhibition of transport by ABC transporters, baicalin could interfere with the absorption and disposition of drugs.

Zhou, Ye, Ruan, and Zeng (2013) studied the influence of a commonly used Chinese medicinal herb Huangqi (*Astragalus membranaceus*, 黄芪) on the P-gp transporter in health volunteers. The results have shown that Huangqi is not a potent modulator of P-gp. In this in vivo study in humans, Huangqi appeared to have inhibitory effect only on elimination rather than absorption of a P-gp substrate. This means concomitantly consume Huangqi may alter the pharmacokinetic behavior of the drugs that are P-gp substrate by increasing the drugs half-life in human bodies.

Ursolic acid is a ubiquitous pentacyclic triterpenoid compound and can be found in many herbal plants such as hawthorn (*Crataegus pinnatifida*, 山楂, Shanzha), Danshen (*S. miltiorrhiza*), gardenia (*Gardenia jasminoides*, 栀子, Zhizi), Nvzhenzi (*Ligustrum lucidum*, 女贞子), Xiakucao (*Prunella vulgaris*, 夏枯草), licorice (*G. uralensis*, 甘草, Gancao), and Lianqiao (Forsythia suspensa, 连翘). Study has found that this compound significantly inhibits the uptake of the popularly prescribed lipid-regulating drug rosuvastatin in to cells, suggesting a potential risk of therapeutic failure in clinical practice if concomitantly use of ursolic acid-containing herbs with statins (Hua, Hua, Nan, Jiang, & Yan, 2014).

## 4. PHARMACODYNAMIC HERB–DRUG INTERACTIONS

While herb–drug interactions in pharmacokinetics have been extensively studied, the pharmacodynamics interactions of herbs and drugs are less researched. The actual and potential pharmacodynamic drug interactions can be very diverse and widely varied (Corrie & Hardman, 2014). Due to the fact that multiple structurally different chemical compounds normally present in a single herb, they may interfere either directly or indirectly with different physiological process via carried action mechanisms such as receptors enzymes, leading to synergic/additive or reduced/ ADRs.

Herb–drug interactions of warfarin have been extensively studies. While abundant information on pharmacokinetic changes caused by different herbs was reported, it has been found that about 80% of all the identifiable herb–warfarin interactions were caused by pharmacodynamics-related mechanisms (Ge, Zhang, & Zuo, 2014; Tsai, Lin, Lu, Chen, & Mahady, 2013). Several pharmacodynamics action mechanisms have been known being involved in the herb–drug interactions, such as *G. biloba* (Yingxingye, 银杏叶) interferes the activity of warfarin by inhibiting the binding of platelet-activating factor to their receptors on platelet (Ge et al., 2014), but Gegen (*Puerariae lobatae*, 葛根) offsets the anticoagulant effects of warfarin by increasing the activity, mRNA, and protein expression of the vitamin K epoxide reductase as well as decreasing the activity of thrombomodulin (Ge et al., 2016).

Another example of herb–drug interaction in pharmacodynamics is the studies on the interaction of a commonly used Chinese medicinal herb Huangqin (*S. baicalensis*, 黄芩) and mefenamic acid, a nonsteroidal antiinflammatory drug (NSAID). Concomitantly administration of Huangqin and mefenamic acid to rat, significant synergistic effects were observed on inhibition of prostaglandin E2 production and on the expression of cyclooxygenase-2 gene. In addition, concurrently use of the herb also showed significant beneficial effects in alleviations of the side effects to stomach caused by mefenamic acid (Fong, Wong, Xie, & Zuo, 2015).

Studies of herb–drug interactions reported in the literature covered a wide categories of drugs. It appears more studies were focused on cardiovascular drugs, immunosuppressants, anticancer drugs, sedatives, and antidepressants (Yang, Hu, Duan, Zhu, & Zhou, 2006). Incidences of herb–drug interactions are higher with those medicines used for chronic diseases in particular the neurological pathology conditions such as depression, epilepsy, and dementia diseases. Some neurological drugs and herbs that may potential have interactions are listed in Table 2. While herb–drug interactions may bring risk to drug safety, but with better understanding the interaction mechanisms, the interactions may be used to improve therapeutic efficacy. A potentially beneficial effect has been reported in a seizures model of rat with a combination use of Chinese medicinal herb *Acorus calamus* (Shuichangpu, 水菖蒲) with antiepileptic drugs, sodium valproate, and carbamazepine. Greater protection as compared to either drug alone was observed, suggesting a potential for add on therapy with the herb. The combined therapy can lower the dose of individual drug and reduce dose-related toxicities (Katyal, Sarangal, & Gupta, 2012).

**Table 2** Herb–Drug Interactions of Some Neurological Drugs

| Drugs | Functions | Herbs or Active Components | Study Design | Interactions | References |
|---|---|---|---|---|---|
| Bupropion | Antidepressant | Huangqin (*Scutellariae baicalensis*, 黄芩)—Baicalin | In vivo | Baicalin remarkably induced CYP2B6-catalyzed bupropion hydroxylation. | Fan, Mao, et al. (2009) |
| Bupropion | Antidepressant | Yinxingye (*Ginkgo biloba*, 银杏叶)—terpene trilactones, flavonols | In vitro | *Ginkgo biloba* and its chemical constituents inhibited CYP2B6.<br>Bupropion hydroxylation was inhibited. | Lau and Chang (2009) |
| Buspirone | Anxiolytics | Gegen (*Pueraria lobata*, 葛根)—puerarin | In vivo, rat; in vitro, rat, human liver microsome | Puerarin, constituent of Gegen, inhibited CYP3A-mediated drug metabolism with potentially clinical significance. Puerarin reduced the elimination of Buspirone by inhibition of the drug's metabolism. | Kim et al. (2014) |
| Caffeine | Stimulant | Baizhi (*Angelicae dahurica*, 白芷) | In vivo, human | Baizhi potently inhibited CYP1A2 activity, metabolism of caffeine was decreased by 90%. | Yi et al. (2009) |
| Carbamazepine | Control of epilepsies | Xiaochaihu Tang (小柴胡汤) extract powder | In vivo, rat | Xiaochaihu Tang extract powder significantly decreased the peak plasma concentration of carbamazepine and lengthened the time to reach the peak concentration. This may be due to the herb decreased the absorption of the drug by delaying the gastric emptying. | Ohnishi et al. (2002) |

*Continued*

**Table 2** Herb–Drug Interactions of Some Neurological Drugs—cont'd

| Drugs | Functions | Herbs or Active Components | Study Design | Interactions | References |
|---|---|---|---|---|---|
| Carbamazepine | Control of epilepsies | Jia-Wei-Xiao-Yao-San (JWXYS, 加味逍遥散) | In vivo, human RCT | JWXYS improved the tolerability during the long-term use of carbamazepine in bipolar disorder, possibly by reducing the plasma concentration of the drug. At the same time, JWXYS showed synergic activities with carbamazepine enhanced the therapeutic effects. | Zhang, Kang, Li, and Tan (2007) |
| Chlorzoxazone | Muscle relaxant | Huangqin (*Scutellariae baicalensis*, 黄芩) | In vivo, human | *Scutellariae baicalensis* induced CYP2E1. Increased the metabolism of chlorzoxazone as a result of induction of CYP2E1. | Yi et al. (2009) |
| Diazepam | Control of epilepsies | Gouteng (*Uncaria tomentosa*, 钩藤) | In vivo | Gouteng enhanced the action of diazepam on spontaneous motor activity. | Quilez, Saenz, and Garcia (2012) |
| Diazepam | Control of epilepsies | St. Jones wart (*Hypericum perforatum*, 圣约翰草/贯叶连翘)—naphthodianthrone | In vivo, mice | St. Jones wart (*Hypericum perforatum*) potentiated the impairment of motor coordination caused by diazepam and prolonged pentobarbital-induced sleeping time. | Raskovic et al. (2014) |
| Lamotrigine | Control of epilepsies | Renshen (*Panax ginseng*, 人参) | Case report | Coadministration of lamotrigine with Ginseng caused drug reaction with eosinophilia and systemic symptoms (DRESS). This was possibly due to inhibition of the uridine diphosphate glucuronosyltransferase 2B7 enzyme by ginseng. | Myers, Watson, and Strock (2015) |

| | | | | | |
|---|---|---|---|---|---|
| Lorazepam | Sedative and analgesic | Xiecao (*Valeriana officinalis*, 缬草), Xipanlian (*Passiflora incarnate*, 西番莲) | Case report | Xiecao and Xipanlian produced additive or synergistic effect to lorazepam, causing adverse drug reactions. This might be due to the herbs increased the inhibitory activity of benzodiazepine drugs through binding to the GABA receptors. | Carrasco et al. (2009) |
| Losartan | Angiotensin-II receptor antagonists | Huangqin (*Scutellariae baicalensis*, 黄芩) | In vivo, human | *Scutellariae baicalensis* inhibited CYP2C9. Reduced the metabolism of losartan due to inhibition of CYP2C9. | Yi et al. (2009) |
| Midazolam | Antiepileptic | Renshen (*Panax ginseng*, 人参) | In vivo, human | *Panax ginseng* induced CYP3A activity. Concomitantly use of the herb significantly reduced the overall bioavailability, maximum plasma concentration, and half-life of the drug midazolam. | Malati et al. (2012) |
| Midazolam | Control of epilepsies | Wuweizi (*Schisandra chinensis*, 五味子)—Schisandra lignans extract | In vivo, rat | Administration of a single dose of Wuweizi increased the oral bioavailability and the peak plasma concentration of midazolam, suggesting an inhibited CYP3A activity. But long-term use of the herb showed significant induction of CYP3A and promoted the in vivo clearance of midazolam. Wuweizi possesses a much stronger inducing than inhibiting effect on CYP3A. Wuweizi showed more intensive regulating effect on intestinal than hepatic CYP3A. | Lai et al. (2009) |

*Continued*

**Table 2** Herb–Drug Interactions of Some Neurological Drugs—cont'd

| Drugs | Functions | Herbs or Active Components | Study Design | Interactions | References |
|---|---|---|---|---|---|
| Midazolam | Control of epilepsies | Danshen (*Salvia miltiorrhiza*, 丹参) | In vivo | Adjuvant use of Danshen with midazolam increased the hypnotic effects of the drug due to reduced metabolism of midazolam through CYP3A inhibition. | Wang, Lee, Zhou, Or, and Yeung (2010) |
| Midazolam | Control of epilepsies | Shenmai injection (生脉注射液) | In vitro | Shenmai injection's SMI could significantly enhance 1-hydroxylation but inhibit 4-hydroxylation of midazolam in human liver microsomes. The herbal constituent panaxytriol showed differential effects on the CYP3A4-mediated metabolism. | Zeng, He, Xia, Zhang, and Xiong (2013) |
| Naproxen | Nonsteroidal antiinflammatory drug | Chuanxinlian (*Andrographis paniculata*, 穿心莲)—andrographolide | In vivo, rat | Chuanxinlian and andrographolide decreased systemic exposure to the drug naproxen by reducing the peak plasma concentration $C_{max}$ and the area under concentration–time curve $AUC_{0-t}$ of naproxen. | Balap, Lohidasan, Sinnathambi, and Mahadik (2017) |
| Paracetamol | Nonopioid analgesics | St John's wort (*Hypericum perforatum*, 圣约翰草/贯叶连翘)—naphthodianthrone | In vivo, mice | *Hypericum perforatum* increased paracetamol plasma concentration. It may be due to the coadministered herb altered the drug's pharmacokinetic properties. | Raskovic et al. (2014) |

| | | | | | |
|---|---|---|---|---|---|
| Paroxetine | Antidepressant | Hongjingtian (*Rhodiola rosea*, 红景天) | Case report | A patient presented the complication related to serotonergic syndrome after the adjuvant use of Rhodiola rosea with paroxetine. The herb may stimulate the distribution of dopamine and serotonin. | Maniscalco, Toffol, Giupponi, and Conca (2015) |
| Phenytoin | Antiepileptic drugs | *Rheum palmatum* (大黄) | In vivo, in vitro | *Rheum palmatum* decreases the oral bioavailability of phenytoin by inducing the activity of the efflux transporter P-gp in the intestine. | Chi, Juang, Chui, Hou, and Chao (2012) |
| Phenytoin | Control of epilepsies | Duxingcai (*Lepidium sativum*, 独行菜), Heizhongcao (*Nigella sativa*, 黑种草), Huluba (*Trigonella foenum-graceum*, 葫芦巴) | In vivo, dog | All the herbs altered the metabolism of phenytoin.<br>Duxingcai increased the maximum plasma concentration ($C_{max}$) and half-life ($t_{1/2}$) but reduced the clearance of phenytoin.<br>Heizhongcao and Huluba reduced area under the curve ($AUC_{0-\infty}$) by 87% and 72%, respectively, suggesting a significant increase in the elimination of phenytoin. | Alkharfy et al. (2013) |
| Phenytoin (Dilantin) | Control of epilepsies | Yinxingye (*Ginkgo biloba*, 银杏叶) | Case report | *Ginkgo biloba* reduced the serum concentration of Dilantin to a subtherapeutic levels possibly by induction of the metabolism enzyme CYP2C9 and CYP2C19. This caused fatal breakthrough seizures. | Kupiec and Raj (2005) |

*Continued*

**Table 2** Herb–Drug Interactions of Some Neurological Drugs—cont'd

| Drugs | Functions | Herbs or Active Components | Study Design | Interactions | References |
|---|---|---|---|---|---|
| Sertraline | Antidepressant (selective serotonin reuptake inhibitor) | Hezi (*Terminalia chebula*, 诃子), Huaite Moyao (*Commiphora wightii*, 怀特没药) | Case report | The chemical components from both of the herbs may have induced the metabolism of sertraline and diminished its efficacy causing recurrent relapse of depression. | Prasad, Tharangani, and Samaranayake (2009) |
| *S*-mephenytoin | Anticonvulsants | Chuanxinlian (*Andrographis paniculata*, 穿心莲), Jixuecao (*Centella asiatica*, 積雪草), Maoxucao (*Orthosiphon stamineus*, 猫须草) | In vitro | All the herbs showed certain degree inhibition to CYP2C19 with an order Maoxucao>Jixuecao>Chuanxinlian. Orthosiphon stamineus's active constituent eupatorine showed potent inhibition to CYP2C19.<br>Metabolism of *S*-mephenytoin was reduced. | Pan et al. (2011) |
| Valproate sodium (Depakote) | Control of epilepsies (anticonvulsants) | Yinxingye (*Ginkgo biloba*, 银杏叶) | Case report | *Ginkgo biloba* reduced the serum concentration of Depakote to a subtherapeutic levels possibly by induction of the metabolism enzyme CYP2C9 and CYP2C19. This caused fatal breakthrough seizures. | Kupiec and Raj (2005) |
| Venlafaxine | Antidepressant | Jixuecao (*Centella asiatica*, 积雪草), Mojiaozao (*Fucus vesiculosus*, 墨角藻) | Case report | Taking venlafaxine caused interstitial pneumonitis and cardiomyopathy. Concomitantly taking *Centella asiatica* and *Fucus vesiculosus* may be one of the potential reasons for this. This adverse drug reaction, as the herbs inhibit CYPs, particularly CYP2D6. | Ferreira, Costa, Dias, Ferreira, and Franco (2014) |

## 5. CONCLUSION

Herb–drug interactions are important issue for clinical therapeutic efficacy of medicine treatment. Herb–drug interactions affect either the pharmacokinetic fates or pharmacodynamics activities of drugs, leading to therapeutic failure or toxicities. For many commonly used Chinese medicinal herbs and some herbal formulas, their pharmacokinetic interactions with some commonly used conventional drugs have been extensively studies. The major mechanisms responsible for most of pharmacokinetic interactions are the modulation of metabolism enzymes activities particularly the CYP isozymes and the membrane transporters activities particularly the efflux transporter P-gp. In comparison with the studies in pharmacokinetics, herb–drug interactions in pharmacodynamics are far more less studied. More attentions and efforts should be put on this area. Herb–drug interactions are complicated, and the clinical outcomes may be a result from the interactions in both pharmacokinetics and pharmacodynamics. Lack of knowledge in the potential adverse herb–drug interactions will put big risk to patients' safety in medical services. On the contrary, some interactions may be therapeutically beneficial. It may be used to help develop new therapeutic strategies in the future.

## REFERENCES

Ajazuddin, A., Alexander, A., Qureshi, A., Kumari, L., Vaishnav, P., Sharma, M., et al. (2014). Role of herbal bioactives as a potential bioavailability enhancer for Active Pharmaceutical Ingredients. *Fitoterapia*, *97*, 1–14.

Alkharfy, K. M., Al-Jenoobi, F. I., Al-Mohizea, A. M., Al-Suwayeh, S. A., Khan, R. M. A., & Ahmad, A. (2013). Effects of Lepidium sativum, Nigella sativa and Trigonella foenum-graceum on phenytoin pharmacokinetics in beagle dogs. *Phytotherapy Research*, *27*(12), 1800–1804.

Al-Khazaali, A., & Arora, R. (2014). P-glycoprotein: A focus on characterizing variability in cardiovascular pharmacotherapeutics. *American Journal of Therapeutics*, *21*(1), 2–9.

Balap, A., Lohidasan, S., Sinnathambi, A., & Mahadik, K. (2017). Herb-drug interaction of Andrographis paniculata (Nees) extract and andrographolide on pharmacokinetic and pharmacodynamic of naproxen in rats. *Journal of Ethnopharmacology*, *195*, 214–221.

Bao, B., Geng, T., Cao, Y., Yao, W., Zhang, L., & Ding, A. (2012). Effects of schizonepetin on activity and mRNA expression of cytochrome P450 enzymes in rats. *International Journal of Molecular Sciences*, *13*(12), 17006–17018.

Brondino, N., De Silvestri, A., Re, S., Lanati, N., Thiemann, P., Verna, A., et al. (2013). A systematic review and meta-analysis of Ginkgo biloba in neuropsychiatric disorders: From ancient tradition to modern-day medicine. *Evidence-Based Complementary and Alternative Medicine*, *2013*, 1–11. http://dx.doi.org/10.1155/2013/915691.

Brown, L., Heyneke, O., Brown, D., Van Wyk, J. P., & Hamman, J. H. (2008). Impact of traditional medicinal plant extracts on anti-retroviral drug absorption. *Journal of Ethnopharmacology, 119*, 588–592.

Budzinski, J. W., Foster, B. C., Vandenhoek, S., & Arnason, J. T. (2000). An in vitro evaluation of human cytochrome P450 3A4 inhibition by selected commercial herbal extracts and tinctures. *Phytomedicine*, 7, 273–282.

Carrasco, M. C., Vallejo, J. R., Pardo de Santayana, M., Peral, D., Martan, M. A., & Altimiras, J. (2009). Interactions of Valeriana officinalis L. and Passiflora incarnate L. in a patient treated with lorazepam. *Phytotherapy Research, 23*(12), 1795–1796.

Caterina, P., Antonello, P., Chiara, G., Chiara, C., Giacomo, L., Antonio, S., et al. (2013). Pharmacokinetic drug-drug interaction and their implication in clinical management. *Journal of Research in Medical Sciences, 18*, 600–609.

Chan, T. (2001). Interaction between warfarin and danshen (*Salvia miltiorrhiza*). *The Annals of Pharmacotherapy, 35*, 501–504.

Chan, A., Liu, W., Lin, S., Leung, H., & Wang, H. (2007). Interaction between traditional Chinese medicine and Western medicine in rats—In-Chen-How and acetaminophen. *Yao Xue Xue Bao, 42*(3), 342–346.

Chen, Y., Liu, W., Chen, B., Fan, L., Han, Y., Wang, G., et al. (2010). Plant polyphenol curcumin significantly affects CYP1A2 and CYP2A6 activity in healthy, male Chinese volunteers. *Annals of Pharmacotherapy, 44*(6), 1038–1045.

Chen, L., Yang, J., Davey, A., Chen, Y., Wang, J., & Liu, X. (2009). Effects of diammonium glycyrrhizinate on the pharmacokinetics of aconitine in rats and the potential mechanism. *Xenobiotica, 39*, 955–963.

Chi, Y. C., Juang, S. H., Chui, W. K., Hou, Y. C., & Chao, P. D. L. (2012). Acute and chronic administrations of Rheum palmatum reduced the bioavailability of phenytoin in rats: A new herb-drug interaction. *Evidence-Based Complementary and Alternative Medicine, 2012*, 1–9. http://dx.doi.org/10.1155/2012/701205.

Colalto, C. (2010). Herbal interactions on absorption of drugs: Mechanisms of action and clinical risk assessment. *Pharmacological Research, 62*(3), 207–227.

Corrie, K., & Hardman, J. (2014). Mechanisms of drug interactions: Pharmacodynamics and pharmacokinetics. *Anaesthesia & Intensive Care Medicine, 15*(7), 305–308.

Deng, Y., Bi, H. C., Zhao, L. Z., He, F., Liu, Y. Q., Yu, J. J., et al. (2008). Induction of cytochrome P450s by terpene trilactones and flavonoids of the Ginkgo biloba extract EGb 761 in rats. *Xenobiotica, 38*(5), 465–481.

Dudhatra, G. B., Mody, S. K., Awale, M. M., Patel, H. B., Modi, C. M., Kumar, A., et al. (2012). A comprehensive review on pharmacotherapeutics of herbal bioenhancers. *Scientific World Journal, 2012*, 1–32. http://dx.doi.org/10.1100/2012/637953.

Eisenberg, D., Davis, R., Ettner, S., Appel, S., Wilkey, S., Van Rompay, M., et al. (1998). Trends in alternative medicine use in the United States, 1990–1997. *JAMA, 280*(18), 1569.

Fan, X., Jiang, Y., Wang, Y., Tan, H., Zeng, H., Wang, Y., et al. (2014). Wuzhi tablet (Schisandra Sphenanthera extract) protects against acetaminophen-induced hepatotoxicity by inhibition of CYP-mediated bioactivation and regulation of NRF2-ARE and p53/p21 pathways. *Drug Metabolism and Disposition, 42*(12), 1982–1990.

Fan, L., Mao, X., Tao, G., Wang, G., Jiang, F., Chen, Y., et al. (2009). Effect of Schisandra chinensis extract and Ginkgo biloba extract on the pharmacokinetics of talinolol in healthy volunteers. *Xenobiotica, 39*(3), 249–254.

Fan, L., Tao, G., Wang, G., Chen, Y., Zhang, W., He, Y., et al. (2009). Effects of Ginkgo biloba extract ingestion on the pharmacokinetics of talinolol in healthy Chinese volunteers. *Annals of Pharmacotherapy, 43*(5), 944–949.

Fan, L., Wang, G., Wang, L., Chen, Y., Zhang, W., Huang, Y., et al. (2007). Herbal medicine Yin Zhi Huang induces CYP3A4-mediated sulfoxidation and CYP2C19-dependent hydroxylation of omeprazole. *Acta Pharmacologica Sinica, 28*(10), 1685–1692.

Fasinu, P., Bouic, P., & Rosenkranz, B. (2012). An overview of the evidence and mechanisms of herb–drug interactions. *Frontiers in Pharmacology, 3*(69), 1–19.

Ferreira, P. G., Costa, S., Dias, N., Ferreira, A. J., & Franco, F. (2014). Simultaneous interstitial pneumonitis and cardiomyopathy induced by venlafaxine. *Jornal Brasileiro de Pneumologia, 40*(3), 313–318.

Fong, S., Wong, Y., Xie, C., & Zuo, Z. (2015). Herb-drug interactions between Scutellariae Radix and mefenamic acid: Simultaneous investigation of pharmacokinetics, anti-inflammatory effect and gastric damage in rats. *Journal of Ethnopharmacology, 170*, 106–116.

Fugh-Berman, A., & Ernst, E. (2001). Herb-drug interactions: Review and assessment of report reliability. *British Journal of Clinical Pharmacology, 52*(5), 587–595.

Gao, J., Shi, Z., Zhu, S., Li, G., Yan, R., & Yao, M. (2013). Influences of processed rhubarbs on the activities of four CYP isozymes and the metabolism of saxagliptin in rats based on probe cocktail and pharmacokinetics approaches. *Journal of Ethnopharmacology, 145*(2), 566–572.

Gaudineau, C., Beckerman, R., Welbourn, S., & Auclair, K. (2004). Inhibition of human P450 enzymes by multiple constituents of the Ginkgo biloba extract. *Biochemical and Biophysical Research Communications, 318*, 1072–1078.

Ge, B., Zhang, Z., & Zuo, Z. (2014). Updates on the clinical evidenced herb warfarin interactions. *Evidence-Based Complementary and Alternative Medicine, 2014*, 1–18. http://dx.doi.org/10.1155/2014/957362.

Ge, B., Zhang, Z., & Zuo, Z. (2016). Radix Puerariae lobatae (Gegen) suppresses the anticoagulation effect of warfarin: A pharmacokinetic and pharmacodynamics study. *Chinese Medicine, 11*(1), 1–17.

Geng, P., Wang, S., Wang, C., Chen, J., Zhang, L., Yang, S., et al. (2015). Evaluation of the impact of Flos Daturae on rat hepatic cytochrome P450 enzymes by cocktail probe drugs. *International Journal of Clinical and Experimental Medicine, 8*(12), 22310–22318.

Graber-Maier, A., Buter, K. B., Aeschlimann, J., Bittel, C., Kreuter, M., Drewe, J., et al. (2010). Effects of Curcuma extracts and curcuminoids on expression of P-glycoprotein and cytochrome P450 3A4 in the intestinal cell culture model LS180. *Planta Medica, 76*, 1866–1870.

Guo, Y., Pope, C., Cheng, X., Zhou, H., & Klaassen, C. (2011). Dose–response of berberine on hepatic cytochromes P450 mRNA expression and activities in mice. *Journal of Ethnopharmacology, 138*(1), 111–118.

Guo, L., Taniguchi, M., Chen, Q., Baba, K., & Yamazoe, Y. (2001). Inhibitory potential of herbal medicines on human cytochrome P450-mediated oxidation: Properties of umbelliferous or citrus crude drugs and their relative prescriptions. *Japanese Journal of Pharmacology, 85*(4), 399–408.

Han, Y., Yu, H., Li, D., Meng, X., Zhou, Z., Yu, Q., et al. (2011). In vitro inhibition of Huanglian [Rhizoma coptidis(L.)] and its six active alkaloids on six cytochrome P450 isoforms in human liver microsomes. *Phytotherapy Research, 25*(11), 1660–1665.

Hao, M., Ba, Q., Yin, J., Li, J., Zhao, Y., & Wang, H. (2011). Deglycosylated ginsenosides are more potent inducers of CYP1A1, CYP1A2 and CYP3A4 expression in HepG2 cells than glycosylated ginsenosides. *Drug Metabolism and Pharmacokinetics, 26*(2), 201–205.

Harris, P., Cooper, K., Relton, C., & Thomas, K. (2012). Prevalence of complementary and alternative medicine (CAM) use by the general population: A systematic review and update. *International Journal of Clinical Practice, 66*(10), 924–939.

He, W., Wu, J., Ning, J., Hou, J., Xin, H., He, Y., et al. (2015). Inhibition of human cytochrome P450 enzymes by licochalcone A, a naturally occurring constituent of licorice. *Toxicology In Vitro, 29*(7), 1569–1576.

Hirota, T., & Ieiri, I. (2015). Drug-drug interactions that interfere with statin metabolism. *Expert Opinion on Drug Metabolism & Toxicology, 11*(9), 1435–1447.

Hou, M., Lu, C., & Tsai, T. (2016). Effects of Jia-Wei-Xiao-Yao-San on the peripheral and lymphatic pharmacokinetics of paclitaxel in rats. *Evidence-Based Complementary and Alternative Medicine, 2016*, 1–10. http://dx.doi.org/10.1155/2016/5614747.

Hu, T., Zhou, X., Wang, L., Or, P., Yeung, J., Kwan, Y., et al. (2015). Effects of tanshinones from Salvia miltiorrhiza on CYP2C19 activity in human liver microsomes: Enzyme kinetic and molecular docking studies. *Chemico-Biological Interactions, 230*, 1–8.

Hua, W., Hua, W., Nan, F., Jiang, W., & Yan, C. (2014). Department of pharmacy, the first affiliated hospital of Nanchang University, Nanchang. *European Journal of Drug Metabolism and Pharmacokinetics, 39*, 221–230.

Huang, G., Chen, S., Tsai, P., Ganzon, J., Lee, C., Shiah, H., et al. (2016). Effects of Dang-Gui-Bu-Xue-Tang, an herbal decoction, on iron uptake in iron-deficient anemia. *Drug Design, Development and Therapy, 10*, 949–957.

Huang, Y., Zheng, S., Xu, Z., & Hou, Y. (2014). Effects of Alismatis rhizome on rat cytochrome P450 enzymes. *Pharmaceutical Biology, 52*(6), 681–687.

Hussain, S. (2011). Patient counselling about herbal-drug interactions. *African Journal of Traditional, Complementary, and Alternative Medicines, 8*(5 Suppl.), 152–163.

Imai, T., Sakai, M., Ohtake, H., Azuma, H., & Otagiri, M. (1999). In vitro and in vivo evaluation of the enhancing activity of glycyrrhizin on the intestinal absorption of drugs. *Pharmaceutical Research, 16*(1), 80–86.

Imai, T., Sakai, M., Ohtake, H., Azuma, H., & Otagiri, M. (2005). Absorption-enhancing effect of glycyrrhizin induced in the presence of capric acid. *International Journal of Pharmaceutics, 294*, 11–21.

Jeong, H., Kong, T., Kwon, S., Hong, S., Yeon, S., Choi, J., et al. (2013). Effect of honokiol on cytochrome P450 and UDP-glucuronosyltransferase enzyme activities in human liver microsomes. *Molecules, 18*(9), 10681–10693.

Jia, W., Du, F., Liu, X., Jiang, R., Xu, F., Yang, J., et al. (2015). Renal tubular secretion of tanshinol: Molecular mechanisms, impact on its systemic exposure, and propensity for dose-related nephrotoxicity and for renal herb-drug interactions. *Drug Metabolism and Disposition, 43*(5), 669–678.

Jiang, R., Dong, J., Li, X., Du, F., Jia, W., Xu, F., et al. (2015). Molecular mechanisms governing different pharmacokinetics of ginsenosides and potential for ginsenoside-perpetrated herb-drug interactions on OATP1B3. *British Journal of Pharmacology, 172*(4), 1059–1073.

Jiang, Y., Fan, X., Wang, Y., Chen, P., Zeng, H., Tan, H., et al. (2015a). Schisandrol B protects against acetaminophen-induced hepatotoxicity by inhibition of CYP-mediated bioactivation and regulation of liver regeneration. *Toxicological Sciences, 143*(1), 107–115.

Jiang, Y., Fan, X., Wang, Y., Tan, H., Chen, P., Zeng, H., et al. (2015b). Hepato-protective effects of six schisandra lignans on acetaminophen-induced liver injury are partially associated with the inhibition of CYP-mediated bioactivation. *Chemico-Biological Interactions, 231*, 83–89.

Jiang, X., Williams, K., Liauw, W., Ammit, A., Roufogalis, B., Duke, C., et al. (2005). Effect of ginkgo and ginger on the pharmacokinetics and pharmacodynamics of warfarin in healthy subjects. *British Journal of Clinical Pharmacology, 59*(4), 425–432.

Jing, X., Peng, Y., Wang, X., & Duane, J. (2015). Effects of Ziziphus jujuba fruit extracts on cytochrome P450 (CYP1A2) activity in rats. *Chinese Journal of Natural Medicines, 13*(8), 588–594.

Jo, J., Kim, S., Nam, W., Seung, E., & Lee, S. (2016). Decreased absorption of midazolam in the stomach due to low pH induced by co-administration of Banha-sasim-tang. *Environmental Health and Toxicology, 9*, 31.

Kalapos-Kovacs, B., Magda, B., Jani, M., Fekete, Z., Szabo, P., Antal, I., et al. (2015). Multiple ABC transporters efflux baicalin. *Phytotherapy Research*, *29*(12), 1987–1990.
Katyal, J., Sarangal, V., & Gupta, Y. (2012). Interaction of hydroalcoholic extract of Acorus calamus Linn. with sodium valproate and carbamazepine. *Indian Journal of Experimental Biology*, *50*(1), 51–55.
Kaur, G., Leslie, E., Tillman, H., Lee, W., Swanlund, D., Karvellas, C., et al. (2015). Detection of ophthalmic acid in serum from acetaminophen-induced acute liver failure patients is more frequent in non-survivors. *PLoS One*, *10*(9), e0139299.
Kesarwani, K., Gupta, R., & Mukerjee, A. (2013). Bioavailability enhancers of herbal origin: An overview. *Asian Pacific Journal of Tropical Biomedicine*, *3*(4), 253–266.
Kim, S. B., Cho, S. S., Cho, H. J., & Yoon, I. S. (2015). Modulation of hepatic cytochrome P450 enzymes by curcumin and its pharmacokinetic consequences in Sprague–Dawley rats. *Pharmacognosy Magazine*, *11*(Suppl. 4), 580–584.
Kim, H., Choi, H., Jeong, T., Jahng, Y., Kim, D., Lee, S., et al. (2013). Selective inhibitory effects of mollugin on CYP1A2 in human liver microsomes. *Food and Chemical Toxicology*, *51*, 33–37.
Kim, S. B., Yoon, I. S., Kim, K. S., Cho, S. J., Kim, Y. S., Cho, H. J., et al. (2014). In vitro and in vivo evaluation of the effect of puerarin on hepatic cytochrome p450-mediated drug metabolism. *Planta Medica*, *80*(7), 561–567.
Kimura, Y., Ito, H., & Hatano, T. (2010). Effects of mace and nutmeg on human cytochrome P450 3A4 and 2C9 activity. *Biological & Pharmaceutical Bulletin*, *33*, 1977–1982.
Kobori, T., Harada, S., Nakamoto, K., & Tokuyama, S. (2014). Mechanisms of P-glycoprotein alteration during anticancer treatment: Role in the pharmacokinetic and pharmacological effects of various substrate drugs. *Journal of Pharmacological Sciences*, *125*(3), 242–254.
Kupiec, T., & Raj, V. (2005). Fatal seizures due to potential herb-drug interactions with ginkgo biloba. *Journal of Analytical Toxicology*, *29*(7), 755–758.
Lai, L., Hao, H., Wang, Q., Zheng, C., Zhou, F., Liu, Y., et al. (2009). Effects of short-term and long-term pretreatment of Schisandra lignans on regulating hepatic and intestinal CYP3A in rats. *Drug Metabolism and Disposition*, *37*(12), 2399–2407.
Lau, A. J., & Chang, T. K. (2009). Inhibition of human CYP2B6-catalyzed bupropion hydroxylation by ginkgo biloba extract: Effect of terpene trilactones and flavonols. *Drug Metabolism and Disposition*, *37*(9), 1931–1937.
Lau, C., Mooiman, K. D., Maas-Bakker, R. F., Beijnen, J. H., Schellens, J. H., & Meijerman, I. (2013). Effect of Chinese herbs on CYP3A4 activity and expression in vitro. *Journal of Ethnopharmacology*, *149*(2), 543–549.
Lee, S., Jang, H., Lee, J., Ma, J., Oh, S., & Kim, S. (2015). Inhibitory effects of Hwang-Ryun-Hae-Dok-Tang on cytochrome P450 in human liver microsomes. *Xenobiotica*, *45*(2), 131–138.
Li, X., Hu, J., Wang, B., Sheng, L., Liu, Z., Yang, S., et al. (2014). Inhibition effects of herbal constituents on P-glycoprotein in vitro and in vivo: Herb-drug interactions mediated via P-gp. *Toxicology and Applied Pharmacology*, *275*(2), 163–175.
Li, Q., & Shu, Y. (2014). Role of solute carriers in response to anticancer drugs. *Molecular and Cellular Therapies*, *2*, 15.
Li, H., Zhang, C., Fan, R., Sun, H., Xie, H., Luo, J., et al. (2016). The effects of Chuanxiong on the pharmacokinetics of warfarin in rats after biliary drainage. *Journal of Ethnopharmacology*, *193*, 117–124.
Li, Y., Zhang, Y., Li, L., Wang, Q., & Wang, N. (2008). Effect of Danggui and Honghua on cytochrome P450 1A2, 2C11, 2E1 and 3A1 mRNA expression in liver of rats. *The American Journal of Chinese Medicine*, *36*(6), 1071–1081.
Liang, Y., Li, S., & Chen, L. (2015). The physiological role of drug transporters. *Protein & Cell*, *6*(5), 334–350.

Lotfi, M., Adib-Hajbaghery, M., Shahsavarloo, Z., & Gandomani, H. (2016). The prevalence of traditional and complementary medicine in the general population in Kashan, Iran, 2014. *European Journal of Integrative Medicine*, *8*(5), 661–669.

Lv, Q., Wang, G., Chen, S., Hu, L., Zhang, X., Ying, G., et al. (2015). In vitro and in vivo inhibitory effects of glycyrrhetinic acid in mice and human cytochrome P450 3A4. *International Journal of Environmental Research and Public Health*, *13*(1), 84.

Maeda, K., Tian, Y., Fujita, T., Ikeda, Y., Kumagai, Y., Kondo, T., et al. (2014). Inhibitory effects of p-aminohippurate and probenecid on the renal clearance of adefovir and benzylpenicillin as probe drugs for organic anion transporter (OAT) 1 and OAT3 in humans. *European Journal of Pharmaceutical Sciences*, *1*, 94–103.

Malagi, K., Kamath, M., & Adiga, S. (2014). Herb-drug interaction in diabetes: A bane or a boon. *Journal of Diabetes Metabolic Disorders & Control*, *1*(3), 00013.

Malati, C., Robertson, S., Hunt, J., Chairez, C., Alfaro, R., Kovacs, J., et al. (2012). Influence of *Panax ginseng* on cytochrome P450 (CYP)3A and P-glycoprotein (P-gp) activity in healthy participants. *The Journal of Clinical Pharmacology*, *52*(6), 932–939.

Maniscalco, I., Toffol, E., Giupponi, G., & Conca, A. (2015). Das Zusammenwirken von Rhodiola rosea (Eisenwurz) und Antidepressiva. *Neuropsychiatrie*, *29*(1), 36–38.

Mazzari, A., & Prieto, J. (2014). Herbal medicines in Brazil: Pharmacokinetic profile and potential herb-drug interactions. *Frontiers in Pharmacology*, *5*, 162.

Meijerman, I., Beijnen, J., & Schellens, J. (2006). Herb-drug interactions in oncology: Focus on mechanisms of induction. *The Oncologist*, *11*(7), 742–752.

Milic, N., Milosevic, N., Golocorbin-Kon, S., Bozic, T., Abenavoli, L., & Borrelli, F. (2014). Warfarin interactions with medicinal herbs. *Natural Product Communications*, *9*, 1211–1216.

Mizuno, K., Katoh, M., Okumura, H., Nakagawa, N., Negishi, T., Hashizume, T., et al. (2009). Metabolic activation of benzodiazepines by CYP3A4. *Drug Metabolism and Disposition*, *37*(2), 345–351.

Mohutsky, M., Anderson, G., Miller, J., & Elmer, G. (2006). Ginkgo biloba: Evaluation of CYP2C9 drug interactions in vitro and in vivo. *American Journal of Therapeutics*, *13*(1), 24–31.

Moses, G., & McGuire, T. (2010). Drug interactions with complementary medicines. *Australian Prescriber*, *33*(6), 177–180.

Myers, A. P., Watson, T. A., & Strock, S. B. (2015). Drug reaction with eosinophilia and systemic symptoms syndrome probably induced by a lamotrigine-ginseng drug interaction. *Pharmacotherapy: The Journal of Human Pharmacology and Drug Therapy*, *35*(3), 9–12.

Neuvonen, P., Niemi, M., & Backman, J. (2006). Drug interactions with lipid-lowering drugs: Mechanisms and clinical relevance. *Clinical Pharmacology and Therapeutics*, *80*(6), 565–581.

Nowack, R. (2008). Review article: Cytochrome P450 enzyme, and transport protein mediated herb-drug interactions in renal transplant patients: Grapefruit juice, St John's Wort - and beyond! *Nephrology (Carlton)*, *13*(4), 337–347.

Obodozie, O. (2012). Pharmacokinetics and drug interactions of herbal medicines: A missing critical step in the phytomedicine/drug development process. In A. Noreddin (Ed.), *Readings in advanced pharmacokinetics—Theory, methods and applications*. InTechOpen, (pp. 127–156). http://dx.doi.org/10.5772/33699.

Oga, E., Sekine, S., Shitara, Y., & Horie, T. (2012). P-glycoprotein mediated efflux in Caco-2 cell monolayers: The influence of herbals on digoxin transport. *Journal of Ethnopharmacology*, *144*(3), 612–617.

Ohnishi, N., Okada, K., Yoshioka, M., Kuroda, K., Nagasawa, K., Takara, K., et al. (2002). Studies on interactions between traditional herbal and western medicines. V. Effects of Sho-saiko-to (Xiao-Cai-hu-Tang) on the pharmacokinetics of carbamazepine in rats. *Biological and Pharmaceutical Bulletin*, *25*(11), 1461–1466.

Or, P., Lam, F., Kwan, Y., Cho, C., Lau, C., Yu, H., et al. (2012). Effects of Radix Astragali and Radix Rehmanniae, the components of an anti-diabetic foot ulcer herbal formula, on metabolism of model CYP1A2, CYP2C9, CYP2D6, CYP2E1 and CYP3A4 probe substrates in pooled human liver microsomes and specific CYP isoforms. *Phytomedicine*, *19*(6), 535–544.

Pan, Y., Abd-Rashid, B. A., Ismail, Z., Ismail, R., Mak, J. W., Pook, P. C. K., et al. (2011). In vitro modulatory effects of Andrographis paniculata, Centella asiatica and Orthosiphon stamineus on cytochrome P450 2C19 (CYP2C19). *Journal of Ethnopharmacology*, *133*(2), 881–887.

Pandit, S., Kanjilal, S., Awasthi, A., Chaudhary, A., Banerjee, D., Bhatt, B., et al. (2016). Evaluation of herb-drug interaction of a polyherbal Ayurvedic formulation through high throughput cytochrome P450 enzyme inhibition assay. *Journal of Ethnopharmacology*, *197*, 165–172. http://dx.doi.org/10.1016/j.jep.2016.07.061.

Pang, C., Sheng, Y., Jiang, P., Wei, H., & Ji, L. (2015). Chlorogenic acid prevents acetaminophen-induced liver injury: The involvement of CYP450 metabolic enzymes and some antioxidant signals. *Journal of Zhejiang University. Science. B*, *16*(7), 602–610.

Prasad, K., Tharangani, P., & Samaranayake, C. (2009). Recurrent relapses of depression in a patient established on sertraline after taking herbal medicinal mixtures—A herb-drug interaction? *Journal of Psychopharmacology*, *23*(2), 216–219.

Qiu, F., Zeng, J., Liu, S., He, M., Zhu, L., Ye, Y., et al. (2014). Effects of danshen ethanol extract on the pharmacokinetics of fexofenadine in healthy volunteers. *Evidence-Based Complementary and Alternative Medicine*, *2014*, 1–5. http://dx.doi.org/10.1155/2014/473213.

Qiu, F., Zhang, R., Sun, J., Jiye, A., Hao, H., Peng, Y., et al. (2008). Inhibitory effects of seven components of danshen extract on catalytic activity of cytochrome P450 enzyme in human liver microsomes. *Drug Metabolism and Disposition*, *36*(7), 1308–1314.

Quilez, A. M., Saenz, M. T., & Garcia, M. D. (2012). Uncaria tomentosa (Willd. Ex. Roem. & Schult.) DC. and eucalyptus globulus Labill. Interactions when administered with diazepam. *Phytotherapy Research*, *26*(3), 458–461.

Ramos-Remus, C., & Raut, A. (2008). Complementary and alternative practices in rheumatology. *Best Practice & Research. Clinical Rheumatology*, *22*(4), 741–757.

Raskovic, A., Cvejic, J., Stilinovic, N., Golocorbin-Kon, S., Vukmirovic, S., Mimica-Dukic, N., et al. (2014). Interaction between different extracts of Hypericum perforatum L. from Serbia and pentobarbital, diazepam and paracetamol. *Molecules*, *19*(4), 3869–3882.

Saengtienchai, A., Ikenaka, Y., Bortey-Sam, N., Jermnark, U., Mizukawa, H., Kawai, Y., et al. (2016). The African hedgehog (Atelerix albiventris): Low phase I and phase II metabolism activities. *Comparative Biochemistry and Physiology, Part C: Toxicology & Pharmacology*, *190*, 38–47.

Sevior, D. K., Hokkanen, J., Tolonen, A., Abass, K., Tursas, L., Pelkonen, O., et al. (2010). Rapid screening of commercially available herbal products for the inhibition of major human hepatic cytochrome P450 enzymes using the N-in-one cocktail. *Xenobiotica*, *40*, 2452–2454.

Sewradj, S., Braver, M., Vermeulen, N., Commandeur, J., Richert, L., & Vos, J. (2016). Inter-donor variability of phase I/phase II metabolism of three reference drugs in cryopreserved primary human hepatocytes in suspension and monolayer. *Toxicology In Vitro*, *33*, 71–79.

Shao, X., Ai, N., Xu, D., & Fan, X. (2016). Exploring the interaction between Salvia miltiorrhiza and human serum albumin: Insights from herb drug interaction reports, computational analysis and experimental studies. *Spectrochimica Acta. Part A, Molecular and Biomolecular Spectroscopy*, *161*, 1–7.

Staines, S. (2011). Herbal medicines: Adverse effects and drug-herb interactions. *Journal of the Malta College of Pharmacy Practice*, *17*, 37–42.

Sun, M., Tang, Y., Ding, T., Liu, M., & Wang, X. (2014). Inhibitory effects of celastrol on rat liver cytochrome P450 1A2, 2C11, 2D6, 2E1 and 3A2 activity. *Fitoterapia*, *92*, 1–8.

Sun, H., Yan, Y., Xu, C., Wan, H., & Liu, D. (2016). Suppression of hepatic Cyp1a2 by total ginsenosides in lipopolysaccharide-treated mice and primary mouse hepatocytes. *Journal of Agricultural and Food Chemistry*, *64*(11), 2386–2393.

Tang, J. C., Zhang, J. N., Wu, Y. T., & Li, Z. X. (2006). Effect of the water extract and ethanol extract from traditional Chinese medicines *Angelica sinensis* (Oliv.) Diels, *Ligusticum* chuanxiong Hort. and *Rheum palmatum* L. on rat liver cytochrome P450 activity. *Phytotherapy Research*, *20*, 1046–1051.

Thu, O. K., Nilsen, O. G., & Hellum, B. (2016). In vitro inhibition of cytochrome P-450 activities and quantification of constituents in a selection of commercial Rhodiola rosea products. *Pharmaceutical Biology*, *54*(12), 3249–3256.

Tian, X., Xu, Z., Li, Z., Ma, Y., Lian, S., Guo, X., et al. (2016). Pharmacokinetics of mangiferin and its metabolite-norathyriol, part 2: Influence of UGT, CYP450, P-gp, and enterobacteria and the potential interaction in Rhizoma Anemarrhenae decoction with timosaponin B2 as the major contributor. *Biofactors*, *42*(5), 545–555.

Tsai, H. H., Lin, H. W., Lu, Y. H., Chen, Y. L., & Mahady, G. B. (2013). A review of potential harmful interactions between anticoagulant/antiplatelet agents and Chinese herbal medicines. *PLoS One*, *8*(5), 64255.

Ueng, Y., Don, M., Peng, H., Wang, S., Wang, J., & Chen, C. (2002). Effects of Wu-chu-yu-tang and its component herbs on drug-metabolizing enzymes. *Japanese Journal of Pharmacology*, *89*(3), 267–273.

Ueng, Y., Tsai, T., Don, M., Chen, R., & Chen, T. (2005). Alteration of the pharmacokinetics of theophylline by rutaecarpine, an alkaloid of the medicinal herb Evodia rutaecarpa, in rats. *Journal of Pharmacy and Pharmacology*, *57*(2), 227–232.

Ueng, Y., Tsai, C., Lo, W., & Yun, C. (2010). Induction of hepatic cytochrome P450s by the herbal medicine Sophora flavescens extract in rats: Impact on the elimination of theophylline. *Drug Metabolism and Pharmacokinetics*, *25*(6), 560–567.

Wang, X., Cheung, C., Lee, W., Or, P., & Yeung, J. (2010). Major tanshinones of Danshen (Salvia miltiorrhiza) exhibit different modes of inhibition on human CYP1A2, CYP2C9, CYP2E1 and CYP3A4 activities in vitro. *Phytomedicine*, *17*(11), 868–875.

Wang, X., Hu, X., Chen, G., Yuan, X., Yang, R., Liang, S., et al. (2015). Effects of vitexin on the pharmacokinetics and mRNA expression of CYP isozymes in rats. *Phytotherapy Research*, *29*(3), 366–372.

Wang, Y., Jiang, Y., Wang, Y., Kang, J., Yu, T., Zhao, H., et al. (2014). Inhibiton of cytochrome P450 isoenzymes and P-gp activity by multiple extracts of Huang-Lian-Jie-Du decoction. *Journal of Ethnopharmacology*, *156*, 175–181.

Wang, X., Lee, W. Y. W., Zhou, X., Or, P. M. Y., & Yeung, J. H. K. (2010). A pharmacodynamic–pharmacokinetic (PD–PK) study on the effects of Danshen (Salvia miltiorrhiza) on midazolam, a model CYP3A probe substrate, in the rat. *Phytomedicine*, *17*(11), 876–883.

Wang, L., & Sweet, D. H. (2013). Competitive inhibition of human organic anion transporters 1 (SLC22A6), 3 (SLC22A8) and 4 (SLC22A11) by major components of the medicinal herb Salvia miltiorrhiza (Danshen). *Drug Metabolism and Pharmacokinetics*, *28*, 220–228.

Wang, W., Tian, D., Zheng, B., Wang, D., Tan, Q., Wang, C., et al. (2015). Peony-glycyrrhiza decoction, an herbal preparation, inhibits clozapine metabolism via cytochrome P450s, but not flavin-containing monooxygenase in in vitro models. *Drug Metabolism and Disposition*, *43*(7), 1147–1153.

Wang, J., van der Heijden, R., Spruit, S., Hankermeier, T., Chan, K., van der Greef, J., et al. (2009). Quality and safety of Chinese herbal medicines guided by a systems biology perspective. *Journal of Ethnopharmacology, 126*(1), 31–41.

Wang, B., Yang, S., Hu, J., & Li, Y. (2014). Multifaceted interaction of the traditional Chinese medicinal herb Schisandra chinensis with cytochrome P450-mediated drug metabolism in rats. *Journal of Ethnopharmacology, 155*(3), 1473–1482.

Wang, X., & Yeung, J. (2012). Investigation of cytochrome P450 1A2 and 3A inhibitory properties of Danshen tincture. *Phytomedicine, 19*(3–4), 348–354.

Wang, X., Zhao, X., Li, D., Lou, Y., Lin, Z., & Zhang, G. (2007). Effects of Ganoderma lucidum polysaccharide on CYP2E1, CYP1A2 and CYP3A activities in BCG-immune hepatic injury in rats. *Biological & Pharmaceutical Bulletin, 30*(9), 1702–1706.

Wanwomolruk, S., & Prachayasittikul, V. (2014). Cytochrome P450 enzyme mediated herbal drug interactions (Part 1). *EXCLI Journal, 13*, 347–391.

Wu, T., Li, X. P., Xu, Y. J., Du, G., & Liu, D. (2013). Ursodeoxycholic acid pretreatment reduces oral bioavailability of the multiple drug resistance-associated protein 2 substrate baicalin in rats. *Planta Medica, 79*(17), 1615–1619.

Wu, X., Ma, J., Ye, Y., & Lin, G. (2016). Transporter modulation by Chinese herbal medicines and its mediated pharmacokinetic herb–drug interactions. *Journal of Chromatography B, 1026*, 236–253.

Yaheya, M., & Isamail, M. (2009). Herb-drug interactions and patient counselling. *International Journal of Pharmacy and Pharmaceutical Sciences, 1*(1), 151–161.

Yamasaki, K., Maruyama, T., Kragh-Hansen, U., & Otagiri, M. (1996). Characterization of site I on human serum albumin: Concept about the structure of a drug binding site. *Biochimica et Biophysica Acta, 1295*, 147–157.

Yan, J., He, X., Feng, S., Zhai, Y., Ma, Y., Liang, S., et al. (2014). Up-regulation on cytochromes P450 in rat mediated by total alkaloid extract from Corydalis yanhusuo. *BMC Complementary and Alternative Medicine, 14*(1), 306–314.

Yang, X. X., Hu, Z. P., Duan, W., Zhu, Y. Z., & Zhou, S. F. (2006). Drug-herb interactions: Eliminating toxicity with hard drug design. *Current Pharmaceutical Design, 12*(35), 4649–4664.

Yang, X., Liao, M., Yang, Z., Guo, J., & Gao, O. (2009). Effect of Panax notoginseng on genes expression of CYP and GST in liver tissues of rats. *Zhongguo Zhong Yao Za Zhi, 34*(18), 2390–2393.

Yao, H., Chang, Y., Uramaru, N., Watanabe, Y., Kitamura, S., Kuo, Y., et al. (2012). Effects of Bu-Zhong-Yi-Qi-Tang on hepatic drug-metabolizing enzymes and plasma tolbutamide concentration in rats. *Journal of Ethnopharmacology, 142*(1), 121–128.

Ye, L., He, X., Kong, L., Liao, Y., Pan, R., Xiao, B., et al. (2014). Identification and characterization of potent CYP2D6 inhibitors in lotus leaves. *Journal of Ethnopharmacology, 153*(1), 190–196.

Yi, S., Cho, J., Lim, K., Kim, K., Kim, J., Kim, B., et al. (2009). Effects of Angelicae tenuissima radix, Angelicae dahuricae radix and Scutellariae radix extracts on cytochrome P450 activities in healthy volunteers. *Basic & Clinical Pharmacology & Toxicology, 105*(4), 249–256.

Yin, O., Tomlinson, B., Waye, M., Chow, A., & Chow, M. (2004). Pharmacogenetics and herb–drug interactions: Experience with Ginkgo biloba and omeprazole. *Pharmacogenetics, 14*, 841–850.

Yu, C., Chan, J., & Sanderson, J. (1997). Chinese herbs and warfarin potentiation by danshen. *Journal of Internal Medicine, 241*, 337–339.

Yu, T., Chen, X., Wang, Y., Zhao, R., & Mao, S. (2014). Modulatory effects of extracts of vinegar-baked Radix Bupleuri and saikosaponins on the activity of cytochrome P450 enzymes in vitro. *Xenobiotica, 44*(10), 861–867.

Zanger, U., & Schwab, M. (2013). Cytochrome P450 enzymes in drug metabolism: Regulation of gene expression, enzyme activities, and impact of genetic variation. *Pharmacology & Therapeutics*, *138*(1), 103–141.

Zeng, C., He, F., Xia, C., Zhang, H., & Xiong, Y. (2013). Identification of the active components in Shenmai injection that differentially affect Cyp3a4-mediated 1′-hydroxylation and 4-hydroxylation of midazolam. *Drug Metabolism and Disposition*, *41*(4), 785–790.

Zhang, F., He, X., Zhai, Y., He, L., Zhang, S., Wang, L., et al. (2015). Mechanism-based inhibition of CYPs and RMs-induced hepatoxicity by rutaecarpine. *Xenobiotica*, *45*(11), 978–989.

Zhang, Z.-J., Kang, W.-H., Li, Q., & Tan, Q.-R. (2007). The beneficial effects of the herbal medicine free and easy wanderer plus (FEWP) for mood disorders: Double-blind, placebo-controlled studies. *Journal of Psychiatric Research*, *41*(10), 828–836.

Zhang, H., Ma, K., Liu, W., Yang, F., Liu, J., & Zhou, H. (2016). Impact of CYP2C19 gene polymorphism on warfarin maintenance doses in patients with non-valvular atrial fibrillation. *Gene*, *591*(1), 80–84.

Zhang, J., Onakpoya, I., Posadzki, P., & Eddouks, M. (2015). The safety of herbal medicine: From prejudice to evidence. *Evidence-Based Complementary and Alternative Medicine*, *2015*, 1–3.

Zhang, Y., Zhang, Y., Guo, Y., Li, W., & Yu, C. (2013). Astragaloside IV inhibited the activity of CYP1A2 in liver microsomes and influenced theophylline pharmacokinetics in rats. *Journal of Pharmacy and Pharmacology*, *65*(1), 149–155.

Zhao, Y., Hellum, B., Liang, A., & Nilsen, O. (2015). Inhibitory mechanisms of human CYPs by three alkaloids isolated from traditional Chinese herbs. *Phytotherapy Research*, *29*(6), 825–834.

Zhao, Y., Liang, A., Zhang, Y., Li, C., Yi, Y., & Nilsen, O. (2016). Impact of tetrahydropalmatine on the pharmacokinetics of probe drugs for CYP1A2, 2D6 and 3A isoenzymes in beagle dogs. *Phytotherapy Research*, *30*(6), 906–914.

Zhou, X., Wang, Y., Hu, T., Or, P., Wong, J., Kwan, Y., et al. (2013). Enzyme kinetic and molecular docking studies for the inhibitions of miltirone on major human cytochrome P450 isozymes. *Phytomedicine*, *20*(3–4), 367–374.

Zhou, Q., Ye, Z., Ruan, Z., & Zeng, S. (2013). Investigation on modulation of human P-gp by multiple doses of Radix Astragali extract granules using fexofenadine as a phenotyping probe. *Journal of Ethnopharmacology*, *146*(3), 744–749.

Zhu, M., Wong, P., & Li, R. (1999a). Effect of oral administration of fennel (Foeniculum vulgare) on ciprofloxacin absorption and disposition in the rat. *The Journal of Pharmacy and Pharmacology*, *51*(12), 1391–1396.

Zhu, M., Wong, P. Y., & Li, R. C. (1999b). Effects of Taraxacum mongolicum on the bioavailability and disposition of ciprofloxacin in rats. *Journal of Pharmaceutical Sciences*, *88*(6), 632–634.

CHAPTER TEN

# Treatment of Chinese Herbal Medicine for Female Infertility

**Dan Jiang*[,1], Lily Li[†], Bai-Yun Zeng[‡]**
*Hallam Institution of TCM in Sheffield UK, Sheffield, United Kingdom
[†]St. Mary's Hospital Paddington, London, United Kingdom
[‡]Neurodegenerative Disease Research Group, Institute of Pharmaceutical Science, Faculty of Life Sciences & Medicine, King's College, London, United Kingdom
[1]Corresponding author: e-mail address: djiang52@hotmail.com

## Contents



## Abstract

Female infertility is when a woman of reproductive age and sexual active, without contraception, cannot get pregnant after a year and more or keeps having miscarriages. Although conventional treatments for infertility such as hormone therapy, in vitro fertilization and many more, helped many female patients with infertility get pregnant during past a few decades, it is far from satisfactory with prolonging treatment time frames and emotional and financial burden. In recent years, more patients with infertile problems are seeking to alternative and complementary medicines to achieve a better outcome. In particular, Chinese herbal medicine (CHM) is increasingly popular for treating infertility due to its effectiveness and complimentary with conventional treatments. However, the mechanisms of action of CHM in treating female infertility are not well

*International Review of Neurobiology*, Volume 135
ISSN 0074-7742
http://dx.doi.org/10.1016/bs.irn.2017.02.011

understood. In this chapter authors reviewed research development of CHM applied in many infertile models and CHM clinical studies in many conditions associated with female infertility, published in past 15 years. The data of review showed that CHM has either specific target mechanisms of action or multitarget mechanisms of action, via regulating relevant hormone levels in female reproductive system, improving ovary function, enhancing uterine receptivity. More studies are warranted to explore the new drugs from CHM and ensure safety, efficacy, and consistency of CHM.

## 1. INTRODUCTION

Infertility in general, is defined as not being able to get pregnant (conceive) after 1 year of unprotected sex (NICE Clinical Guideline, 2013; WHO Infertility, 2013). Fertility decreases with age. Estimated infertility from 782 couples recruited from seven European centres is 8% for women aged 19–26 years, 13%–14% for women aged 27–34 years, and 18% for women aged 35–39 years (Dunson, 2004). Common reasons causing female infertility are tubal blockage, pelvic inflammatory diseases caused by infections such as tuberculosis, uterine problems, previous tubal ligation, endometriosis, and advanced maternal age (Corbett & Morin-Papunen, 2013; Matzuk & Lamb, 2008; Vercellini, Viganò, Somigliana, & Fedele, 2014). Up to 20% of infertile couples have unexplained infertility. In these cases, abnormalities are likely to be present but not detected by current methods, or only functional disorders are apparent (Gurunath, Pandian, Anderson, & Bhattacharya, 2011; Pandian, Gibreel, & Bhattacharya, 2015). Conventional treatments of infertility generally involve the use of fertility medication, e.g., follicle-stimulating hormone (FSH), human chorionic gonadotropin, gonadotropin-releasing hormone analogues, ovarian stimulating medication such as clomiphene citrate, human menopausal gonadotropin, and many more (Williams, Mortada, & Porter, 2016), and surgery including tuboplasty, salpingectomy, oophorectomy, unilateral or bilateral salpingo-oophorectomy (Lang, Dunaway, & Roniger, 1990; Sotrel, 2009), or assisted reproductive technologies such as in vitro fertilization (IVF) and related techniques (Pandian et al., 2015). These therapies can treat some causes of infertility, but results are not always satisfactory. Although the overall live birth rate per embryo transfer has increased from 19.2% in 2002 to 23.3% in 2013 (Chambers et al., 2016), the result still falls short of the expectations of the infertile woman/couple. Furthermore, side effects from hormonal treatment are not without risk, such as over stimulated ovaries (Rashidi,

Najmi, & Mobasseri, 2015) or psychiatric disorders (Freeman, Toth, & Cohen, 2013).

Due to the lack of a satisfactory outcome from conventional medical treatments for many women with infertility, they are increasingly seeking to alternative therapies to improve their results. During past 15 years, Chinese herbal medicine (CHM), as a holistic approach to many conditions, has been increasingly used to treat infertility in many Western countries (Coulson & Jenkins, 2005; Smith et al., 2010; Stankiewicz, Smith, Alvino, & Norman, 2007). Further, CHM treatment increased fertility rate and played an important role in supporting conventional infertility treatments (Ried, 2015; Ried & Stuart, 2011; Zhou & Qu, 2009). However, the mechanisms of action of CHM in treating infertility are not well understood. In this chapter we reviewed recent research development of CHM treatment with female infertility from both basic research and clinical studies, in particular, focusing on the underlying mechanisms of action of CHM.

## 2. EFFECT OF CHM ON FEMALE INFERTILITY MODELS

There are many studies which aim to elucidate the mechanism of CHM in treating female infertility in animal models of common diseases associated with infertility.

### 2.1 Promoting Endometrial Receptivity in the Uterus

It is believed that a receptive uterine state is essential for successful conceive either naturally or using an assisted pregnancy technique. Poor endometrial receptivity is one of the most common reasons for infertility and prevents embryo implantation into the uterus. Many studies showed that CHM promoted a receptive uterine state (Guo, Wang, & Li, 2011). In a mouse model of infertility, effect of extracts from a CHM remedy (Zhuyun recipe, 助孕) on improving endometrial receptivity in mice with embryonic implantation dysfunction and ovulation stimulation was conducted (Yu, Yang, & Liang, 2011). Female pregnant kunming mice were randomly divided into six groups: A control group; B ovulation stimulation (OS) group; C OS + CHM formula group; D embryo implantation dysfunction (EID) group; E, EID + CHM formula; and F CHM-only group. Uterine samples were collected at gestation day 4 and detected with immunohistochemistry and real-time PCR analyses for expression of endometrial leukemia inhibitory factor (LIF) and integrin β3 subunit, markers of endometrial receptivity (Bhatt, Brunet, & Stewart, 1991; Illera et al., 2000). It was found, at the end of study, that OS group and EID group showed a significant decrease in pregnancy rate and the lower level expression of both endometrial LIF and integrin β3 subunit during the implantation window compared with control group. The OS + CHM formula and EID + CHM formula groups showed a higher pregnant rate and higher level of endometrial LIF and integrin β3 subunit expression compared to the OS and EID groups. The number of implanted embryos in EID group was lower than in the control group; in contrast, the number of implanted embryos in EID + CHM formula group was significantly higher than in the EID group and control group. However, no significant difference was found in the measured indices between the CHM-only group and control group. The studies showed that CHM helped reverse the expression of endometrial LIF and integrin β3 subunit, improved the uterine receptivity in mice, and increased pregnancy rate and the number of embryonic implantations (Yu et al., 2011). In another study, Bu Shen Huo Xue Decoction (BSHXD, 补肾活血饮), prescribed to aid in preparing the endometrium for implantation in China, was recently investigated for its mechanism of actions in a rat model with controlled ovarian hyperstimulation (COH) treatment (Gong, Lou, Lu, Huang, & Jin, 2015). BSHXD treatment markedly increased the number of live births and enhanced the implantation capacity in the rat model by restoring protein expression of endometrial LIF, and without affecting expression of Ang-2 protein, which tends to destabilizing quiescent endothelium (Fiedler et al., 2006), compared with COH alone group (Gong et al., 2015). This suggested that BSHXD treatment could help patients with an impaired endometrium following COH treatment.

## 2.2 Improving Ovarian Function

Premature ovarian failure (POF) and insufficient storage function of the ovaries are both common reasons for female infertility and result from a variety of hormonal disorders (Matzuk & Lamb, 2008). Established formulas of CHM have been found to effectively stimulate ovaries and raise the low hormone level. Li and Luo (2010) used rat model where ovulation was inhibited by hydroxyl urea, which matched kidney deficiency and dampness accumulation patterns in TCM theory. They showed that Luo's established recipe (罗氏促排卵汤) increased ovulation, modified serum levels of FSH, LH, and estrogen, and promoted pregnancy rate when compared to the control groups (Li & Luo, 2010). In another study, Fu et al. (2012) investigated the underlying mechanism of therapeutic effect of CHM formula Tongmai

Danshen Tablet (TMDST, 通脉丹参片) in tripterygium glycosides-induced POF rat model. They found that TMDST treatment significantly restored normal estrous cycle and promoted near or above normal ovarian index such as, serum estradiol and, serum inhibin B level in comparison with those of normal controls. Further, TMDST significantly elevated expression of estrogen receptor, vascular endothelial growth factor (VEGFR), and VEGFR-2, markedly lowered intracellular TNF-α and Caspase-3, improved ovarian vascular structure, thinner vascular wall, and larger vascular lumen compared with control groups (Fu et al., 2012). This suggested that therapeutic effect of TMDST in treating POF was through improving angiogenesis and antiapoptosis pathway.

## 2.3 Preventing Spontaneous Abortion

Spontaneous abortion also known as miscarriage is a common complication of pregnancy that affected about 5% of child-bearing women (Rai & Regan, 2006). CHM such as Zhuyun No. 3, an established formula for preventing miscarriage, showed that it could help prevent miscarriage. Study of mechanism behind Zhuyun No. 3 (ZY3, 助孕3号) was conducted in rat models of abortion using hydroxyurea and mifepristone, to mimic kidney deficiency and spleen deficiency, respectively (according to TCM theory), to understand the underlying mechanisms of ZY3 (Luo et al., 2003). It was found that ZY3 formula significantly increased the expression of progesterone receptors in deciduas of models of both types compared with respective control groups. The water extract and drug serum of ZY3 lowered the potency of uterus contractility in vivo and in vitro isolated uterine smooth muscle, respectively (Luo et al., 2003). The study showed that by reinforcing kidney and strengthening Pi, ZY3 protected pregnancy via elevating the target cell progesterone receptors and relaxing the uterine smooth muscle.

## 2.4 Reregulating Luteum Inhibition and Immune Function

Transforming growth factor beta 1 (TGF-β1) plays the important role in promoting the embryo embedding at the surface of uterus via modulating immune tolerance (Ayatollahi, Geramizadeh, & Samsami, 2005). Alterations in TGF-β1 were linked to pregnancy failure. Zhao, Luo, Ni, and Zhou (2008) studied the effect of CHM formula Zhuyun No. 3 (ZY3, 助孕3号) on serum TGF-β1 in luteum inhibition abortion rat model and explored its mechanisms in preventing abortion from the perspective of immune regulation. The rat model was produced by feeding pregnancy rats with hydroxyurea for 7 days, then feeding with mifepristone for 1 day. Results showed that treatment with Zhuyun 3 dose-dependently increased in the serum levels of TGF-β1 compared with control groups (Zhao et al., 2008), indicating that Zhuyun 3 participated immune regulation via modifying TGF-β1expression which might lead to prevent abortion.

Taken together, CHM formulas mentioned earlier increased fertility rates in the infertility models by improving uterine embedding conditions and rebalance hormonal levels in the uterus.

# 3. CHM INCREASES CLINICAL PREGNANCY RATE

## 3.1 Improving Fallopian Tube Obstruction

Blocked fallopian tubes are unable to let the ovum and sperm converge, thus making fertilization impossible. Study of CHM formula for fallopian tube obstruction was conducted (Jiang, Liang, & Liu, 2006) in which patients with salpingostomy were treated with CHM formula Bu Shen Huo Xue decoction (BSHXD, 补肾活血饮), starting from the 5th day of menstrual cycle for 14 days, with salpingostomy alone as a control. Three-month treatment was taken as one therapeutic course and total treatment lasted for four courses. It was demonstrated at the end of treatment that the condition of follicular development, thickness of endometrium, and level of serum estradiol in the preovulatory phase were all markedly improved in both BSHXD plus salpingostomy and salpingostomy alone groups compared with respective baseline evaluation. However, the improvement was significantly superior in BSHXD plus salpingostomy group compared with that in salpingostomy group. Moreover, the pregnancy rate was also higher in the former than in the latter (Jiang et al., 2006). An earlier study with fallopian tube obstruction reported that the fallopian tube patency rate was 86.7% and the pregnant rate 85.0% in the CHM-Western medicine (WM) group, respectively, while 66.7% and 63.3% in the CHM alone group, respectively, and 53.3% and 50.0% in the WM alone group, respectively, in comparison with their respective baseline assessments. The levels of serum C-reactive protein and interleukin-1 beta were all lowered after three courses of treatment, and the effect was more evident in the CHM-WM group, indicating rebalance of immune system might be one of factors enhancing pregnant rate (Kang, Xia, & He, 2001). However, there was no mention about the names and treatment regimes of either CHM formula or Western medicine.

## 3.2 Improving Ovarian Function

Lin'erlai (LEL, 琳儿来) prescription, an established CHM recipe for treating infertility, was used to treat patients with anovulatory infertility, in comparison with patients control group treated with clomiphene, a nonsteroidal fertility medicine that stimulates ovulation (Huang, 2007). The total effective rate in the LEL group was 96.7%, which was significantly higher than 53.3% in the control group. The abortion rate was 10.0% in the LEL group, which was significantly lower than 54.6% in the control group. The symptom scores decreased significantly in the LEL group but remained unchanged in control group. The study demonstrated that LEL had a better effect than clomiphene in treating anovulatory infertility (Huang, 2007). Further, Tan et al. (2012) reported a meta-analysis where 1659 participants were involved in 15 studies showing the efficacy of CHM treatment of infertility caused by anovulation as compared to clomiphene treatment. Analysis data indicated that CHM significantly improved the cervical mucus score, increased the ovulation rate leading to higher the pregnancy rate, and reduced the miscarriage rate compared to clomiphene. No significant adverse effects were identified from the use of CHM in this review (Tan et al., 2012). Similarly, a Cochrane review found that the addition of CHM to clomiphene is associated with improved clinical pregnancy outcomes compared with clomiphene alone (Zhang et al., 2010).

Effect of CHM formula on improving ovulation failure and on incidence rate of luteinized unruptured follicle syndrome (LUF) and luteal phase defect (LPD), by tonifying kidney and activating blood (according to TCM theory), was assessed in patients with ovulation disorder (Zhou & Li, 2006). CHM formula (for details, see Table 1) was adjusted at different stages of menstrual cycle. At the end of treatment, it was revealed that CHM formula treatment significantly increased ovulation rate, markedly decreased the occurrence rates of LUF and LPD, raised the conception rate compared with control group (Zhou & Li, 2006). Similar therapy, using several of herbal formulas at different stages of the menstrual cycle, according to the cyclical hormonal changes, improved ovarian function and enhanced pregnant rate in patients with infertility in Western countries (Jiang, 2014).

## 3.3 Improving Polycystic Ovary Syndrome

Polycystic ovary syndrome (PCOS) is one of the most common causes of female infertility, resulting in abnormal follicular development and ovulation dysfunction partially due to imbalance of steroid hormone releasing

**Table 1** Details of CHM Formulas Administered in Different Stages of Menstrual Cycle in Zhou and Li (2006)

| Periodic Stage | Treating Principle | Chinese Herbal Formula Elements |
|---|---|---|
| During menstruation | To move Qi's stagnation and activity blood for supporting follicle's growth | Cangzhu/Xanthium Sibiricum part |
| | | Xiangfu/Cyperus Rotundus L. |
| | | Mudanpi/Paeonia Suffruticosa Andr. |
| | | Danshen/*Salviae miltiorrhizae* Bge. |
| | | Chishao/Paeonia Lactiflora Pall. |
| | | Wulingzhi |
| | | Zelan/Lycopus Lucidus Turcz var. |
| | | Hirtus Regel |
| | | Zishiying/ |
| | | Shanzha/Crataegus Pinnatifida Bge. |
| | | Yimucao/Leonurus Japonicus Hoult |
| | | Fuling/Poria Cocos (Schw.) Wolf |
| Post of menstruation | To nourish Yin and blood for promoting essence (egg)'s growing up and bursting out | Danggui/Angelica Sinensis (Oliv.) Diels |
| | | Chishao/Paeonia Lactiflora Pall. |
| | | Shudihuang/Radix *Glutinosa Libosch* |
| | | Mudanpi/Paeonia Suffruticosa Andr. |
| | | Fuling/Poria Cocos (Schw.) Wolf |
| | | Shanyao/Dioscorea Opposita Thunb. |
| | | Shanyurou/Cornus Officinalis Sieb et. |
| | | Zucc |
| | | Zhizi/Gardenia Jasminoides Ellis |
| | | Chuanduan/Dipsacus Asperoides C.Y. |
| | | Cheng et |
| | | Tusizi/Cuscuta Glabra Roxb |
| | | Fupenzi/Rubus Chingi Hu |
| Ovulation | To tonify kidney Yang and to nourish Yin essence and blood for preparing a pregnancy | Danggui/Angelica Sinensis (Oliv.) Diels |
| | | Danshen/*Salviae miltiorrhizae* Bge. |
| | | Chishao/Paeonia Lactiflora Pall. |
| | | Zelan/Lycopus Lucidus Turcz var. |

**Table 1** Details of CHM Formulas Administered in Different Stages of Menstrual Cycle in Zhou and Li (2006)—cont'd

| Periodic Stage | Treating Principle | Chinese Herbal Formula Elements |
|---|---|---|
| | | Hirtus Regel |
| | | Chongweizi/Leonurus Heterophyllus |
| | | Sweet |
| | | Honghua/Carpesium Tinctorins L. |
| | | Xiangfu/Cyperus Rotundus L. |
| Before of menstruation | To move liver Qi stagnation and nourish blood for feeding embryo and preventing a miscarriage | Chaihu/Bupleurum Chinense DC. |
| | | Xiangfu/Cyperus Rotundus L. |
| | | Aiye/Artemisisia Argyi Levl. et Vant. |
| | | Xuduan/Dipsacus Asperoides C.Y. |
| | | Cheng et |
| | | Duzhong/Eucommia Ulmoides Oliv. |
| | | Ejiao/Asini Gelatinum, Comi Asini Colla |
| | | Sangjisheng/Taxillus Chinensis (DC.) Danser |

(Corbett & Morin-Papunen, 2013; Williams et al., 2016). Recently, some CHM formulas showed beneficial effect on PCOS. For example, CHM formula Erzhi Tiangui Granule (ETG, 二至天癸颗粒) was found to significantly improve oocyte quality of patients with PCOS, via regulating the anti-Müllerian hormone levels in the serum and in the follicular fluid, adjusting the androgen level, improving the pathophysiological changes of PCOS, and activating the ovarian microenvironment (Lian & Zhao, 2012). Another CHM formula Bushen Huatan formula (BHF, 补肾化痰汤) was reported to ameliorate insulin resistance and also to improve the function of ovarian ovulation in patients with PCOS (Lee et al., 2006). In another report, a metabolomics approach was used to investigate the mechanism of action of CHM formula BHF on improving PCOS (Lu et al., 2016). Following BHF intervention for three menstrual cycles, body weight, Body Mass Index (BMI), waist circumference, and Hip circumference of all patients with PCOS significantly decreased compared to baseline assessment. Then, patients were divided into normoinsulinemic (NI) group and hyperinsulinemic (HI) group for serum metabolic studies. There were significant changes in serum metabolic following treatment compared with baseline in

NI group and HI group, respectively. In NI group, significant phospholipid metabolism improvement was observed. In HI group, changes in amino acid metabolism were observed (Lu et al., 2016). The results suggested that BHF was effective in treating patients with PCOS by reduced inflammatory reaction and oxidative stress in both NI group and HI group.

## 3.4 Improving Mycoplasma Infection

Female mycoplasma genitalium infection can cause infertility. Relevant tests showed an increased in cytokines (IL-1beta, IL-2, TNF-α) in the peripheral blood and in the cervical mucus, and positive culture of mycoplasma from genital tracts of infertile women with mycoplasma infection (Lis, Rowhani-Rahbar, & Manhart, 2015). Recently, a randomized study was conducted to evaluate antimycoplasma effect of CHM formula Xiaozhi decoction in 72 patients with positive genital mycoplasma cultures (Lou, Yang, & Liu, 2012). The patients were randomly divided into a CHM group and a Western medicine group. The TCM group was treated with Xiaozhi decoction twice every day for 6 weeks. The Western medicine group was treated with azithromycin for 3 days consecutively and underwent six courses of treatment. At the end of treatment, CHM group showed a significant decrease in serum cytokines such as IL-1beta, IL-2, TNF-α, and in the cervical mucus compared with baseline and with end value of Western medicine group. This suggested that Xiaozhi decoction could be used to treat infertile women with mycoplasma infection (Lou et al., 2012).

In a series of studies of CHM treatment for female infertility, Ried et al. first (Ried & Stuart, 2011) identified 8 randomized controlled trials (RCTs), 13 cohort studies, 3 case series, and 6 case studies involving 1851 women with infertility. Meta-analysis of the RCTs suggested a 3.5 greater likelihood of achieving pregnancy with CHM therapy over a 4-month period compared with clomiphene therapy alone. They suggested that management of female infertility with CHM could improve pregnancy rates twofold compared with clomiphene alone (Ried & Stuart, 2011). Then, they found that the quality of the menstrual cycle, integral to TCM diagnosis, appears to be fundamental to successful treatment of female infertility (Ried & Alfred, 2013). Recently, the same group published an updated review of the same topic, where they collected data from 40 RCTs involving 4247 women with infertility into a systematic review. Meta-analysis study suggested a 1.74 higher probability of achieving pregnancy with CHM therapy than with clomiphene alone. In addition, the authors concluded that fertility indicators such as biphasic basal body temperature, ovulation rates, cervical mucus score, and appropriate thickness of the endometrial lining are positively influenced by CHM therapy, implicating an alleviating physiological effect for viable pregnancy (Ried, 2015).

# 4. ENHANCING THE SUCCESS RATE OF IVF AND EMBRYO TRANSFER

## 4.1 Promoting Uterus Receptivity

ETG (二至天癸颗粒) not only was used to treat PCOS, improving ovarian function, but also used to improve efficacy of in vitro fertilization–embryo transfer (IVF-ET). In a series studies, Lian, Teng, and Zhang (2007) and Lian et al. (2013) assessed the effect and mechanisms of ETG on infertile women who undergo IVF-ET. First, in a randomized study, ETG treatment distinctively increased the amount of oocyte, elevated the quality of embryo, and raised the successful rate of IVF-ET, and this was correlated to the increase of leukemia inhibitory factor levels in follicular fluid compared with Western drug group (Lian et al., 2007). Then, a randomized, double-blinded, and placebo-controlled clinical trial was conducted to explore the effects of ETG on DNA methyltransferases (DNMT) 1 protein expression in endometrium of infertile women who were to undergo IVF-ET. DNMT 1protein has regulatory effects on genes associated with shedding of endometrium in the menstrual cycles and was decreased in patients with infertility (Yamagata et al., 2009). Treatment with ETG significantly improved syndrome scores, decreased dosage, and duration of gonadotropin compared with placebo treatment group. Further, the high-quality oocyte and embryo rates, and clinical pregnancy rate were all higher in the ETG group than those in placebo group. The DNMT1 protein expression in the endometrium was much more abundant in the ETG group than that in the placebo group, which might lead to enhanced endometrial receptivity and improvement in clinical pregnancy rate (Lian et al., 2013).

## 4.2 Improving Pregnant Rate in Patients With Failure of IVF Treatment

Recently, a study exploring the effect of traditional Chinese comprehensive therapy (TCCT), consisting of both Chinese herbs and ear acupuncture, on promoting gestation in patients with previously failure IVF-ET was conducted (Xu et al., 2015). Sixty-seven patients were enrolled in this study and divided into two groups: a treatment group with 35 patients and a control group with 32 patients. The treatment group was given TCCT for 3 months, and then administered IVF-ET or natural pregnancy was awaited. The control group was administered IVF-ET without TCCT 3 months after the previous IVF-ET or natural pregnancy attempt. The natural pregnancy rate of the two groups was calculated after treatment. After treatment with TCCT, seven patients in the treatment group became pregnant, while there were no successful conceptions in the control group, showing that TCCT treatment group had a significantly higher conception rate (20%) than the control group (0%). The study suggested that TCCT could increase the number of fertilized eggs, the fertilization rate, pregnancy rate, and clinical pregnancy rate after another IVF-ET treatment and could promote natural pregnancy rate in patients with previously failed IVF-ET (Xu et al., 2015). A systematic review and meta-analysis of RCTs were conducted to evaluate the effectiveness of CHM with concurrent IVF vs IVF alone on the outcomes of IVF (Cao et al., 2013). Twenty trials involving 1721 women were included in the study, and three trials were evaluated as having an unclear risk of bias. The remaining trials were evaluated as having a high risk of bias. Despite the consideration the authors found that the combination of CHM and IVF significantly increased clinical pregnancy rates and ongoing pregnancy rates. The study indicated that combination of IVF and CHM in the included trials improved IVF success, however, due to the high risk of bias observed with the trials; the significant differences found were felt to be likely inaccurate. The authors also mentioned that no conclusion could be made with respect to the reproductive toxicity of CHM (Cao et al., 2013).

## 5. CONCLUSION

The quality of the environment in the body is considered as important as the quality of eggs, sperm, and embryo for a successful pregnancy. The fundamental principle of TCM therapy, in particular CHM, is to restore balance within the body, which affects hormonal regulation of the menstrual cycle, and provides a physiological environment to facilitate conception, implantation, and maintenance of a viable pregnancy. Studies reviewed here showed CHM helped infertile female patients in many ways to achieve a successful pregnancy. Further, CHM can complement with Western medicine and fertility treatment to improve pregnancy rates while reducing treatment time frames and emotional and financial burden. Although CHM in particular its formulas has been increasingly used to treat female infertility, more studies concerning its efficacy, consistency, and safety are still needed.

## ACKNOWLEDGMENT

This work was partly supported by ATCM (The Association of Traditional Chinese Medicine and Acupuncture UK). Authors thank for the IT support from King's College London.

## REFERENCES

Ayatollahi, M., Geramizadeh, B., & Samsami, A. (2005). Transforming growth factor beta-1 influence on fetal allografts during pregnancy. *Transplantation Proceedings, 37*, 4603–4604.

Bhatt, H., Brunet, L. J., & Stewart, C. L. (1991). Uterine expression of leukemia inhibitory factor coincides with the onset of blastocyst implantation. *Proceedings of the National Academy of Sciences of United States of America, 88*, 11408–11412.

Cao, H., Han, M., Ng, E. H., Wu, X., Flower, A., Lewith, G., et al. (2013). Can Chinese herbal medicine improve outcomes of in vitro fertilization? A systematic review and meta-analysis of randomized controlled trials. *PloS One, 12*, e81650.

Chambers, G. M., Wand, H., Macaldowie, A., Chapman, M. G., Farquhar, C. M., Bowman, M., et al. (2016). Population trends and live birth rates associated with common ART treatment strategies. *Human Reproduction, 31*, 2632–2641.

Corbett, S., & Morin-Papunen, L. (2013). The polycystic ovary syndrome and recent human evolution. *Molecular and Cellular Endocrinology, 373*, 39–50.

Coulson, C., & Jenkins, J. (2005). Complementary and alternative medicine utilisation in NHS and private clinic settings: A United Kingdom survey of 400 infertility patients. *Journal of Experimental & Clinical Assisted Reproduction, 2*, 5.

Dunson, D. B. (2004). Increased infertility with age in men and women. *Obstetrics and Gynaecology, 103*, 51–56.

Fiedler, U., Reiss, Y., Scharpfenecker, M., Grunow, V., Koidl, S., Thurston, G., et al. (2006). Angiopoietin-2 sensitizes endothelial cells to TNF-alpha and has a crucial role in the induction of inflammation. *Nature Medicine, 12*, 235–239.

Freeman, M. P., Toth, T. L., & Cohen, L. S. (2013). Assisted reproduction and risk of depressive relapse: Considerations for treatment. *Annals of Clinical Psychiatry, 25*, 283–288.

Fu, Y., Zhao, Z., Wu, Y., Wu, K., Xu, X., Liu, Y., et al. (2012). Therapeutic mechanisms of Tongmai Dasheng tablet on tripterygium glycosides induced rat model for premature ovarian failure. *Journal of Ethnopharmacology, 139*, 26–33.

Gong, X., Lou, J. Y., Lu, Q. D., Huang, H. T., & Jin, Z. (2015). Bu Shen Huo Xue decoction restores endometrial leukemia-inhibitory factor but not Angiopoietin-2 expression, and improves uterine receptivity in the controlled ovarian stimulation rat model. *Experimental and Therapeutic Medicine, 9*, 751–756.

Guo, J., Wang, L. N., & Li, D. (2011). Exploring the effects of Chinese medicine in improving uterine endometrial blood flow for increasing the successful rate of in vitro fertilization and embryo transfer. *Zhong Xi Yi Jie He Xue Bao, 9*, 1301–1306.

Gurunath, S., Pandian, Z., Anderson, R. A., & Bhattacharya, S. (2011). Defining infertility—A systematic review of prevalence studies. *Human Reproduction Update, 17*, 575–588.

Huang, Y. (2007). Clinical observation on anovulatory infertility of blood and essence asthenia type treated by Lin'erlai Prescription. *Chinese Journal of Integrated Traditional and Western Medicine, 27*, 1028–1030.

Illera, M. J., Juan, L., Stewart, C. L., Cullinan, E., Ruman, J., & Lessey, B. A. (2000). Effect of peritoneal fluid from women with endometriosis on implantation in the mouse model. *Fertility and Sterility, 74*, 41–48.

Jiang, D. (2014). Treating female infertility by strengthening the natural menstrual cycle with acupuncture and Chinese herbal medicine. *The European Journal of Oriental Medicine*, *7*, 28–35.

Jiang, Y. C., Liang, R. N., & Liu, R. T. (2006). Effects of combined therapy of salpingostomy and Bushen Huoxue drugs on fallopian tube obstructive infertility. *Chinese Journal of Integrated Traditional and Chinese Medicine*, *26*, 325–328.

Kang, J. L., Xia, W., & He, Q. Y. (2001). Clinical study on treatment of oviduct obstruction by integrative traditional Chinese and western medicine. *Chinese Journal of Integrated Traditional and Western Medicine*, *21*, 416–418.

Lang, E. K., Dunaway, H. E., Jr., & Roniger, W. E. (1990). Selective osteal salpingography and transvaginal catheter dilatation in the diagnosis and treatment of fallopian tube obstruction. *American Journal of Roentgenology*, *154*, 735–740.

Lee, Y. S., Kim, W. S., Kim, K. H., Yoon, M. J., Cho, H. J., & Shen, Y. (2006). Berberine, a natural plant product, activates AMP-activated protein kinase with beneficial metabolic effects in diabetic and insulin-resistant states. *Diabetes*, *55*, 2256–2264.

Li, S., & Luo, S. (2010). Study on kidney tonifying of CHM to the rat models with ovulated inhibition. *Chinese Journal of Basic Medicine in Traditional Chinese Medicine*, *16*, 567–569.

Lian, F., Teng, Y. L., & Zhang, J. W. (2007). Clinical study on effect of Erzhi Tiangui granule in improving the quality of oocytes and leukemia inhibitory factor in follicular fluid of women undergoing in vitro fertilization and embryo transfer. *Chinese Journal of Integrated Traditional and Western Medicine*, *27*, 976–979.

Lian, F., Wang, R. X., Feng-Mei, M., Zhen-Gao, S., Li-Hong, W., & Lei, S. (2013). Effects of Chinese medicines for tonifying the kidney on DNMT1 protein expression in endometrium of infertile women during implantation period. *Journal of Alternative and Complementary Medicine*, *19*, 353–359.

Lian, F., & Zhao, S. (2012). Effects of Shen invigorating and Chong-channel regulating method on anti-Müllerian hormone and oocyte quality in polycystic ovarian syndrome patients. *Chinese Journal of Integrated Traditional and Western Medicine*, *32*, 9–12.

Lis, R., Rowhani-Rahbar, A., & Manhart, L. E. (2015). Mycoplasma genitalium infection and female reproductive tract disease: A meta-analysis. *Clinical Infectious Diseases*, *61*, 418–426.

Lou, Y. F., Yang, X. F., & Liu, L. C. (2012). Study on changes in cytokines of infertile women with mycoplasma infection and intervention with traditional Chinese medicines. *China Journal of Chinese Materia Medica*, *37*, 3158–3160.

Lu, C., Zhao, X., Li, Y., Li, Y., Yuan, C., Xu, F., et al. (2016). Serum metabolomics study of traditional Chinese medicine formula intervention to polycystic ovary syndrome. *Journal of Pharmaceutical and Biomedical Analysis*, *20*, 127–133.

Luo, S. P., Liang, G. Z., Zhang, Y. Z., Xu, L. M., Deng, G. P., & Ouyang, H. Q. (2003). Study of mechanism on Zhuyun No. 3 in preventing and treating spontaneous abortion. *Chinese Journal of Integrated Traditional and Western Medicine*, *23*, 522–525.

Matzuk, M. M., & Lamb, D. J. (2008). The biology of infertility: Research advances and clinical challenges. *Nature Medicine*, *14*, 1197–1213.

NICE Clinical Guideline CG156. (2013). *Fertility: Assessment and treatment for people with fertility problems*. London, United Kingdom: National Institute for Health and Care Excellence.

Pandian, Z., Gibreel, A., & Bhattacharya, S. (2015). In vitro fertilisation for unexplained subfertility. *The Cochrane Database of Systematic Reviews*, *11*, CD003357 http://dx.doi.org/10.1002/14651858.

Rai, R., & Regan, L. (2006). Recurrent miscarriage. *Lancet*, *368*, 601–611.

Rashidi, M., Najmi, Z., & Mobasseri, A. (2015). *Advantages of recombinant follicle-stimulating hormone over human menopausal gonadotropin in intrauterine insemination: A randomized clinical trial in polycystic ovary syndrome-associated infertility*. Gynaecological and Obstetric Investigation. PMID: 26228499.

Ried, K. (2015). Chinese herbal medicine for female infertility: An updated meta-analysis. *Complement Therapy Medicine, 23*, 116–128.

Ried, K., & Alfred, A. (2013). Quality of life, coping strategies and support needs of women seeking Traditional Chinese Medicine for infertility and viable pregnancy in Australia: A mixed methods approach. *BMC Women's Health, 13*, 17–19.

Ried, K., & Stuart, K. (2011). Efficacy of traditional Chinese herbal medicine in the management of female infertility: A systematic review. *Complement Therapy Medicine, 19*, 319–321.

Smith, J. F., Eisenberg, M. L., Millstein, S. G., Nachtigall, R. D., Shindel, A. W., Wing, H., et al. (2010). The use of complementary and alternative fertility treatment in couples seeking fertility care: Data from a prospective cohort in the United States. *Fertility and Sterility, 93*, 2169–2174.

Sotrel, G. (2009). Is surgical repair of the fallopian tubes ever appropriate? *Review Obstetrics and Gynaecology, 2*(3), 176–185.

Stankiewicz, M., Smith, C., Alvino, H., & Norman, R. (2007). The use of complementary medicine and therapies by patients attending a reproductive medicine unit in South Australia: A prospective survey. *Australian and New Zealand Journal of Obstetrics and Gynaecology, 47*, 145–149.

Tan, L., Tong, Y., Sze, S. C., Xu, M., Shi, Y., & Song, X. Y. (2012). Chinese herbal medicine for infertility with anovulation: A systematic review. *Journal of Alternative and Complementary Medicine, 18*, 1087–1100.

Vercellini, P., Viganò, P., Somigliana, E., & Fedele, L. (2014). Endometriosis: Pathogenesis and treatment. *Nature Review Endocrinology, 10*, 261–275.

Williams, T., Mortada, R., & Porter, S. (2016). Diagnosis and treatment of polycystic ovary syndrome. *American Family Physician, 94*, 106–113.

World Health Organization. (2013). *Health topics: Infertility*. Available http://www.who.int/topics/infertility/en/.

Xu, X., Jia, W., Cheng, S., Zhong, Y., Xia, W., & Zeng, Q. (2015). Effect of traditional Chinese comprehensive therapy on gestation in patients with previously failed in vitro fertilization and embryo transfer. *Journal of Traditional Chinese Medicine, 35*, 428–433.

Yamagata, Y., Asada, H., Tamura, I., Lee, L., Maekawa, R., Taniguchi, K., et al. (2009). DNA methyltransferase expression in the human endometrium: Downregulation by progesterone and estrogen. *Human Reproduction, 24*, 1126–1132.

Yu, N., Yang, J., & Liang, T. (2011). Extracts from a traditional Chinese herbal remedy (Zhuyun recipe) improve endometrial receptivity in mice with embryonic implantation dysfunction and ovulation stimulation. *Journal of Ethnopharmacology, 137*, 389–395.

Zhang, J., Li, T., Zhou, L., Tang, L., Xu, L., Wu, T., et al. (2010). Chinese herbal medicine for subfertile women with polycystic ovarian syndrome. *Cochrane Database of Systematic Reviews, 9*, CD007535 http://dx.doi.org/10.1002/14651858.

Zhao, H. Y., Luo, S. P., Ni, Y. C., & Zhou, Y. (2008). Effects of Zhuyun No. 3 serum TGF-β1 in abortion rats of kidney deficiency and luteum inhibition. *Shanghai Journal of Traditional Chinese Medicine, 42*, 71–73.

Zhou, J., & Li, D. J. (2006). Tonifying kidney and activating blood method in treating ovulation failure of 32 cases: B ultrasonic evaluation. *Chinese Journal of Clinical Rehabilitation, 10*, 114–116.

Zhou, J., & Qu, F. (2009). Treating gynaecological disorders with traditional Chinese medicine: A review. *African Journal of Traditional, Complementary, and Alternative Medicines, 6*, 494–517.

CHAPTER ELEVEN

# Phytochemical Modulation of Apoptosis and Autophagy: Strategies to Overcome Chemoresistance in Leukemic Stem Cells in the Bone Marrow Microenvironment

**Helen C. Owen[1,2], Sandra Appiah[1,2], Noor Hasan, Lucy Ghali, Ghada Elayat, Celia Bell**
Faculty of Science and Technology, Middlesex University, The Burroughs, Hendon, London, United Kingdom
[2]Corresponding authors: e-mail addresses: H.Owen@mdx.ac.uk; S.Appiah@mdx.ac.uk

## Contents



[1] Joint first authors.

*International Review of Neurobiology*, Volume 135
ISSN 0074-7742
http://dx.doi.org/10.1016/bs.irn.2017.02.012

## Abstract

Advances in scientific research and targeted treatment regimes have improved survival rates for many cancers over the past few decades. However, for some types of leukemia, including acute lymphoblastic and acute myeloid leukemia, mortality rates have continued to rise, with chemoresistance in leukemic stem cells (LSCs) being a major contributing factor. Most cancer drug therapies act by inducing apoptosis in dividing cells but are ineffective in targeting quiescent LSCs. Niches in the bone marrow, known as leukemic niches, behave as "sanctuaries" where LSCs acquire drug resistance. This review explores the role of the bone marrow environment in the maintenance of LSCs and its contribution to chemoresistance and considers current research on the potential use of phytochemicals to overcome chemoresistance through the modulation of signaling pathways involved in the survival and death of leukemic clonal cells and/or leukemic stem cells. Phytochemicals from traditional Chinese medicine, namely baicalein, chrysin, wogonin (constituents of *Scutellaria baicalensis*; huáng qín; 黄芩), curcumin (a constituent of *Curcuma longa*, jiāng huáng, 姜黄), and resveratrol (a constituent of *Polygonum cuspidatum*; hŭ zhàng, 虎杖) have been shown to induce apoptosis in leukemic cell lines, with curcumin and resveratrol also causing cell death via the induction of autophagy (a nonapoptotic pathway). In order to be effective in eliminating LSCs, it is important to target signaling pathways (such as Wnt/β-catenin, Notch, and Hedgehog). Resveratrol has been reported to induce apoptosis in leukemic cells through the inhibition of the Notch and Sonic hedgehog signaling pathways, therefore showing potential to affect LSCs. While these findings are of interest, there is a lack of reported research on the modulatory effect of phytochemicals on the autophagic cell death pathway in leukemia, and on the signaling pathways involved in the maintenance of LSCs, highlighting the need for further work in these areas.


## 1. INTRODUCTION

Leukemia is the 10th most common cause of cancer death worldwide, resulting in an estimated 265,000 deaths in 2012 (3% of the total cancer deaths) (Cancer Research UK, 2016). While positive advances in treatment for some of the more common cancers such as breast, lung, and prostate cancer have resulted in a sharp decline in the death rate since the 1970s, this trend has not been seen in all types of leukemia. This is largely attributed to chemoresistance, a frequent and serious problem in the treatment of patients with leukemia, in particular those with acute myeloid leukemia (AML). AML is the most common leukemia in adults but has the lowest survival rate and the mortality rates from AML continue to rise (Cancer Research UK, 2016). Unlike AML, the survival rate of those suffering from chronic myeloid leukemia (CML) improved significantly with the introduction of tyrosine kinase inhibitors (TKIs) in 2001

(Kantarjian et al., 2012). Indeed, these drugs revolutionized therapy for CML and inspired targeted therapy for other forms of cancer. However, a growing number of chemoresistant cases of CML are now appearing due in part to the difficulty of patients in adhering to the treatment regime and/or the cancer cells developing resistance against the prolonged use of chemotherapeutics (Assouline & Lipton, 2011; Iqbal & Iqbal, 2014).

Our understanding of the evolution and pathogenesis of chemoresistance in hematological cancers has benefited greatly from recent advances in genomics, hematopoietic stem cell (HSC) biology, and studies using in vivo models. Consequently, there is now great interest surrounding the development of novel, less toxic, targeted therapies that could be used alongside chemotherapy to reduce drug resistance. Response to chemotherapy depends on multiple factors including age, cytogenetic, and molecular abnormalities as well as organ function at diagnosis (Tsuji et al., 2016). However, these factors alone do not fully explain the observed patterns of resistance and other factors have been implicated. It is postulated that the microenvironment in which adult bone marrow stem cells are maintained in healthy individuals may play a role, with leukemia cells targeting the microenvironment to make it more supportive of cancer progression (see Section 3). It is also understood that a subpopulation of stem cells, known as leukemic stem cells, evade the cytotoxicity of some established chemotherapy drugs and can cause drug resistance after prolonged use (Assouline & Lipton, 2011; Ho et al., 2016; Iqbal & Iqbal, 2014). More recently, the role of autophagy, and its interplay with apoptosis, has been explored as an important mechanism contributing to chemoresistance. Autophagy is increasingly being recognized as a pathway that could have a substantial impact on the outcome of innovative therapeutic strategies (Li et al., 2016; Nikoletopoulou, Markaki, Palikaras, & Tavernarakis, 2013; Ouyang et al., 2012).

A further area of interest is the potential of combining current treatments using chemotherapeutic drugs with the use of phytochemicals. Several chemotherapy drugs have been derived from natural products including vincristine, an approved drug for the treatment of acute leukemia (National Cancer Institute, 2016), which is a constituent of the plant species *Vinca rosea* L. (Apocynaceae). There is a need to develop more novel chemotherapy drugs with selective toxicity to leukemic cells and which can also target leukemic stem cells. This review explores the role of the bone marrow microenvironment (BMM) in the maintenance of leukemic stem cells and its contribution to chemoresistance and explores the potential of the use of phytochemicals (baicalein, chrysin, curcumin, resveratrol, and wogonin) alongside current chemotherapy options to help overcome chemoresistance.

## 2. CURRENT TREATMENT STRATEGIES FOR MYELOID LEUKEMIA AND THE DEVELOPMENT OF CHEMORESISTANCE

Leukemia describes a heterogeneous group of acquired clonal disorders caused by abnormal differentiation of HSCs to white blood cells (WBCs) (Khwaja et al., 2016). Hematopoiesis is a continuous process in which HSCs differentiate and develop into mature and specialized blood cells in the bone marrow (Table 1). Normal HSCs, like other stem cells, are undifferentiated, long-lived cells capable of asymmetric division, facilitating both self-renewal (clonal expansion) and the generation of differentiated progeny (clonal extinction) (Pina & Enver, 2007).

Leukemias present as either acute or chronic diseases and are characterized by the exponential proliferation of myeloid or lymphoid cells. Acute leukemia is a rapidly progressive group of disorders caused by the accumulation of immature WBCs in the bone marrow and blood. Chronic leukemia is marked by the hyperproliferation of neoplastic myeloid or lymphoid progenitors that retain the capacity for terminal differentiation at the initial stages of the condition, with the latter stages of the condition characterized by excessive proliferation of immature WBCs (Passegué, Jamieson, Ailles, & Weissman, 2003). The current gold standard for the treatment of leukemias involves three phases of therapy termed "induction," "consolidation," and "complete remission." Induction therapy is aimed at substantially reducing the population of leukemic cells using cytotoxic agents and typically involves treatment with cytarabine and anthracyclines. This is followed by consolidation therapy, which aims to eliminate any residual disease and prevent relapse. In the final phase, "complete remission," the patient is given maintenance doses of a chemotherapy drug to prevent expansion of any remaining abnormal immature WBCs (blasts cells) (Burnett, Wetzler, & Löwenberg, 2011). Cytarabine and anthracycline act mainly on dividing cells and it has been shown that leukemia stem cells (LSCs) exist in a quiescent nondividing state and therefore frequently resist these drugs (Ho et al., 2016). Targeting the treatment for CML varies depending on the patient's condition, response to the drug, and CML stage; chronic, accelerated, or blast crisis phase (American Cancer Society, 2015).

**Table 1** Hematopoietic Stem Cell Differentiation in Bone Marrow; Function and Surface Markers Identified From Mature, Specialized Normal Blood Cell Population

| Cell Types (Reference Range) | | | Function[a,b] | Surface Markers for Normal Cells |
|---|---|---|---|---|
| Common myeloid progenitor (excessive proliferation of the blood cells leads to myeloid leukemia) | Myeloblasts 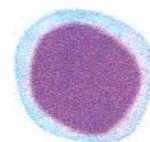 | Neutrophils  (2.0–7.0 × $10^9$/L) | Participation in innate inflammatory response | CD15, CD31, CD66, CXCR1, CXCR2, $CD14^{low}$, CD11b[c] |
| | | | Phagocytosis (bacteria, yeast), and removal of debris | |
| | | Monocytes 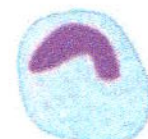 (0.2–1.0 × $10^9$/L) | Cytokine release (interleukins) | CD14, CD33[d] |
| | | | Cooperation with lymphocytes in generating antibodies | |
| | | | Expression of tissue factor | |
| | | | Participation in inflammation, phagocytosis, and debris removal | |
| | | Eosinophils 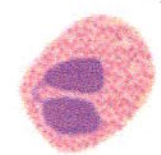 (0.02–0.5 × $10^9$/L) | Protect against parasitic infections | CD11b, CD125, CCR1, CCR3[c] |
| | | | Allergic response, release of histamine | |
| | | Basophiles 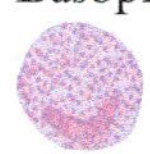 (0.02–0.1 × $10^9$/L) | Hypersensitivity reaction | CD123, CCR3, CRTH2, CD63, CD203C[b] |
| | | | Heparin and histamine release | |

*Continued*

**Table 1** Hematopoietic Stem Cell Differentiation in Bone Marrow; Function and Surface Markers Identified From Mature, Specialized Normal Blood Cell Population—cont'd

| Cell Types (Reference Range) | | | Function[a,b] | Surface Markers for Normal Cells |
|---|---|---|---|---|
| Common lymphoid progenitor (excessive proliferation of the blood cells leads to lymphoid leukemia) | Lymphoblasts 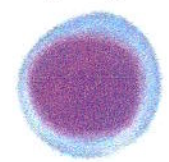 | T-lymphocytes 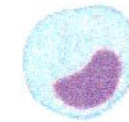 | Participation in adaptive immune response, cooperation in antibody production | CD3, CD4, CD8[d] |
| | | | Destruction of cells infected by viruses | |
| | | B-lymphocytes  | Participation in adaptive humoral immune response, antibody generation | CD19, CD20[d] |
| | | | Memory cells, secondary antibody response | |
| | | Natural killer cells 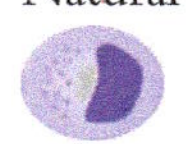 | Participation in innate inflammatory response | CD8, CD16, CD56[d] |
| | | | Cytotoxic respond to viral-infected and malignant cells | |

The hematopoietic stem cell (HSC) is a common progenitor of a variety of blood cell types. HSCs differentiate into a common myeloid progenitor (CMP) or common lymphoid progenitor (CLP). CMPs develop into myeloblasts (which further develop into one of four distinctive types of white blood cells: neutrophil, monocyte, eosinophil, or basophil) or erythrocytes, thrombocytes, and mast cells. CLPs develop into lymphoblasts with further differentiation into either T-lymphocytes or B-lymphocytes, or natural killer cells. Abnormal hematopoietic differentiation of HSCs to WBCs causes either acute or chronic disease of myeloid or lymphoid origin depending on the affected progenitor cell and the rate of disease progression. Hematopoietic stem cells: normal markers include $LKS^+$, $CD34^-$, $CD150^+$, $CD48^-$) and leukemic stem cell markers include $CD34^+$, $CD38^-$ (Lane, Scadden, & Gilliland, 2009).

Lymphocytes reference range ($1.0–3.0 \times 10^9$/L) and white blood cells (reference range: $4.0–10.0 \times 10^9$/L).

[a]Moore, Knight, and Blann (2010).

[b]Metcalfe et al. (2016).

[c]Murdoch, Muthana, Coffelt, and Lewis (2008).

[d]BD Bioscience (2010).

Chemotherapy drugs approved by the US Food and Drug Administration (FDA) for the treatment of myeloid leukemia (National Cancer Institute, 2016) are summarized in Table 2. Imatinib, a first-generation TKI, has revolutionized CML treatment (Assouline & Lipton, 2011). However, there has been a rise in the number of CML patients developing resistance to this TKI, which is not resolved by increasing the dose of imatinib as some patients suffer side effects and may require an alternative treatment regime. More effective second- and third-generation TKI drugs such as nilotinib, dasatinib, bosutinib, and ponatinib (Table 2) may tackle imatinib resistance (Assouline & Lipton, 2011; Jabbour, Kantarjian, & Cortes, 2015).

Although most leukemia patients experience remission following treatment, the emergence of chemoresistance and resultant clinical relapses contribute to unsatisfactory clinical outcomes for many patients. Relapsed disease is generally much less responsive to treatment and despite salvage chemotherapy most patients eventually succumb to their disease. Following chemotherapy, it is probable that some subpopulations of leukemic stem cells develop chemoresistance. These may be capable of repopulating blood, thereby reinitiating the disease. This process may play an important role in the development of chemoresistance in leukemias. Research is ongoing to identify molecules and pathways that are preferentially expressed in these stem cells and to determine the role of the BMM in supporting the growth and differentiation of the stem cells with the aim of developing strategies to improve clinical outcomes and reduce chemoresistance (e.g., Tabe & Konopleva, 2014).

## 3. CHEMORESISTANCE IN LEUKEMIA STEM CELLS MEDIATED BY THE BMM

Niches within the BMM are responsible for controlling the number, quiescence, proliferation, multilineage differentiation capacity, and self-renewal of HSCs (Fig. 1). The potential for limitless self-renewal is one of the hallmarks of cancer (Hanahan & Weinberg, 2000), although it is recognized that this capacity is often restricted to a subpopulation of tumor cells, known as cancer stem cells, in the bone marrow (Lapidot et al., 1994). Leukemias are maintained by a subset of self-renewing LSCs, which have been implicated in progression, recurrence, and therapeutic resistance (Horton & Huntly, 2012). LSCs have the capacity to alter the BMM, creating leukemic niches (Bakker, Qattan, Mutti, Demonacos, & Krstic-Demonacos, 2016) (Fig. 1). The BMM has a complex and dynamic

**Table 2** Approved Chemotherapy Drugs for the Treatment of Myeloid Leukemia

| | Acute Myeloid Leukemia (AML) | | Both AML and CML | | Chronic Myeloid Leukemia (CML) |
|---|---|---|---|---|---|
| Synthetic | *Thioguanine*<br>• Guanosine analogue antimetabolite<br>• As remission induction and consolidation therapy | Synthetic | *Cyclophosphamide*<br>• Alkylating agent<br>• Approved to treat (alone or with other drugs) acute lymphocytic leukemia (ALL) (in children), chronic lymphocytic leukemia (CLL), AML, and CML | | *Mechlorethamine hydrochloride*<br>• The hydrochloride salt of mechlorethamine<br>• Treatment of chronic leukemia types |
| Synthetic | *Mitoxantrone hydrochloride*<br>• Hydrochloride salt of an anthracenedione<br>• Used in combined therapy | Synthetic | *Cytarabine (ara-C; cytosine arabinoside)*<br>• Analogue of cytidine (a nucleoside)<br>• Treat AML, CML, ALL in combined therapy | | *Busulfan*<br>• A derivative of dimethanesulfonate<br>• Treat CML as a palliative treatment and prepare patients for stem cell transplant |
| Natural source | *Daunorubicin hydrochloride*<br>• Hydrochloride salt of an anthracycline (derived from *Streptomyces* bacterium; *S. peucetius* var. *caesius*, a soil bacteria)<br>• Treatment of ALL (adults and children) and AML (adults) | Naturalsource; Anthracycline (*Streptomyces* bacterium) | *Doxorubicin hydrochloride*<br>• Hydrochloride salt of an anthracycline, closely related to daunorubicin<br>• Approved to treat ALL, AML, lymphoma, and other cancer types | Synthetic; Tyrosine kinase inhibitors; First generation | *Imatinib mesylate*<br>• A mesylate salt, a tyrosine kinase inhibitor<br>• Treat Philadelphia (Ph) chromosome positive cancer (CML and ALL) |
| Natural source | *Arsenic trioxide ($As_2O_3$)*<br>• Naturally occurring as minerals (arsenolite and cloudetite)<br>• Treat acute promyelocytic leukemia (when other treatments have been ineffective or with recurred cancer) | Naturalsource; Anthracycline (*Streptomyces* bacterium) | | Synthetic; Tyrosine kinase inhibitors; Second generation | *Nilotinib*<br>• Aminopyrimidine-derivative Bcr-Abl tyrosine kinase inhibitor; binding mode is energetically more favorable than that of imatinib<br>• Treat Ph chromosome positive (CML and ALL) adult patients |
| Natural source | *Vincristine sulfate*<br>• Sulfate salt of a natural alkaloid (plant, *Vinca rosea Linn*)<br>• Treatment of acute leukemia types | Naturalsource; Anthracycline (*Streptomyces* bacterium) | *Idarubicin hydrochloride*<br>• Hydrochloride salt of an anthracycline (derivative of daunorubicin)<br>• Penetrates cell membrane more effectively than other anthracycline drugs<br>• Approved to treat AML (with other drugs) | Synthetic; Tyrosine kinase inhibitors; Second generation | *Dasatinib*<br>• Inhibitor of protein-tyrosine kinases<br>• Treat Ph chromosome positive (CML and ALL) patients with intolerance to other drugs |
| Combination drug | *ADE*: Combination of cytarabine (ara-C), daunorubicin, and etoposide phosphate<br>Approved to treat AML (in adults and children)<br>[Etoposide, semisynthetic derivate of podophyllotoxin (constituent of mandrake root, *Podophyllum peltatum*)] | | | Synthetic; Tyrosine kinase inhibitors; Second generation | *Bosutinib*<br>• A quinolone derivative and dual kinase inhibitor<br>• Approved for CML with positive Ph chromosome (when other treatments have been ineffective) |
| | | | | Synthetic; Tyrosine kinase inhibitors; Third generation | *Ponatinib hydrochloride*<br>• A multitargeted receptor tyrosine kinase inhibitor (TKI)<br>• Treat T315I mutation (chemoresistance to other TKI) in Ph chromosome positive patients |

| | |
|---|---|
| Natural source | *Hydroxyurea (hydroxycarbamate)*<br>• A monohydroxyl-substituted urea antimetabolite<br>• Approved to treat CML (refractory), and squamous cell carcinoma |
| | *Omacetaxine mepesuccinate*<br>• A semisynthetic formulation of the cytotoxic plant alkaloid homoharringtonine isolated from the evergreen tree *Cephalotaxus harringtonia*<br>• Approved to treat CML that is unresponsive to TKIs. |

Source of information: National Cancer Institute (2016)
Chemotherapy drugs, approved by the FDA for the treatment of myeloid leukemia are either chemically synthesized or derived from natural sources

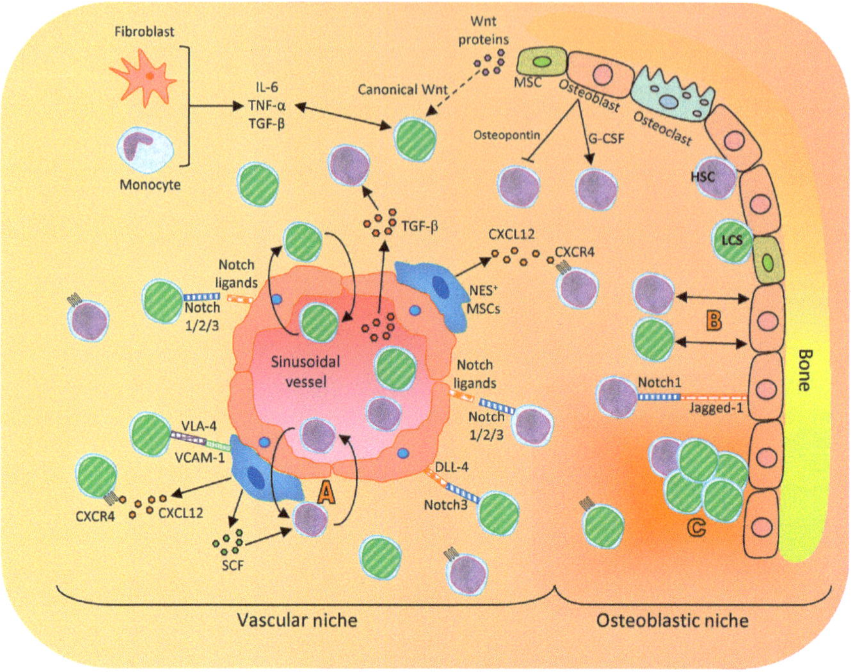


**Fig. 1** The bone marrow microenvironment and the mechanisms involved in maintaining HSCs and LSCs. The bone marrow consists of two submicroenvironments, the osteoblastic (endosteal) niche and the vascular (sinusoidal) niche where stem cells are protected from environmental stress and maintain self-renewal and multilineage differentiation capacity. The osteoblastic niche is mainly involved in maintaining HSC quiescence, and the vascular niche promotes HSC proliferation and differentiation. The subtle interactions within the BMM between HSCs, progenitor cells, stromal cells, and cytokines/chemokines are essential for the maintenance of normal hematopoiesis. Certain signals preserve HSCs in their immature state, while other signals are necessary for promoting the maturation of specific subtypes of blood cells. The signals responsible for inducing blood cell development have been well characterized. Survival and proliferation signals within the BMM preferentially support growth of leukemic cells by supplying the necessary cell contact-mediated signals and cytokines to LSCs as described in Section 2. (A) The sinusoids within the vascular niche of the bone marrow have thin walls, allowing cell trafficking into and out of the bone marrow; (B) there is cross talk between stem cells and the bone marrow environment to promote maintenance and survival of HSCs/LSCs; (C) LSCs alter the bone marrow niche, creating "leukemic" niches, an environment more favorable to the cancer. *Sources of information: Konopleva, M., Tabe, Y., Zeng, Z., & Andreeff, M. (2009). Therapeutic targeting of microenvironmental interactions in leukemia: Mechanisms and approaches.* Drug Resistance Updates, 12*(4), 103–113; Calvi, L. M., & Link, D. C. (2014). Cellular complexity of the bone marrow hematopoietic stem cell niche.* Calcified Tissue International, 94*(1), 112–124; Bakker, E., Qattan, M., Mutti, L., Demonacos, C., & Krstic-Demonacos, M. (2016). The role of microenvironment and immunity in drug response in leukemia.*

three-dimensional architecture made up of two submicroenvironments known as the osteoblastic (endosteal) niche and the vascular (sinusoidal) niche (Fig. 1). Important factors for the maintenance of HSCs by the osteoblastic niche are matrix metalloproteinase 9, cathepsin K, and growth factors (Bakker et al., 2016). The osteoblastic niche is mainly involved in maintaining HSC quiescence, and the vascular niche promotes HSC proliferation and differentiation (Konopleva, Tabe, Zeng, & Andreeff, 2009). As illustrated in Fig. 1, osteoblasts can both stimulate and negatively regulate HSC proliferation through granulocyte colony-stimulating factor, Jagged-1/Notch-1, and osteopontin signaling (Chiarini et al., 2015). The vascular niche, comprising bone marrow vasculature, produces and secretes various cytokines. Transforming growth factor-β (TGF-β) signaling maintains HSCs in their dormant state (Calvi & Link, 2014), while Notch ligands and stem cell factor secreted by endothelial cells sustain HSC self-renewal and proliferation. Delta-like-4, a Notch ligand, is expressed on endothelial cells and specifically activates Notch 3 signaling in LSCs, thus promoting tumor escape from dormancy (Chiarini et al., 2015). Chemokine C-X-C motif ligand 12 (CXCL12) expression in nestin$^+$ mesenchymal stem cells (NES$^+$ MSCs) and its receptor C-X-C receptor 4 (CXCR4) regulate HSC release into the bloodstream (Chiarini et al., 2015; Konopleva et al., 2009). Interactions between very late antigen-4 (expressed by HSCs and LSCs) and vascular cell adhesion molecule-1 are another mechanism involved in retaining these cells in the bone marrow niches and have also been shown to be crucial for the persistence of minimal residual disease (Bakker et al., 2016). Interleukin-6 (IL-6), tumor necrosis factor-α (TNF-α), and TGF-β secreted by the fibroblasts and monocytes can increase proliferation of LSCs (Cogle et al., 2015). The canonical Wnt pathway is involved in protective effects toward LSCs; MSCs secrete Wnt proteins that are involved in Wnt signaling activation in LSCs, therefore sustaining their proliferation and cell cycle progression (Calvi & Link, 2014; Chiarini et al., 2015).

---

Biochimica et Biophysica Acta, 1863, *414–426; Chiarini, F., Lonetti, A., Evangelisti, C., Buontempo, F., Orsini, E., Evangelisti, C., et al. (2015). Advances in understanding the acute lymphoblastic leukemia bone marrow microenvironment: From biology to therapeutic targeting.* Biochimica et Biophysica Acta, 1863, *449–463; Cogle, C. R., Saki, N., Khodadi, E., Li, J., Shahjahani, M., & Azizidoost, S. (2015). Bone marrow niche in the myelodysplastic syndromes.* Leukemia Research, 39*(10), 1020–1027; Morris, V. (2013). Defining the bone marrow microenvironment.* Science Spotlight, 3*(12); Rankin, E. B., Narla, A., Park, J. K., Lin, S., & Sakamoto, K. M. (2015). Biology of the bone marrow microenvironment and myelodysplastic syndromes.* Molecular Genetics and Metabolism, 116*(1), 24–28.*

Individual cancer genomes are genetically heterogeneous and this is most likely to reflect heterogeneity at the level of LSCs. Indeed, transplantation of primary leukemia cells into immunodeficient mice revealed variable competitive regeneration of subclones in patterns that reflected the diversity within the primary tumor (Notta et al., 2011). In the face of overwhelming leukemic cell heterogeneity, researchers are currently looking at targeted combined therapeutic strategies from natural sources to overcome the limitations of individual drugs. Several studies have shown that certain plant-derived compounds from traditional Chinese medicine have the ability to improve the efficiency of antileukemic agents and reduce chemoresistance as discussed in Section 4.

# 4. PHYTOCHEMICAL MODULATION OF LEUKEMIC CELL DEATH

Phytochemicals from traditional Chinese medicine, namely baicalein, chrysin, wogonin (constituents of *Scutellaria baicalensis*; huáng qín; 黄芩), curcumin (a constituent of *Curcuma longa*, jiāng huáng, 姜黄), and resveratrol (a constituent of *Polygonum cuspidatum*; hǔ zhàng, 虎杖), were selected for evaluation of their ability to modulate leukemic cell death via apoptosis (Section 4.1) and autophagy (Section 4.2), as well as their potential to affect signaling pathways involved in the maintenance of LSCs (Section 4.3).

## 4.1 Phytochemical Modulation of Apoptotic Cell Death

Programmed cell death can be induced via different mechanisms including apoptotic and nonapoptotic pathways (autophagy, anoikis, and necroptosis). Apoptosis is the primary mode of cell death presenting characteristic morphology (Jia, Li, Qin, & Liang, 2009; Rao et al., 2011) and is regulated by intrinsic and extrinsic signaling pathways (Panno, 2005). These death pathways activate the enzyme caspase, a cysteine protease, that cleaves proteins at aspartic acid residues (Sellers & Fisher, 1999) and trigger a cascade of proteolytic events, ultimately leading to cell death (Ashkenazi, 2008).

The intrinsic pathway is mediated by intracellular (Li et al., 2016) or extracellular stimuli (Ouyang et al., 2012), which result in mitochondrial perturbation and mitochondrial membrane degradation with resulting release of cytochrome *c* (Ouyang et al., 2012). Cytochrome *c* interacts with Apaf-1 and procaspase 9 to form an apoptosome, resulting in the activation of initiator caspase 9, which in turn triggers the downstream effector caspases (3, 6, and 7), leading to apoptosis (Li et al., 2016; Ouyang et al., 2012).

Therapies that modulate caspase activity are therefore worthy candidates as chemotherapy drugs. Baicalein (Li et al., 2004), chrysin (Woo, Jeong, Park, & Kwon, 2004), curcumin (Jia et al., 2009; Rao et al., 2011; Shakor et al., 2014; Yang et al., 2012), and resveratrol (Puissant et al., 2008) have all been shown to upregulate the production of caspase 3, leading to the induction of apoptosis in leukemic cell lines (Fig. 2). Jia et al. (2009) have also reported that curcumin can induce cell death (by both apoptosis and autophagy) via reduction in mitochondrial membrane potential, which causes the release of cytochrome *c*, leading to caspase-dependent killing. Apoptosis induced through mitochondrial perturbation is also a result of transcriptional and posttranslational activation of the Bcl-2 family of proteins, Bax, Bak, Bad, BH3, and t-Bid (proapoptotic molecules) as well as Bcl-2, Bcl-$X_L$, and Mcl-1 (antiapoptotic molecules) (Li et al., 2016; Nikoletopoulou et al., 2013). An imbalance in the ratio of pro- and antiapoptotic molecules determines cell survival or death (Adams & Cory, 2007; Nikoletopoulou et al., 2013). Of the compounds of interest in this review, curcumin demonstrated apoptotic effects by upregulating t-Bid (Jia et al., 2009) and Bax (Guo et al., 2015; Shakor et al., 2014) and downregulating Bcl-2 (Guo et al., 2015; Jia et al., 2009; Rao et al., 2011; Shakor et al., 2014; Weng et al., 2015), while in leukemic cell lines Wogonin was found to reduce Bcl-2 and Mcl-1 expression (Huang et al., 2010; Polier et al., 2015). In a mouse model of lymphocytic leukemia, as well as in human AML and ALL, resveratrol has been found to cause a reduction in Bcl-2 expression and upregulate Bax (Li et al., 2007, 2010), again suggesting proapoptotic actions.

The extrinsic apoptotic pathway involves ligation of death ligands to cell surface death receptors including the tumor necrosis factor superfamily (Li et al., 2016). The binding of death ligands to their receptors forms a death-inducing signaling complex, which includes Fas-associated death domain-containing protein and procaspases 8 and 10 (Li et al., 2016). Activated initiator caspases 8 and 10 cleave and activate effector caspases 3, 6, and 7 (Sellers & Fisher, 1999). Additionally, the initiator caspases 8 and 10 stimulate Bid, which affects mitochondrial permeability (Li et al., 2016). Curcumin (Yang et al., 2012), baicalein (Chen et al., 2013), and resveratrol (Puissant et al., 2008) have been shown to induce apoptosis via the extrinsic pathway in addition to stimulating the intrinsic pathway.

The nuclear transcription factor p53 (tumor suppressor gene), which responds to signals such as DNA damage, oxidative stress, and ischemia, is commonly mutated in human cancers (Nikoletopoulou et al., 2013).

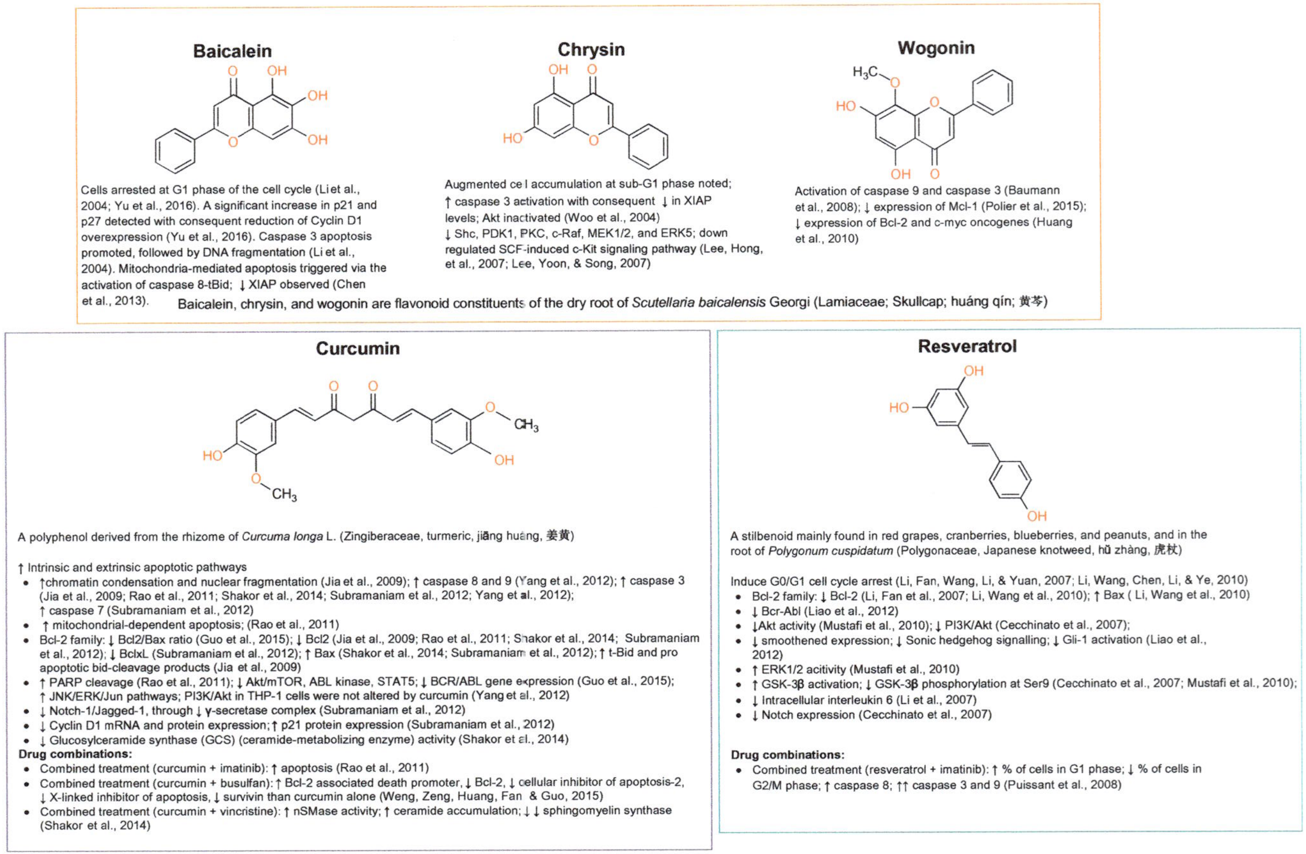


**Fig. 2** Modulation of apoptotic pathway in leukemia cells by baicalein, chrysin, wogonin, curcumin, and resveratrol (Baumann et al., 2008; Lee, Yoon, et al., 2007; Mustafi, Chakraborty, & Raha, 2010; Subramaniam et al., 2012; Yu et al., 2016).

Cytoplasmic p53 is involved in regulating cell cycle progression (at G1/S and G2/M checkpoints) (Sellers & Fisher, 1999) and causing apoptosis of cells by inducing the transcription of the death receptor components and directly interacting with the Bcl-2 family members. Increased mitochondrial outer membrane permeabilization by the upregulation of p53 leads to cell death (Nikoletopoulou et al., 2013). Some cell processes including cell survival, proliferation, growth, and metabolism are regulated by phosphatidylinositol-3-kinase/Akt and the mammalian target of the rapamycin (mTOR) signaling pathways (Ersahin, Tuncbag, & Cetin-Atalay, 2015). However, various genetic and epigenetic mechanisms can cause its deregulation in some cancers (Ouyang et al., 2012), resulting in angiogenesis and metastasis (Ersahin et al., 2015). This pathway has been shown to increase the activity of nuclear factor-κB (associated with cell survival), but negatively regulates the activity of p53, thus making a significant contribution to chemoresistance (Grandage, Gale, Linch, & Khwaja, 2005).

In addition to inducing leukemic cell death as single agents, some phytochemicals when combined with approved chemotherapy drugs can lead to enhanced activity through either synergistic or additive effects. Reports indicate that there is enhanced antileukemic activity of imatinib when combined with either curcumin (Rao et al., 2011) or resveratrol (Puissant et al., 2008). Curcumin has also shown promising apoptotic effects in combination with busulfan (Weng et al., 2015) and vincristine (Shakor et al., 2014).

## 4.2 Phytochemical Modulation of Autophagic Cell Death

The primary cell death pathway targeted by chemotherapeutic compounds is apoptosis. However, recently autophagy (also known as programmed cell death II) has been linked to the acquisition of resistance in cancer to chemotherapy in different cancers in some cell populations (Ojha, Jha, Singh, & Bhattacharyya, 2014). However, the therapeutic potential of autophagy modulation in LSCs remains to be thoroughly investigated. Autophagy is also known to play an important role in the regulation of numerous physiological functions, including hematopoiesis (Leveque et al., 2013), and dysregulation of autophagy may contribute to the development of several disorders, including acute leukemias (Liu, Chen, Atkinson, Claxton, & Wang, 2016). Findings suggest that autophagy can act either as a chemoresistance mechanism or have tumor suppressive functions, with the outcome depending on the environment surrounding the tumor (Sehgal et al., 2015). It has been shown that treatment of leukemic cell lines with curcumin can induce autophagy and result in apoptosis, thus stimulating cell death rather than the promotion of cell survival (Guo et al., 2015; Jia et al., 2009; Wu et al., 2011).

Autophagy functions to degrade and recycle cellular components such as aged proteins and organelles that can then be used to generate ATP and provide essential building blocks during nutrient and/or oxygen deprivation in order to maintain cell survival. Over the past 15 years, many regulators and autophagy-related genes (ATGs) have been identified. Autophagy begins with the activation of the initiation complex, a serine/threonine kinase complex containing ULK1 and ULK2, ATG13, IP2000, and ATG101 (Fig. 3). Following adenosine monophosphate-activated protein kinase (AMPK) activation or inhibition of mTOR (i.e., during chemotherapy), this complex becomes active. The subsequent formation of a phagophore forms a complex with VPS15, UVRAG, AMBRA1, Rubicon, and Beclin1 and is inhibited by the proapoptotic protein Bcl-2 through interactions with Beclin1. Misfolded proteins and organelles conjugate with ubiquitin and are localized to the phagophore membrane by the ubiquitin-binding protein p62/SQSTM1. The next step, autophagosome maturation, is regulated by the ATG/(LC3)–PE conjugation system (Feng, Yao, & Klionsky, 2015). The conversion of LC3-I to LC3-II during this conjugation process

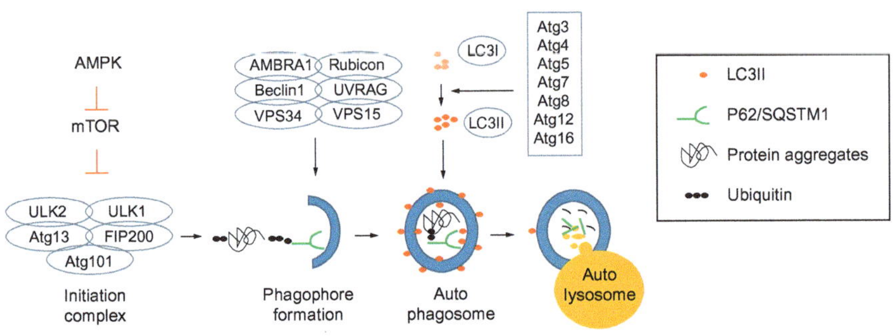


**Fig. 3** The regulation of autophagy. Autophagy begins with the activation of the initiation complex, which becomes active following AMP-activated protein kinase (AMPK) activation or inhibition of mTOR. Following this, a phagophore is formed, and misfolded proteins and organelles conjugate with ubiquitin and are localized to the phagophore membrane by the ubiquitin-binding protein p62/SQSTM1. The next step, autophagosome maturation, is regulated by the ATG12/ATG5/ATG16 and ATG8 (LC3)–PE conjugation system. The final step of the autophagy process involves fusion of the autophagosome with lysosomes, which release enzymes to break down waste materials and consequently release amino acids, which are then recycled to provide energy for the cell.

results in LC3-II binding to the surface of the newly formed autophagosome, which now engulfs the misfolded proteins and organelles. The final step of the autophagy process involves fusion of the autophagosome with lysosomes, which release pH-sensitive enzymes into the autophagosome and break down the waste materials, freeing up amino acids which can then be recycled to provide energy for the cell (Fig. 3).

Autophagy was first linked to cancer with the identification of BECLIN1 as a mammalian autophagy-promoting gene and the finding that Beclin-1 protein expression was frequently low in human breast epithelial carcinoma cells, but expressed at high levels in normal breast epithelia (Liang et al., 1999). Since then, a number of studies have shown that the role of autophagy differs depending on the stage of cancer development. Initially, autophagy likely has a preventive effect against cancer, but as the cancer develops, the cancer cells could utilize autophagy for their own cytoprotection, probably due to the role of this process in improving nutrient and/or oxygen supply to rapidly dividing cancer cell populations (Janku, McConkey, Hong, & Kurzrock, 2011). Indeed, the aberrant expression of BECLIN1 has been reported in many types of cancer including hepatocellular, colorecetal, nasopharyngeal, endometrial, and gastric (Giatromanolaki et al., 2011; Koukourakis et al., 2010; Shi, Ding, Zhou, Qiu, & Fan, 2009; Wan et al., 2010; Xia, Wang, Zhao, & Song, 2013) and has been shown to correlate with poor prognosis.

Recently, autophagy has been shown to play a role in the chemoresistance of CML stem cells to TKIs. Treatment of CML cells in vitro with TKI results in induction of autophagy, while inhibiting autophagy using the antimalarial drug hydroxychloroquine (HCQ) has been shown to sensitize mature and immature CML stem cells to TKI treatment (Helgason et al., 2013). Furthermore, cell death following treatment with TKIs is increased by knocking down key signaling molecules in the autophagy pathway (ATG7 or ATG5) in CML cell lines or $CD34^{+}38^{-}$ CML stem cells (Bellodi et al., 2009); therefore, it is plausible that downregulating autophagy can thus sensitize cancer cells to treatment. Although there are no specific inhibitors of autophagy clinically available at present, it is possible to inhibit autophagy with HCQ, an antimalarial drug, which has previously been approved for the treatment of autoimmune disorders such as rheumatoid arthritis (Helgason, Karvela, & Holyoake, 2011). There have been over 40 clinical trials using HCQ in cancer patients, either alone or in combination with cytotoxic agents (Jiang & Mizushima, 2014; Yang & Chen, 2011). However, although HCQ has been shown to block autophagy in vitro, it is still unclear whether autophagy can be blocked with HCQ in the clinical setting. There is therefore a need for the development of more specific autophagy modulators for clinical use (inducers or inhibitors depending on the specific cancer microenvironment).

Investigation into the interaction of phytochemicals with respect to regulation of autophagy is ongoing. Resveratrol-treated CML and ALL cell lines (Jiao et al., 2013; Puissant et al., 2010), and curcumin-treated myeloid leukemia cell lines (Jia et al., 2009; Wu et al., 2011) exhibited increased levels of LC3-II with an upregulation of beclin 1, followed by the formation of autophagosomes, suggestive of induced autophagy. When treated with an autophagy inhibitor, phytochemical-induced cytotoxicity was not evident, indicating that the induction of autophagy had an antiproliferative effect on the cell lines. The other three phytochemicals discussed in this chapter, baicalein, chrysin, and wogonin, have also been reported to induce autophagy in cancers of epithelial origin (Aryal et al., 2014; Chow, Chen, Liang, Huang, & Wang, 2012; Xuan, Zhang, Wang, Fu, & Zhang, 2016). However, the induction of autophagy by baicalein in hepatocellular carcinoma was reported to confer cell survival rather than induce cell death (Wang et al., 2014). Further work is required to evaluate the effect of these compounds on blood cancers and to evaluate the role of autophagy in cell survival, cell death, and chemoresistance.

## 4.3 Phytochemical Modulation of Cell Death via Wnt/β-Catenin, Notch, and Hedgehog Signaling Pathways

Understanding the mechanisms by which chemoresistance can occur is important when developing novel therapeutic approaches to treating cancer. Most of the work undertaken to date on the effects of the phytochemicals discussed in this review has been conducted on cancer cell lines and not on LSCs. While the precise mechanisms involved in chemoresistance remain unclear, there are several compelling studies where LSCs have been shown to reinitiate the embryonic signaling pathways Wnt/β-catenin, Notch, and Hedgehog (Bakker et al., 2016; Etet, Vecchio, & Kamdje, 2012; Hu & Li, 2016), leading to increased self-renewal of LSCs and reduced drug efficacy, resulting in chemoresistance (Roy, Ray, & Jana, 2013). Inhibition of the Wnt/β-catenin (Sinclair, Latif, & Holyoake, 2013) and Hedgehog pathways (Xu et al., 2015) and activation of the Notch pathway (Lobry et al., 2013) have been shown to induce cell death and lead to a reduction in LSCs. The treatment of human acute T-lymphoblastic leukemia cell line (T-ALL), Jurkat, with baicalein displayed increased cell apoptosis via the downregulation of mRNA and protein levels of β-catenin (Liu, Liu, et al., 2016). When treated with resveratrol, the human T-ALL cell line, MOLT-4, exhibited an inhibition of the Notch signaling pathway and activation of apoptosis (Cecchinato et al., 2007). Promising results from a study on lung cancer cell lines indicate that curcumin-induced autophagic cell death via downregulating Hedgehog signaling pathway resulted in a decrease in Smoothened (a positive regulatory protein) and Gli1 (glioma-associated oncogene transcription gene) (Xu et al., 2015). The CML cell line, K562, when treated with resveratrol, exhibited increased cell apoptosis through a reduction in Smoothened and Gli1 activation, which in turn resulted in the downregulation of Bcr-Abl via the inhibition of the Sonic hedgehog signaling pathway (Liao et al., 2012).

## 5. APPROACHES TO SUPPORT FUTURE RESEARCH IN OVERCOMING LEUKEMIA CHEMORESISTANCE

Over the past 50 years, our understanding of the evolution and pathogenesis of chemoresistance during leukemia has benefited greatly from recent advances in hematopoietic stem cell biology and cell death pathways. In this section, we describe methodologies and investigations into alternative regulatory systems required to identify the key mechanisms involved in leukemia chemoresistance to help develop novel therapeutics that can continue to impact mortality rates in patients.

### 5.1 Necroptosis-Regulated Leukemic Cell Death

Autophagy and apoptosis are clearly important modulators of chemoresistance in LSCs in the BMM; however, recent advances in the field suggest that an alternative cell death pathway may also play a role. Necrosis was originally considered to be an accidental and unregulated cell death. However, accumulating evidence has shown that necrosis can be induced and proceeds in a regulated manner similar to apoptosis, although through a caspase-independent mechanism. This regulated necrosis is termed necroptosis, and interestingly, it has recently been shown to be induced by chemotherapeutic agents in cancers such as neuroblastoma (Nicolai, Pieraccioli, Peschiaroli, Melino, & Raschellà, 2015), suggesting a novel therapeutic target that should be further investigated for chemoresistant cancers. Fu et al. (2013) investigated the necroptotic effect of shikonin, a type of Chinese herbal medicine purified from *Lithospermum erythrorhizon* (the red-root gromwell, zǐcǎo, 紫草), on murine osteocarcinoma cell lines. Their results showed an increase in receptor-interacting proteins (RIP-1 and -3), the well-known initiators of necroptosis, therefore indicating the possibility of using natural compounds to induce necroptosis where apoptotic cell death fails to occur. To date very few studies have been published on phytochemical modulators of necroptosis.

## 5.2 Leukemia Cell Culture Using 3D Models to Investigate Chemoresistance

Tumors develop through dynamic processes where cancer cells interact with one another and with their surrounding matrix in a 3D structure (Alemany-Ribes & Semino, 2014). Numerous biomimetic 3D models have recently been developed to study HSC expansion and differentiation (Owen & Shoichet, 2010). An interesting approach to 3D leukemia models is the ex vivo expansion of leukemic cells isolated from patients. A 2D and 3D culture of osteoblast niche with bioderived bone as a scaffold, containing osteoblasts derived from differentiated MSCs of CML patients, showed the reconstructed osteoblast niche maintained LSCs from the bone marrow of CML patients for a longer period compared with 2D culture (Hou et al., 2009). Furthermore, Aljitawi et al. (2014) investigated the cytotoxic and apoptotic effects of chemotherapy drugs doxorubicin and cytarabine on leukemia cell lines, HL60, Kasumi-1, MV411, and human bone marrow MSCs, in 2D and 3D coculture conditions, and found that the 3D cultured leukemia cells were more resistant to drug-induced apoptosis compared to cells cultured in 2D or in suspension (Aljitawi et al., 2014).

While the phytochemicals reported in this chapter, baicalein, chrysin, curcumin, resveratrol, and wogonin, show strong evidence of inducing leukemia cell death, it is important to stress that these studies were performed on cells cultured in 2D/suspension conditions, which does not take into account the cell–extracellular matrix interactions nor the various cell populations and structures present in the BMM. It has been well established that the microenvironment plays a central role in regulating tumor development and behavior, such as drug resistance. Odot et al. (2004) evaluated the effect of curcumin on murine melanoma cells, B16-R, when cultured in 2D monolayer or 3D spheroids. The $IC_{50}$ values for monolayer culture and spheroid culture were 27 and 175 μM, respectively, therefore showing that spheroids, which are more in vivo-like solid tumors, are more resistant to the therapeutic effects of curcumin than monolayer cells. Abuelba et al. (2015) investigated the effect of curcumin on cell viability of breast adenocarcinoma cell line, MDA-MB-231, when cultured in 2D monolayer or 3D spheroid models. Treatment with curcumin concentration of 15 μM for 48 h inhibited tumor development and proliferation with a viability of 65.7% in 2D culture, while an increased tumor cell viability of 82.15% was observed following spheroidal culture. The comparison of 2D cell culture with 3D spheroid cell culture of human hepatocellular carcinoma cells, HepG2, treated with resveratrol indicated 50% more live cells with intact cell membrane integrity in the 3D culture compared to 2D culture; therefore the efficacy of resveratrol is decreased in 3D models compared to 2D cell culture (Patra, Peng, Liao, Lee, & Tung, 2016). These studies show promising evidence that 3D tumor cell culture systems would be the ideal environment for in vitro assays to explore the antileukemic effects of phytochemicals combined with chemotherapy drugs. Applying this method of investigation in drug discovery, cancer stem cell research, and cancer cell biology is more likely to increase the success rate of any subsequent in vivo study to overcome chemoresistance.

## 5.3 The Use of Chemotherapy Drugs With Nanomaterials in Treating Chemoresistance

The combination of chemotherapeutic drugs with nanomaterials is also a recently emerging field, which has been shown to selectively target cancer stem cells. Compared to other approaches such as combinatorial therapy of chemotherapy drugs with inhibitors of chemoresistance pathways, delivery of existing chemotherapy drugs with nanomaterials such as nanodiamonds has recently been shown to provide a broader range of protection in a package that is both safer and more effective in a leukemia cell line (Man et al., 2014). Delivery of chemotherapy drugs by nanodiamonds not only enables enhanced killing of chemoresistant cancer stem cells (Ho, Wang, & Chow, 2015) but may be a useful alternative to alleviate the toxic side effects of standard chemotherapy drugs for leukemia patients.

As a therapeutic agent, curcumin is able to pass through cellular membranes due to its lipophilic properties. However, when administered orally, the hydrophobic nature of the compound also leads to low bioavailability of the drug. Strategies to increase the bioavailability of curcumin and therapeutic selectivity include encapsulating the compound in lipid nanoemulsion (Onodera et al., 2015) and curcumin-loaded magnetic nanoparticles (Yallapu et al., 2013). Coformulation with imatinib in poly(lactide-*co*-glycolide) nanoparticles (Acharya & Sahoo, 2016) and doxorubicin–curcumin composite nanoparticle formulation (Pramanik et al., 2012) has also demonstrated enhanced anticancer effects from combinational studies. These promising results indicate that nanotechnology could be an effective strategy to help overcome cancer drug resistance. More studies are required to investigate the effect of encapsulated phytochemicals on leukemia using in vivo studies.

## 5.4 Epigenetic Profiling of Leukemia Cells to Overcome Chemoresistance

Epigenetic regulation of gene expression is crucial for both cancer stem cell biology and chemoresistance. Epigenetic mechanisms regulate gene expression in the absence of alterations in DNA sequences. These mechanisms include DNA methylation, histone modification, and RNA-mediated targeting among others, and they regulate many biological processes that are fundamental to cancer. Epigenetic mechanisms are thought to be regulators of the cancer stem cell state, since cancer stem cells are assumed to be genetically identical to their differentiated progeny (Magee, Piskounova, & Morrison, 2012). A growing body of evidence suggests that certain epigenetic mechanisms regulate the biological properties of cancer stem cells, with inhibition of DNA methylation suppressing cancer stem cell properties in a number of hematological malignancies (Tsai et al., 2012). Interestingly, the study of histone deacetylases (HDACs) has also supported the hypothesis that epigenetic mechanisms may regulate chemotherapy resistance in LSCs. HDACs are chromatin-remodeling enzymes that have gained recent attention as therapeutic targets in cancer (Bolden, Peart, & Johnstone, 2006). It has been reported that HDAC inhibition suppresses residual CML stem cells following imatinib therapy in vivo. Interestingly, combined exposure to imatinib and an HDAC inhibitor suppressed the residual bone marrow cancer stem cell burden associated with imatinib therapy. In a later study, the same group demonstrated that selective sirtuin HDAC inhibition suppressed CML stem cell engraftment and growth in vivo in a p53-dependent mechanism (Li et al., 2012). Interestingly, curcumin has been shown to mediate inhibition of HDAC activity in multiple in vitro cancer models, and recent investigations provide evidence that HDAC inhibitors may provide a potent strategy for cancer treatment (Lee, Hong, et al., 2007). Furthermore, it has been demonstrated that promoter hypoacetylation of several histones is curcumin mediated and correlates with gene silencing (Pollack, Sapkota, & Boss, 2009). In leukemia cells, curcumin treatment induces global hypomethylation through the inhibition of DNA methyltransferases (Fu & Kurzrock, 2010); however, the role of other phytochemicals in the epigenetic regulation of leukemia resistance remains unclear.

Together, these studies suggest that epigenetic mechanisms are important regulators of LSCs, and that curcumin-induced HDAC inhibition in particular may improve the efficacy of chemotherapy. LSC epigenomic profiling may help to dissect specific chemoresistance pathways further and could have a significant clinical impact for patient stratification and the design of therapeutic regimens.

## 6. CONCLUSIONS

Our understanding of the evolution and pathogenesis of chemoresistance in hematological cancers has benefited greatly from recent advances in genomics, hematopoietic stem cell biology, and studies using in vivo models. It is clear that there are a number of factors, which contribute to the development of chemoresistance to the current drug therapies available for leukemia. Autophagy and apoptosis appear to play a central role, and these mechanisms are targets for research into novel therapies. Most cancer drug therapies target dividing cells and a treatment strategy is to induce apoptosis in these cells. Since cancer cells seem to develop resistance to apoptosis-induced cell death, we recommend more studies to be conducted on nonapoptotic cell death pathways such as autophagy and necroptosis. Other strategies such as targeting quiescent LSCs and the microenvironment that supports them, and understanding the signaling pathways involved in LSC maintenance and proliferation are also essential in understanding this complex cellular phenomenon. While there is preliminary evidence to support the use of phytochemicals, in particular curcumin and resveratrol, to overcome chemoresistance through the modulation of signaling pathways involved in the survival and death of leukemic clonal cells LSCs, further research is required. Future work should include the development of 3D models to investigate the ability of curcumin and resveratrol to modulate apoptotic and autophagic death of acute leukemia cells and LSCs in the BMM.

## ACKNOWLEDGMENT

We thank Milan Vu for help with graphics and editing the manuscript.

# REFERENCES

Abuelba, H., Cotrutz, C. E., Stoica, B. A., Stoica, L., Olinici, D., & Petreuş, T. (2015). In vitro evaluation of curcumin effects on breast adenocarcinoma 2D and 3D cell cultures. *Romanian Journal of Morphology and Embryology*, *56*(1), 71–76.

Acharya, S., & Sahoo, S. K. (2016). Exploitation of redox discrepancy in leukemia cells by a reactive oxygen species nanoscavenger for inducing cytotoxicity in imatinib resistant cells. *Journal of Colloid and Interface Science*, *467*, 180–191.

Adams, J. M., & Cory, S. (2007). The Bcl-2 apoptotic switch in cancer development and therapy. *Oncogene*, *26*(9), 1324–1337.

Alemany-Ribes, M., & Semino, C. E. (2014). Bioengineering 3D environments for cancer models. *Advanced Drug Delivery Reviews*, *79*, 40–49.

Aljitawi, O. S., Li, D., Xiao, Y., Zhang, D., Ramachandran, K., Stehno-Bittel, L., et al. (2014). A novel three-dimensional stromal-based model for in vitro chemotherapy sensitivity testing of leukemia cells. *Leukemia & Lymphoma*, *55*(2), 378–391.

American Cancer Society. (2015). *Treating chronic myeloid leukemia by phase*. http://www.cancer.org/cancer/leukemia-chronicmyeloidcml/detailedguide/leukemia-chronic-myeloid-myelogenous-treating-treating-by-phase. Last accessed 20 October 2016.

Aryal, P., Kim, K., Park, P. H., Ham, S., Cho, J., & Song, K. (2014). Baicalein induces autophagic cell death through AMPK/ULK1 activation and downregulation of mTORC1 complex components in human cancer cells. *The FEBS Journal*, *281*(20), 4644–4658.

Ashkenazi, A. (2008). Targeting the extrinsic apoptosis pathway in cancer. *Cytokine & Growth Factor Reviews*, *19*(3), 325–331.

Assouline, S., & Lipton, J. H. (2011). Monitoring response and resistance to treatment in chronic myeloid leukemia. *Current Oncology*, *18*(2), 71–83.

Bakker, E., Qattan, M., Mutti, L., Demonacos, C., & Krstic-Demonacos, M. (2016). The role of microenvironment and immunity in drug response in leukemia. *Biochimica et Biophysica Acta*, *1863*, 414–426.

Baumann, S., Fas, S. C., Giaisi, M., Müller, W. W., Merling, A., Gülow, K., et al. (2008). Wogonin preferentially kills malignant lymphocytes and suppresses T-cell tumor growth by inducing PLCγ1- and Ca2+-dependent apoptosis. *Blood*, *111*(4), 2354–2363.

BD Bioscience. (2010). *CD marker handbook*. https://www.bdbiosciences.com/documents/cd_marker_handbook.pdf. Last accessed 6 October 2016.

Bellodi, C., Lidonnici, M. R., Hamilton, A., Helgason, G. V., Soliera, A. R., Ronchetti, M., et al. (2009). Targeting autophagy potentiates tyrosine kinase inhibitor-induced cell death in Philadelphia chromosome-positive cells, including primary CML stem cells. *The Journal of Clinical Investigation*, *119*(5), 1109–1123.

Bolden, J. E., Peart, M. J., & Johnstone, R. W. (2006). Anticancer activities of histone deacetylase inhibitors. *Nature Reviews. Drug Discovery*, *5*(9), 769–784.

Burnett, A., Wetzler, M., & Löwenberg, B. (2011). Therapeutic advances in acute myeloid leukemia. *Journal of Clinical Oncology*, *29*(5), 487–494.

Calvi, L. M., & Link, D. C. (2014). Cellular complexity of the bone marrow hematopoietic stem cell niche. *Calcified Tissue International*, *94*(1), 112–124.

Cancer Research UK. (2016). *Leukaemia (all subtypes combined) statistics*. http://www.cancerresearchuk.org/health-professional/cancer-statistics/statistics-by-cancer-type/leukaemia#dgg72fy48VWYawLQ.99. Last accessed 6 November 2016.

Cecchinato, V., Chiaramonte, R., Nizzardo, M., Cristofaro, B., Basile, A., Sherbet, G. V., et al. (2007). Resveratrol-induced apoptosis in human T-cell acute lymphoblastic leukaemia MOLT-4 cells. *Biochemical Pharmacology*, *74*(11), 1568–1574.

Chen, Y. J., Wu, C. S., Shieh, J. J., Wu, J. H., Chen, H. Y., Chung, T. W., et al. (2013). Baicalein triggers mitochondria-mediated apoptosis and enhances the antileukemic effect of vincristine in childhood acute lymphoblastic leukemia CCRF-CEM cells. *Evidence-based Complementary and Alternative Medicine*, *2013*, 124747.

Chiarini, F., Lonetti, A., Evangelisti, C., Buontempo, F., Orsini, E., Evangelisti, C., et al. (2015). Advances in understanding the acute lymphoblastic leukemia bone marrow microenvironment: From biology to therapeutic targeting. *Biochimica et Biophysica Acta, 1863*, 449–463.

Chow, S. E., Chen, Y. W., Liang, C. A., Huang, Y. K., & Wang, J. S. (2012). Wogonin induces cross-regulation between autophagy and apoptosis via a variety of Akt pathway in human nasopharyngeal carcinoma cells. *Journal of Cellular Biochemistry, 113*(11), 3476–3485.

Cogle, C. R., Saki, N., Khodadi, E., Li, J., Shahjahani, M., & Azizidoost, S. (2015). Bone marrow niche in the myelodysplastic syndromes. *Leukemia Research, 39*(10), 1020–1027.

Ersahin, T., Tuncbag, N., & Cetin-Atalay, R. (2015). The PI3K/AKT/mTOR interactive pathway. *Molecular BioSystems, 11*(7), 1946–1954.

Etet, P. S., Vecchio, L., & Kamdje, A. N. (2012). Signaling pathways in chronic myeloid leukemia and leukemic stem cell maintenance: Key role of stromal microenvironment. *Cellular Signalling, 24*(9), 1883–1888.

Feng, Y., Yao, Z., & Klionsky, D. J. (2015). How to control self-digestion: Transcriptional, post-transcriptional, and post-translational regulation of autophagy. *Trends in Cell Biology, 25*(6), 354–363.

Fu, Z., Deng, B., Liao, Y., Shan, L., Yin, F., Wang, Z., et al. (2013). The anti-tumor effect of shikonin on osteosarcoma by inducing RIP1 and RIP3 dependent necroptosis. *BMC Cancer, 13*(1), 1.

Fu, S., & Kurzrock, R. (2010). Development of curcumin as an epigenetic agent. *Cancer, 116*(20), 4670–4676.

Giatromanolaki, A., Koukourakis, M. I., Koutsopoulos, A., Chloropoulou, P., Liberis, V., & Sivridis, E. (2011). High Beclin 1 expression defines a poor prognosis in endometrial adenocarcinomas. *Gynecologic Oncology, 123*(1), 147–151.

Grandage, V. L., Gale, R. E., Linch, D. C., & Khwaja, A. (2005). PI3-kinase/Akt is constitutively active in primary acute myeloid leukaemia cells and regulates survival and chemoresistance via NF-kB, MAPkinase and p53 pathways. *Leukemia, 19*(4), 586–594.

Guo, Y., Li, Y., Shan, Q., He, G., Lin, J., & Gong, Y. (2015). Curcumin potentiates the antileukemia effects of imatinib by downregulation of the AKT/mTOR pathway and BCR/ABL gene expression in Ph+ acute lymphoblastic leukemia. *The International Journal of Biochemistry & Cell Biology, 65*, 1–11.

Hanahan, D., & Weinberg, R. A. (2000). The hallmarks of cancer. *Cell, 100*(1), 57–70.

Helgason, G. V., Karvela, M., & Holyoake, T. L. (2011). Kill one bird with two stones: Potential efficacy of BCR-ABL and autophagy inhibition in CML. *Blood, 118*(8), 2035–2043.

Helgason, G. V., Mukhopadhyay, A., Karvela, M., Salomoni, P., Calabretta, B., & Holyoake, T. L. (2013). Autophagy in chronic myeloid leukaemia: Stem cell survival and implication in therapy. *Current Cancer Drug Targets, 13*(7), 724–734.

Ho, T. C., LaMere, M., Stevens, B. M., Ashton, J. M., Myers, J. R., O'Dwyer, K. M., et al. (2016). Evolution of acute myelogenous leukemia stem cell properties after treatment and progression. *Blood, 128*(13), 1671–1678.

Ho, D., Wang, C. K., & Chow, E. K. (2015). Nanodiamonds: The intersection of nanotechnology, drug development, and personalized medicine. *Science Advances, 1*(7), e1500439.

Horton, S. J., & Huntly, B. J. (2012). Recent advances in acute myeloid leukemia stem cell biology. *Haematologica, 97*(7), 966–974.

Hou, L., Liu, T., Tan, J., Meng, W., Deng, L., Yu, H., et al. (2009). Long-term culture of leukemic bone marrow primary cells in biomimetic osteoblast niche. *International Journal of Hematology, 90*(3), 281–291.

Hu, Y., & Li, S. (2016). Survival regulation of leukemia stem cells. *Cellular and Molecular Life Sciences, 73*(5), 1039–1050.

Huang, S., Wang, C., Yang, R., Chu, C., Wu, H., & Pang, J. S. (2010). Wogonin, an active compound in Scutellaria baicalensis, induces apoptosis and reduces telomerase activity in the HL-60 leukemia cells. *Phytomedicine*, *17*(1), 47–54.

Iqbal, N., & Iqbal, N. (2014). Imatinib: A breakthrough of targeted therapy in cancer. *Chemotherapy Research and Practice*, *2014*, 357027.

Jabbour, E., Kantarjian, H., & Cortes, J. (2015). Use of second- and third-generation tyrosine kinase inhibitors in the treatment of chronic myeloid leukemia: An evolving treatment paradigm. *Clinical Lymphoma, Myeloma & Leukemia*, *15*(6), 323–334.

Janku, F., McConkey, D. J., Hong, D. S., & Kurzrock, R. (2011). Autophagy as a target for anticancer therapy. *Nature Reviews. Clinical Oncology*, *8*(9), 528–539.

Jia, Y. L., Li, J., Qin, Z. H., & Liang, Z. Q. (2009). Autophagic and apoptotic mechanisms of curcumin-induced death in K562 cells. *Journal of Asian Natural Products Research*, *11*(11), 918–928.

Jiang, P., & Mizushima, N. (2014). Autophagy and human diseases. *Cell Research*, *24*, 69–79.

Jiao, G., Yan, L., Qiang, L., Xia, G., Ling, G., Gui, M., et al. (2013). Resveratrol induces apoptosis and autophagy in T-cell acute lymphoblastic leukemia cells by inhibiting Akt/mTOR and activating p38-MAPK. *Biomedical and Environmental Sciences*, *26*(11), 902–911.

Kantarjian, H., O'Brien, S., Jabbour, E., Garcia-Manero, G., Quintas-Cardama, A., Shan, J., et al. (2012). Improved survival in chronic myeloid leukemia since the introduction of imatinib therapy: A single-institution historical experience. *Blood*, *119*(9), 1981–1987.

Khwaja, A., Bjorkholm, M., Gale, R. E., Levine, R. L., Jordan, C. T., Ehninger, G., et al. (2016). Acute myeloid leukaemia. *Nature Reviews. Disease Primers*, *10*(2), 16010.

Konopleva, M., Tabe, Y., Zeng, Z., & Andreeff, M. (2009). Therapeutic targeting of microenvironmental interactions in leukemia: Mechanisms and approaches. *Drug Resistance Updates*, *12*(4), 103–113.

Koukourakis, M. I., Giatromanolaki, A., Sivridis, E., Pitiakoudis, M., Gatter, K. C., & Harris, A. L. (2010). Beclin 1 over- and underexpression in colorectal cancer: Distinct patterns relate to prognosis and tumour hypoxia. *British Journal of Cancer*, *103*(8), 1209–1214.

Lane, S. W., Scadden, D. T., & Gilliland, D. G. (2009). The leukemic stem cell niche: Current concepts and therapeutic opportunities. *Blood*, *114*(6), 1150–1157.

Lapidot, T., Sirard, C., Vormoor, J., Murdoch, B., Hoang, T., Caceres-Cortes, J., et al. (1994). A cell initiating human acute myeloid leukaemia after transplantation into SCID mice. *Nature*, *367*(6464), 645–648.

Lee, Y. H., Hong, S. W., Jun, W., Cho, H. Y., Kim, H. C., Jung, M. G., et al. (2007). Anti-histone acetyltransferase activity from allspice extracts inhibits androgen receptor-dependent prostate cancer cell growth. *Bioscience, Biotechnology, and Biochemistry*, *71*(11), 2712–2719.

Lee, S. J., Yoon, J., & Song, K. S. (2007). Chrysin inhibited stem cell factor (SCF)/c-kit complex-induced cell proliferation in human myeloid leukemia cells. *Biochemical Pharmacology*, *74*(2), 215–225.

Leveque, L., Vu, T., Kuns, R. D., Teal, B. E., Lor, M., Lineburg, K. E., et al. (2013). Autophagy is required for long-term hematopoietic stem cell (HSC) function and G-CSF-induced HSC mobilization. *Blood*, *122*(21), 892.

Li, T., Fan, G. X., Wang, W., Li, T., & Yuan, Y. K. (2007). Resveratrol induces apoptosis, influences IL-6 and exerts immunomodulatory effect on mouse lymphocytic leukemia both in vitro and in vivo. *International Immunopharmacology*, 7(9), 1221–1231.

Li, Y. C., Tyan, Y. S., Kuo, H. M., Chang, W. C., Hsia, T. C., & Chung, J. G. (2004). Baicalein induced in vitro apoptosis undergo caspases activity in human promyelocytic leukemia HL-60 cells. *Food and Chemical Toxicology*, *42*(1), 37–43.

Li, T., Wang, W., Chen, H., Li, T., & Ye, L. (2010). Evaluation of anti-leukemia effect of resveratrol by modulating SATA3 signaling. *International Immunopharmacology, 10*(1), 18–25.

Li, L., Wang, L., Li, L., Wang, Z., Ho, Y., McDonald, T., et al. (2012). Activation of p53 by SIRT1 inhibition enhances elimination of CML leukemia stem cells in combination with imatinib. *Cancer Cell, 21*(2), 266–281.

Li, J., Yang, Z., Li, Y., Xia, J., Li, D., Li, H., et al. (2016). Cell apoptosis, autophagy and necroptosis in osteosarcoma treatment. *Oncotarget,* 7(28), 44763–44778.

Liang, X. H., Jackson, S., Seaman, M., Brown, K., Kempkes, B., Hibshoosh, H., et al. (1999). Induction of autophagy and inhibition of tumorigenesis by beclin 1. *Nature, 402*(6762), 672–676.

Liao, H. F., Su, Y. C., Zheng, Z. Y., Cai, C. J., Hou, M. H., Chao, K. C., et al. (2012). Sonic hedgehog signaling regulates Bcr-Abl expression in human chronic myeloid leukemia cells. *Biomedicine & Pharmacotherapy, 66*(5), 378–383.

Liu, Q., Chen, L., Atkinson, J. M., Claxton, D. F., & Wang, H. G. (2016). Atg5-dependent autophagy contributes to the development of acute myeloid leukemia in an MLL-AF9-driven mouse model. *Cell Death & Disease,* 7(9), e2361.

Liu, X., Liu, S., Chen, J., He, L., Meng, X., & Liu, S. (2016). Baicalein suppresses the proliferation of acute T-lymphoblastic leukemia Jurkat cells by inhibiting the Wnt/β-catenin signaling. *Annals of Hematology, 95*(11), 1787–1793.

Lobry, C., Ntziachristos, P., Ndiaye-Lobry, D., Oh, P., Cimmino, L., Zhu, N., et al. (2013). Notch pathway activation targets AML-initiating cell homeostasis and differentiation. *The Journal of Experimental Medicine, 210*(2), 301–319.

Magee, J. A., Piskounova, E., & Morrison, S. J. (2012). Cancer stem cells: Impact, heterogeneity, and uncertainty. *Cancer Cell, 21*(3), 283–296.

Man, H. B., Kim, H., Kim, H., Robinson, E., Liu, W. K., Chow, E. K., et al. (2014). Synthesis of nanodiamond-daunorubicin conjugates to overcome multidrug chemoresistance in leukemia. *Nanomedicine, 10*(2), 359–369.

Metcalfe, D. D., Pawankar, R., Ackerman, S. J., Akin, C., Clayton, F., Falcone, F. H., et al. (2016). Biomarkers of the involvement of mast cells, basophils and eosinophils in asthma and allergic diseases. *World Allergy Organization Journal, 9*(7), 7.

Moore, G. W., Knight, G., & Blann, D. A. (2010). *Haematology*. New York, USA: Oxford University Press Inc.

Murdoch, C., Muthana, M., Coffelt, S. B., & Lewis, C. E. (2008). The role of myeloid cells in the promotion of tumour angiogenesis. *Nature Reviews. Cancer, 8*(8), 618–631.

Mustafi, S. B., Chakraborty, P. K., & Raha, S. (2010). Modulation of Akt and ERK1/2 pathways by resveratrol in chronic myelogenous leukemia (CML) cells results in the downregulation of Hsp70. *PloS One, 5*(1), e8719.

National Cancer Institute. (2016). *Drugs approved for leukemia.* https://www.cancer.gov/about-cancer/treatment/drugs/leukemia. Last accessed 20 October 2016.

Nicolai, S., Pieraccioli, M., Peschiaroli, A., Melino, G., & Raschellà, G. (2015). Neuroblastoma: Oncogenic mechanisms and therapeutic exploitation of necroptosis. *Cell Death & Disease, 6,* e2010.

Nikoletopoulou, V., Markaki, M., Palikaras, K., & Tavernarakis, N. (2013). Crosstalk between apoptosis, necrosis and autophagy. *Biochimica et Biophysica Acta, 1833*(12), 3448–3459.

Notta, F., Mullighan, C. G., Wang, J. C., Poeppl, A., Doulatov, S., Phillips, L. A., et al. (2011). Evolution of human BCR-ABL1 lymphoblastic leukaemia-initiating cells. *Nature, 469*(7330), 362–367.

Odot, J., Albert, P., Carlier, A., Tarpin, M., Devy, J., & Madoulet, C. (2004). In vitro and in vivo anti-tumoral effect of curcumin against melanoma cells. *International Journal of Cancer, 111*(3), 381–387.

Ojha, R., Jha, V., Singh, S. K., & Bhattacharyya, S. (2014). Autophagy inhibition suppresses the tumorigenic potential of cancer stem cell enriched side population in bladder cancer. *Biochimica et Biophysica Acta, 1842*(11), 2073–2086.

Onodera, T., Kuriyama, I., Andoh, T., Ichikawa, H., Sakamoto, Y., Lee-Hiraiwa, E., et al. (2015). Influence of particle size on the in vitro and in vivo anti-inflammatory and anti-allergic activities of a curcumin lipid nanoemulsion. *Journal of Molecular Medicine, 35*(6), 1720–1728.

Ouyang, L., Shi, Z., Zhao, S., Wang, F. T., Zhou, T. T., Liu, B., et al. (2012). Programmed cell death pathways in cancer: A review of apoptosis, autophagy and programmed necrosis. *Cell Proliferation, 45*(6), 487–498.

Owen, S. C., & Shoichet, M. S. (2010). Design of three-dimensional biomimetic scaffolds. *Journal of Biomedical Materials Research. Part A, 94*(4), 1321–1331.

Panno, J. (2005). *Cancer: The role of genes, lifestyle and environment.* USA: Facts on File, Inc.

Passegué, E., Jamieson, C. H. M., Ailles, L. E., & Weissman, I. L. (2003). Normal and leukemic hematopoiesis: Are leukemias a stem cell disorder or a reacquisition of stem cell characteristics? *Proceedings of the National Academy of Sciences of the United States of America, 100,* 11842–11849.

Patra, B., Peng, C. C., Liao, W. H., Lee, C. H., & Tung, Y. C. (2016). Drug testing and flow cytometry analysis on a large number of uniform sized tumor spheroids using a microfluidic device. *Scientific Reports, 6,* 21061.

Pina, C., & Enver, T. (2007). Differential contributions of haematopoietic stem cells to foetal and adult haematopoiesis: Insights from functional analysis of transcriptional regulators. *Oncogene, 26*(47), 6750–6765.

Polier, G., Giaisi, M., Köhler, R., Müller, W. W., Lutz, C., Buss, E. C., et al. (2015). Targeting CDK9 by wogonin and related natural flavones potentiates the anti-cancer efficacy of the Bcl-2 family inhibitor ABT-263. *International Journal of Cancer, 136,* 688–698.

Pollack, B. P., Sapkota, B., & Boss, J. M. (2009). Ultraviolet radiation-induced transcription is associated with gene-specific histone acetylation. *Photochemistry and Photobiology, 85*(3), 652–662.

Pramanik, D., Campbell, N. R., Das, S., Gupta, S., Chenna, V., Bisht, S., et al. (2012). A composite polymer nanoparticle overcomes multidrug resistance and ameliorates doxorubicin-associated cardiomyopathy. *Oncotarget, 3*(6), 640–650.

Puissant, A., Grosso, S., Jacquel, A., Belhacene, N., Colosetti, P., Cassuto, J. P., et al. (2008). Imatinib mesylate-resistant human chronic myelogenous leukemia cell lines exhibit high sensitivity to the phytoalexin resveratrol. *The FASEB Journal, 22*(6), 1894–1904.

Puissant, A., Robert, G., Fenouille, N., Luciano, F., Cassuto, J. P., Raynaud, S., et al. (2010). Resveratrol promotes autophagic cell death in chronic myelogenous leukemia cells via JNK-mediated p62/SQSTM1 expression and AMPK activation. *Cancer Research, 70*(3), 1042–1052.

Rao, J., Xu, D. R., Zheng, F. M., Long, Z. J., Huang, S. S., Wu, X., et al. (2011). Curcumin reduces expression of Bcl-2, leading to apoptosis in daunorubicin-insensitive CD34+ acute myeloid leukemia cell lines and primary sorted CD34+ acute myeloid leukemia cells. *Journal of Translational Medicine, 9*(1), 1.

Roy, A., Ray, M., & Jana, K. (2013). Cancer stem cells, Wnt, hedgehog and notch signaling, the role of dietary phytochemicals: New insights for cancer therapy. *Translational Medicine, 3.* e125. http://dx.doi.org/10.4172/2161-1025.1000e125.

Sehgal, A. R., Konig, H., Johnson, D. E., Tang, D., Amaravadi, R. K., Boyiadzis, M., et al. (2015). You eat what you are: Autophagy inhibition as a therapeutic strategy in leukemia. *Leukemia, 29*(3), 517–525.

Sellers, W. R., & Fisher, D. E. (1999). Apoptosis and cancer drug targeting. *The Journal of Clinical Investigation, 104*(12), 1655–1661.

Shakor, A. B. A., Atia, M., Ismail, I. A., Alshehri, A., El-Refaey, H., Kwiatkowska, K., et al. (2014). Curcumin induces apoptosis of multidrug-resistant human leukemia HL60 cells by complex pathways leading to ceramide accumulation. *Biochimica et Biophysica Acta, 1841*(12), 1672–1682.

Shi, Y. H., Ding, Z. B., Zhou, J., Qiu, S. J., & Fan, J. (2009). Prognostic significance of Beclin 1-dependent apoptotic activity in hepatocellular carcinoma. *Autophagy, 5*(3), 380–382.

Sinclair, A., Latif, A. L., & Holyoake, T. L. (2013). Targeting survival pathways in chronic myeloid leukaemia stem cells. *British Journal of Pharmacology, 169*(8), 1693–1707.

Subramaniam, D., Ponnurangam, S., Ramamoorthy, P., Standing, D., Battafarano, R. J., Anant, S., et al. (2012). Curcumin induces cell death in esophageal cancer cells through modulating notch signaling. *PLoS One*, 7(2), e30590.

Tabe, Y., & Konopleva, M. (2014). Advances in understanding the leukaemia microenvironment. *British Journal of Haematology, 164*(6), 767–778.

Tsai, H. C., Li, H., Van-Neste, L., Cai, Y., Robert, C., Rassool, F. V., et al. (2012). Transient low doses of DNA-demethylating agents exert durable antitumor effects on hematological and epithelial tumor cells. *Cancer Cell, 21*(3), 430–446.

Tsuji, D., Ikeda, M., Yamamoto, K., Nakamorim, H., Kim, Y. I., Kawasaki, Y., et al. (2016). Drug-related genetic polymorphisms affecting severe chemotherapy-induced neutropenia in breast cancer patients: A hospital-based observational study. *Medicine (Baltimore), 95*(44), e5151.

Wan, X. B., Fan, X. J., Chen, M. Y., Xiang, J., Huang, P. Y., Guo, L., et al. (2010). Elevated Beclin 1 expression is correlated with HIF-1$\alpha$ in predicting poor prognosis of nasopharyngeal carcinoma. *Autophagy, 6*(3), 395–404.

Wang, Z., Jiang, C., Chen, W., Zhang, G., Luo, D., Cao, Y., et al. (2014). Baicalein induces apoptosis and autophagy via endoplasmic reticulum stress in hepatocellular carcinoma cells. *BioMed Research International, 2014*, 732516. http://dx.doi.org/10.1155/2014/732516. 13 pages.

Weng, G., Zeng, Y., Huang, J., Fan, J., & Guo, K. (2015). Curcumin enhanced busulfan-induced apoptosis through downregulating the expression of survivin in leukemia stem-like KG1a cells. *BioMed Research International, 2015*, 630397. 16 pages.

Woo, K. J., Jeong, Y., Park, J., & Kwon, T. K. (2004). Chrysin-induced apoptosis is mediated through caspase activation and akt inactivation in U937 leukemia cells. *Biochemical and Biophysical Research Communications, 325*(4), 1215–1222.

Wu, J. C., Lai, C. S., Badmaev, V., Nagabhushanam, K., Ho, C. T., & Pan, M. H. (2011). Tetrahydrocurcumin, a major metabolite of curcumin, induced autophagic cell death through coordinative modulation of PI3K/Akt-mTOR and MAPK signaling pathways in human leukemia HL-60 cells. *Molecular Nutrition & Food Research, 55*(11), 1646–1654.

Xia, P., Wang, J. J., Zhao, B. B., & Song, C. L. (2013). The role of beclin-1 expression in patients with gastric cancer: A meta-analysis. *Tumour Biology, 34*(6), 3303–3307.

Xu, J. H., Yang, H. P., Zhou, X. D., Wang, H. J., Liang, G., & Tang, C. L. (2015). Autophagy accompanied with bisdemethoxycurcumin-induced apoptosis in non-small cell lung cancer cells. *Biomedical and Environmental Sciences, 28*(2), 105–115.

Xuan, H. Z., Zhang, J. H., Wang, Y. H., Fu, C. L., & Zhang, W. (2016). Anti-tumor activity evaluation of novel chrysin–organotin compound in MCF-7 cells. *Bioorganic & Medicinal Chemistry Letters, 26*(2), 570–574.

Yallapu, M. M., Ebeling, M. C., Khan, S., Sundram, V., Chauhan, N., Gupta, B. K., et al. (2013). Novel curcumin-loaded magnetic nanoparticles for pancreatic cancer treatment. *Molecular Cancer Therapeutics, 12*(8), 1471–1480.

Yang, C. W., Chang, C. L., Lee, H. C., Chi, C. W., Pan, J. P., & Yang, W. C. (2012). Curcumin induces the apoptosis of human monocytic leukemia THP-1 cells via the activation of JNK/ERK pathways. *BMC Complementary and Alternative Medicine, 12*(1), 1.

Yang, P. M., & Chen, C. C. (2011). Life or death? Autophagy in anticancer therapies with statins and histone deacetylase inhibitors. *Autophagy*, 7(1), 107–108.
Yu, C., Zeng, J., Yan, Z., Ma, Z., Liu, S., & Huang, Z. (2016). Baicalein antagonizes acute megakaryoblastic leukemia in vitro and in vivo by inducing cell cycle arrest. *Cell & Bioscience*, *6*(20), 1–9.

CHAPTER TWELVE

# Chinese Herbal Medicine for the Treatment of Drug Addiction

**Weili Zhu*, Yinan Zhang*, Yingjie Huang*, Lin Lu†,1**

*National Institute on Drug Dependence and Beijing Key Laboratory of Drug Dependence, Peking University, Beijing, China

†Institute of Mental Health, Peking University Sixth Hospital, and Key Laboratory of Mental Health, Beijing, China

[1]Corresponding author: e-mail address: linlu@bjmu.edu.cn

## Contents



## Abstract

This chapter summarizes recent developments in preclinical and clinical research on Chinese herbal medicines and their neurochemical mechanism of action for the treatment of drug addiction. We searched Chinese and English scientific literature and selected several kinds of Chinese herbal medicines that have beneficial effects on drug addiction. *Ginseng* (*Renshen*) may be clinically useful for the prevention of opioid abuse and dependence. *Rhizoma Corydalis* (*Yanhusuo*) may be used to prevent relapse to chronic drug dependence. Alkaloids of *Uncaria rhynchophylla* (*Gouteng*) appear to have positive effects on methamphetamine and ketamine addiction. Both *Salvia miltiorrhiza* (*Danshen*) and *Radix Pueraiae* (*Gegen*) have beneficial inhibitory effects on alcohol intake. Sinomenine has been shown to have preventive and curative effects on opioid dependence. L-Stepholidine, an alkaloid extract of the Chinese herb *Stephania intermedia* (*Rulan*), attenuated the acquisition, maintenance, and reacquisition of morphine-induced conditioned place preference and antagonized the heroin-induced reinstatement of heroin seeking. Traditional Chinese herbal medicines may be used to complement current treatments for drug addiction, including

*International Review of Neurobiology*, Volume 135
ISSN 0074-7742
http://dx.doi.org/10.1016/bs.irn.2017.02.013

withdrawal and relapse. As the molecular mechanisms of action of traditional Chinese herbal medicines are elucidated, further advances in their use for the treatment of drug addiction are promising.

# 1. INTRODUCTION

Drug addiction is a major public health problem that affects an increasing number of people and causes serious burdens to society and the economy. The burden of mental, neurological, and substance use disorders is estimated to increase by 10% in China between 2013 and 2025 (Charlson, Baxter, Cheng, Shidhaye, & Whiteford, 2016). Drug-seeking and -taking behaviors are the main characteristics of drug addiction (Feltenstein & See, 2008). Although various therapies for drug abuse and dependence have been developed, including pharmacological, psychological, and sociological interventions, their efficacy is still limited.

Chinese herbal medicines have been used in China for more than 2000 years as the main therapeutic tool, and such treatments for drug addiction have been used for more than 200 years (Lu et al., 2009). In the last two decades, an increasing number of clinical trials and animal studies have shown that Chinese herbal medicines have efficacy against the acquisition and relapse of drug addiction and withdrawal syndromes.

This chapter discusses the role of Chinese herbal medicines and their main constituents in the treatment of drug addiction in both rodents and humans. We also consider ways in which improving the clinical efficacy of Chinese herbal medicines may increase positive outcomes in drug addiction and related psychiatric disorders.

## 1.1 *Ginseng* (*Renshen*, 人参)

*Ginseng* is one of the most widely used herbal medicines in the world. It has been used for more than 2000 years, mainly in China, Korea, and Japan. *Ginseng* belongs to the *Panax* genus under the family Araliaceae and refers to a group of 11 species of perennial plants. *Ginseng* that is extracted from the roots of these plants is commercially available (Rastogi, Santiago-Moreno, & Dore, 2014). *Ginseng* is composed of many chemical constituents, including triterpene saponins, polysaccharides, peptidoglycans, nitrogen-containing compounds, fatty acids, carbohydrates, and phenolic compounds. It also contains essential oil-containing polyacetylenes and sesquiterpenes (Sticher, 1998). *Ginseng*'s major active components are ginsenosides, which are a form of triterpene glycosides (saponins) (Rastogi et al., 2014). One hundred fifty ginsenosides have been isolated from *ginseng*, 40 of which have been found in *Panax Ginseng* alone (Christensen, 2009), mainly Rb1, Rb2, Rc, Rd, Rg1, Rg2, Rh1, and Re (Attele, Wu, & Yuan, 1999). *Ginseng* affects various aspects of neurodevelopmental, neurodegenerative, and neuropsychiatric disorders (Kim, Kim, & Shin, 2013). *Ginseng* may be clinically useful for the prevention of opioid and psychostimulant abuse and dependence (Tokuyama & Takahashi, 2001). *Panax Ginseng* also decreases alcohol consumption. Most traditional Chinese herbal plants can reduce alcohol absorption from the gastrointestinal tract (Carai et al., 2000). In rats that exhibit cocaine-induced behavioral sensitization, Lee et al. (2008) found that *ginseng* total saponins (GTSs) significantly inhibited the chronic cocaine-induced increase in locomotor activity, associated with a reduction of dopamine (DA) release and postsynaptic neuronal activity, suggesting that GTSs may be used for the treatment of cocaine addiction. Opioid-induced hyperalgesia is characterized by nociceptive sensitization that is caused by the cessation of chronic opioid use. Li, Tang, et al. (2014) investigated the effects of ginsenosides Re, Rg1, and Rb1 on opioid-induced hyperalgesia and found that Re (300 mg/kg) inhibited opioid-induced hyperalgesia in both a thermal sensitivity test and an acetic acid-induced writhing test. Rg1 and Rb1 ginsenosides did not prevent opioid-induced hyperalgesia in either test. In the acetic acid-induced writhing test, Rg1 had a tendency to aggravate opioid-induced hyperalgesia. These results suggest that the ginsenoside Re, rather than Rg1 and Rb1, may reverse opioid-induced hyperalgesia. Lee et al. (2011) investigated the effects of wild *ginseng* on morphine addiction. Compared with *Panax Ginseng*, wild *ginseng* (*Panax quinquefolius*) significantly inhibited morphine-induced increases in Fos expression in the nucleus accumbens (NAc) and tyrosine hydroxylase (TH) expression in the ventral tegmental area (VTA) and locomotor activity. The inhibitory effect of wild *ginseng* on behavioral sensitization after repeated exposure to morphine was closely associated with reductions of DA biosynthesis and postsynaptic neuronal activity, suggesting that wild *ginseng* may inhibit the behavioral effects of morphine, possibly by modulating the central dopaminergic system. The results indicate that wild *ginseng* may be useful for preventing and treating morphine addiction. Kim, Shim, Chung, and Lee (2006) investigated whether the effects of *ginseng* saponins occur by inhibiting the psychomotor stimulant-induced enhancement of dopaminergic transmission. These authors found that GTSs decreased nicotine-induced Fos protein expression in the NAc and striatum, suggesting an inhibitory effect on the nicotine-induced enhancement of dopaminergic transmission. Altogether, these data suggest that GTSs attenuate the nicotine-induced enhancement of dopaminergic transmission and may be useful as a therapeutic agent for nicotine addiction.

## 1.2 *Rhizoma Corydalis* (*Yanhusuo*, 延胡索)

*Rhizoma Corydalis* is a perennial herb of the Papaveraceae family. The dried tuber of *Corydalis yanhusuo W.T. Wang* is used in traditional Chinese medicine for pain relief, anticancer effects, and the enhancement of vital energy (Gao, He, Li, & Wang, 2009). *C. yanhusuo* is one of the top 10 most frequently used herbs in drug addiction (Min et al., 2007). To date, approximately 60 alkaloids from dried *C. yanhusuo* bulbs have been identified, which are known as its major pharmacologically active compounds (Sun, Liu, Lin, Miao, & Lin, 2014). Alkaloids of *C. yanhusuo*, such as L-tetrahydropalmatine (L-THP), act on the neurotransmitter DA in the central nervous system and are effective for the treatment of opiate addiction (Chang & Lin, 2001; Yue et al., 2012). Both L-THP and *C. yanhusuo* normalized the levels of DA, glutamate NR2B, and $D_2$ receptor expression that were altered by chronic morphine treatment in reward-related circuits (Luo et al., 2012; Yu et al., 2013). L-THP has high affinity for $D_1$ and $D_2$ receptors and lower affinity for $D_3$ receptors (Jin, 1987; Marcenac, Jin, & Gonon, 1986). Pharmacological studies have demonstrated that L-THP is a natural antagonist of DA $D_1$ and $D_2$ receptors and is useful for the treatment of drug addiction by antagonizing DA receptors. In an animal model of cocaine self-administration, Mantsch et al. reported that L-THP did not inhibit natural reward-related behavior but decreased drug-induced self-administration behavior and relapse in rats (Mantsch et al., 2010). A heroin self-administration study found that L-THP inhibited heroin-induced self-administration and cue-induced relapse (Yue et al., 2012). L-THP also prevented morphine- and methamphetamine (METH)-induced conditioned place preference (CPP) (Luo et al., 2012; Su et al., 2013; Yu et al., 2013) and had a dose-dependent inhibitory effect on the formation, development, and expression of METH-induced behavioral sensitization in mice (Zhao et al., 2014). The underlying molecular mechanism of action of L-THP on METH-induced locomotor sensitization might be related to the inhibition of extracellular signal-regulated kinase 1/2 phosphorylation in the NAc and caudate putamen (Zhao et al., 2014). A recent study also showed that a combination of L-THP and low-dose naltrexone reduced drug-seeking behavior during reinstatement and upregulated the levels of both plasma β-endorphin and hypothalamic proopiomelanocortin (Sushchyk, Xi, & Wang, 2016). A randomized, double-blind, placebo-controlled study in cocaine abusers found that 3.5-day intake of L-THP was safe and well tolerated, with no significant side effects, suggesting that oral L-THP may be promising for the treatment of cocaine use disorder (Hassan et al., 2017). L-isocorypalmine (L-ICP), a monodemethylated analog of L-THP, was isolated from *C. yanhusuo* and shown to have high affinity for $D_1$ and $D_5$ receptors and moderate affinity for $D_2$, $D_3$, and $D_4$ receptors (Xu et al., 2013). Pretreatment with L-ICP or *C. yanhusuo* before cocaine administration for 6 days inhibited cocaine-induced CPP; L-ICP alone did not cause any place preference (Xu et al., 2013).

## 1.3 *Salvia miltiorrhiza* (*Danshen*, 丹参)

*Danshen* consists of the dried roots of the Lamiaceae plant *Salvia miltiorrhiza Bge*. It has been widely used in traditional Chinese medicine for the treatment of cardiovascular disease, hemorrhage, hepatitis, miscarriage, edema, menstrual disorders, insomnia (Zhou, Zuo, & Chow, 2005), and substance dependence (Mah, Tang, Liauw, Nagel, & Schneider, 1998). More than 30 kinds of tashinone compounds have been isolated from *S. miltiorrhiza*, many of which have various biological activities, including antioxidant, antiplatelet, antitumor, and antiviral effects (Du & Zhang, 2004; Lam, Yeung, Cheung, & Or, 2006; Lin et al., 2006; Ryu, Lee, & Choi, 1997; Wong et al., 2010; Wu et al., 2007; Zhang, An, Zhang, & Pu, 2010; Zhang & Wang, 2006). The main active components of *Danshen* that may play a role in inhibiting alcohol intake are tanshinones I and II, cryptotanshinone, and miltirone (Brunetti et al., 2003). Miltirone, in contrast to diazepam (a full γ-aminobutyric acid [GABA] receptor agonist), shows affinity for central benzodiazepine receptors in the presence of 100 mM GABA, indicating that it might act as a partial agonist (Imanshahidi & Hosseinzadeh, 2006; Lee et al., 1991). Miltirone produces muscle relaxation, sedation, dependence, and a withdrawal syndrome in mice (Lee et al., 1991). Miltirone (1–10 μM) partially inhibited the increase in mRNA levels of the α4 subunit of the GABA receptor that was induced by ethanol withdrawal in cultured hippocampal neurons (Mostallino et al., 2004). *Danshen* has been shown to significantly increase rat striatal $K^+$-stimulated DA release. Protein kinase C (PKC) inhibitors and a mitogen-activated protein kinase (MAPK) inhibitor blocked the ability of

*Danshen* to elicit DA release. *Danshen* may stimulate DA release, and *Danshen*-induced increases in MAPK and PKC appear to be critical for the enhancement in transporter-mediated DA release (Chung et al., 2006). A study in selectively bred Sardinian alcohol-preferring (sP) rats demonstrated the ability of an *S. miltiorrhiza* extract that contained 13% tanshinone IIA, administered orally, to significantly decrease spontaneous alcohol intake and maintenance (Colombo et al., 1999; Vacca et al., 2003). Treatment with *S. miltiorrhiza* extract (0, 50, 100, and 200 mg/kg, i.g.) significantly reduced lever responding for alcohol, the amount self-administered alcohol, and break points for alcohol responding (Maccioni et al., 2014). IDN 5082, a standardized extract of *S. miltiorrhiza*, dose dependently suppressed the acquisition of alcohol-drinking behavior (Brunetti et al., 2003). Serra et al. found that IDN 5082 prevented development of the alcohol deprivation effect in sP rats, indicating that IDN 5082 may have antirelapse properties (Serra et al., 2003). Additionally, *S. miltiorrhiza* also showed potential efficacy against alcoholic liver disease. Tanshinone IIA ameliorated alcoholic liver disease by reducing lipopolysaccharide- and ethanol-induced Kupffer cell sensitization (Yin et al., 2008). Cryptotanshinone can also inhibit alcoholic liver disease by blocking hepatic cell death and fatty acid synthesis (Yin et al., 2009). Salvianolic acid B, isolated from *S. miltiorrhiza*, relieved acute ethanol-induced hepatocyte apoptosis through sirtuin 1-mediated deacetylation of the p53 transcription factor pathway (Li, Lu, et al., 2014). Altogether, these results suggest that *S. miltiorrhiza* may be promising for the treatment of alcoholism and possibly treat and prevent alcoholic liver disease.

## 1.4 *Uncaria rhynchophylla* (*Gouteng*, 钩藤)

*Uncaria rhynchophylla* (*Gouteng*) is a traditional Chinese medicine that belongs to the *Rubiaceae* genus. Thirty-four *U. rhynchophylla* plants have been identified worldwide, mainly distributed in torrid zones. Fourteen types of gambir plants have been found in southern China, including in Yunnan, Guangxi, Guizhou, Sicuan, Jiangxi, and Fujian provinces (Shi, Yu, Chen, & Xu, 2003). The total alkaloid content of *U. rhynchophylla* is ~0.2%, of which rhynchophylline (Rhy) comprises 28%–50% and isorhynchophylline (Isorhy) comprises 15%. It also contains trace components of hirsutine, hirsuteine, corynantheine, dihydrocorynantheine, isocorynoxeine, akuammigine, geissoschijine, and methylethe (Phillipson & Hemingway, 1975; Yamanaka, Kimizuka, Aimi, Sakai, & Haginiwa, 1983).

Gambir plants are prescribed in traditional Chinese medicine for the treatment of cardiovascular disease, lightheadedness, convulsions, numbness, and hypertension. A study in adult zebrafish found that total alkaloids of *U. rhynchophylla* significantly inhibited METH-induced CPP and reduced the expression of TH, NR2B, and glutamate receptor 2 (GluR2), suggesting that alkaloids of *U. rhynchophylla* might have beneficial effects on METH addiction.

## 1.5 *Radix Puerariae* (*Gegen*, 葛根)

*Radix Pueraiae*, also called Kudzu, is the dried root of leguminous *Pueraria lobata* plants. The tea of *Radix Pueraiae* was first recorded in the Chinese materia medica Shen Nong Ben Cao Jing (Pharmacopeia of Shen Nong, ~200 BC) and used as an antipyretic, antidiarrhetic, diaphoretic, and antiemetic agent. As early as 1000 years ago, *Radix Pueraia*e was used to treat alcohol-related problems, such as alcohol intoxication (Sun Simiao, ~600 AD) and alcohol abuse (Li Dongyuan, ~1200 AD). *Radix Pueraiae* has been used in traditional Chinese medicine for the treatment of fevers, gastrointestinal disorders, muscle aches, allergies, skin problems, high blood pressure, and chronic alcoholism (Abascal & Yarnell, 2007; Lee, Lee, & Jeong, 1999). Increasing evidence shows that *Radix Pueraiae* may be beneficial for the treatment of alcohol abuse. A crude extract of *Radix Pueraiae* that included two major constituents, daidzein and daidzin, significantly suppressed free-choice ethanol intake in Golden hamsters (Keung & Vallee, 1993a). Daidzin is a potent, selective, and reversible inhibitor of mitochondrial acetaldehyde dehydrogenase (ALDH-2), which is essential for the oxidation of acetaldehyde derived from ethanol metabolism (Keung & Vallee, 1993b). Daidzin suppressed ethanol intake in hamsters by inhibiting the formation of its reactive intermediates 5-hydroxyindole-3-acetic acid and 3,4-dihydroxyphenylacetic acid from the metabolism of respective amines (Keung & Vallee, 1998). Puerarin, one of the major constituents of *Radix Pueraiae*, reduced ethanol intake in alcohol-preferring rats (Benlhabib, Baker, Keyler, & Singh, 2004). Puerarin from kudzu roots inhibited addiction-related abnormalities. Polyenylphosphatidylcholine from soy and curcumin from turmeric suppressed inflammation-related abnormalities in alcohol-drinking rats that were subjected to lipopolysaccharide-induced pleurisy, suggesting that a mixture of Puerarin and either polyenylphosphatidylcholine or curcumin may be a therapeutic alternative for alcohol-related disorders

(Singh, Jiang, Benlhabib, & Gupta, 2007). Treatment with *Radix Pueraiae* dose dependently inhibited the alcohol-induced suppression of Fos expression in the hippocampus, suggesting that *Radix Pueraiae* may alleviate the alcohol-induced disruption of hippocampal function (Jang et al., 2003). *Radix Pueraiae* also increased the number of bromodeoxyuridine- and NADPH-d-positive cells that was decreased by alcohol administration (Jang et al., 2002). Pretreatment with puerarin reversed the changes in $GABA_A$ receptor expression and increased ADH activity in alcoholism models (Zhang, Li, et al., 2010). Moreover, the number of neurons decreased in the hippocampal dentate gyrus in puerarin-treated mice. Puerarin treatment decreased glutamate and GABA levels and increased glutamate, GABA, tumor necrosis factor α, and interleukin 1β levels in mice that were subjected to chronic alcohol poisoning, reflecting normalization of these neurotransmitters and immune factors (Cui et al., 2015). Both puerarin and daidzein have high affinity for benzodiazepine receptors and inhibit [$^3$H]flunitrazepam binding to membranes in the cortex, cerebellum, and hippocampus (Shen, Witt, Nielsen, & Sterner, 1996).

## 1.6 *Caulis sinomenii* (*Qingfengteng*, 清风藤)

*Caulis sinomenii* is the dry rattan of the leaves of the ivy plants Menispermaceae and *Sinomenium acutum var. Cinereum*. Its main effects include dispelling wind, removing obstructions in the meridians, analgesia, and sedation (Mo, An, & Zhou, 2006). Sinomenine is the primary active ingredient of *C. sinomenii*, whose chemical structure is similar to morphine. Sinomenine is known as one of the strongest histamine-releasing agents in plants. Sinomenine also has significant analgesic, sedative, and anti-inflammatory effects and lowers blood pressure (Liu et al., 2005; Zhao et al., 2005). As the main ingredient of *C. sinomenii*, sinomenine has been shown to prevent and cure opioid dependence (Chen et al., 2005). Animal studies have shown that sinomenine has inhibitory effects on morphine withdrawal symptoms and physical dependence in mice, rats, and guinea pigs (Lu et al., 2004). Morphine-dependent mice exhibit a significant reduction of body weight, whereas sinomenine-treated mice do not present significant weight loss. Sinomenine was shown to relieve withdrawal symptoms and attenuate weight loss in morphine-dependent mice (Wang, Mo, & Shao, 2003). Mice that were treated with sinomenine for 5 days did not exhibit obvious naloxone-precipitated morphine withdrawal symptoms. Although sinomenine has a similar molecular structure as morphine, it does not cause similar somatic withdrawal symptoms (Ding, Li, & Jiang, 2000). In vivo and in vitro studies showed that morphine addiction increases neuronal $Ca^{2+}$ concentrations in rats and mice. In the spinal cord in morphine-dependent mice, the mRNA levels, protein levels, and phosphorylation of calmodulin-dependent protein kinase II$\alpha$ significantly increase and participate in morphine tolerance (Feng, Lu, Huang, Zhu, & Yao, 2005; Liang, Li, & Clark, 2004). Wang et al. reported that sinomenine normalized the levels of norepinephrine (NE), DA, serotonin (5-hydroxytryptamine [5-HT]), and other monoamine neurotransmitters in the brain in morphine-dependent mice after withdrawal (Wang, Mo, Zhu, & Wen, 2002). Excitatory neurotransmitters, such as NE, DA, and 5-HT, are involved in the formation of morphine dependence and tolerance (Koyuncuoglu, Genc, Gungor, Eroglu, & Sagduyu, 1977). Pretreatment with sinomenine reduced cyclic adenosine monophosphate (cAMP) levels in the nervous system in morphine-tolerant mice, prevented the formation of morphine-induced CPP in mice, and prevented the formation of psychological dependence (Wang, Mo, & Zhang, 2002). Central histamine may be involved in the formation of morphine dependence and withdrawal (Mo, Leung, & Yung, 2005). DA activates adenylate cyclase and catalyzes the generation of cAMP, and cAMP levels increase in the nervous system in morphine-dependent mice (Chi, Li, Liu, Zhu, & Gu, 2003). Cyclic guanosine monophosphate that is produced by activation of the nitric oxide/neuronal nitric oxide synthase (NO/nNOS) system inhibits the activity of phosphodiesterase 3, which catalyzes cAMP metabolism, thus increasing cAMP levels in the cerebellum (Li et al., 2003). Sinomenine may attenuate morphine addiction and significantly alleviate morphine withdrawal symptoms. The mechanism of action of sinomenine may be associated with regulation of the NO/nNOS system in the cerebellum and spinal cord (Liu, Zheng, Yang, Yi, & Hu, 2007).

## 1.7 *Stephania intermedia* (*Rulan*, 汝兰)

L-Stepholidine (L-SPD) is an alkaloid extract of the Chinese herb *Stephania intermedia*. It acts as a $D_1$ receptor partial agonist and $D_2$ receptor full antagonist (Jin, Zhu, & Fu, 2002; Zou, Liu, & Jin, 1997). Previous data showed that L-SPD produced antipsychotic-like effects in two animal models of schizophrenia: prepulse inhibition and the paw test in rats (Ellenbroek,

Zhang, & Jin, 2006). Using in vivo single-unit recording in rats, Gao et al. found that L-SPD increased the firing rate of VTA DA neurons and induced neuronal excitation with dual $D_2$ receptor antagonist/5-$HT_{1A}$ receptor agonist properties, suggesting that L-SPD may have potential antipsychotic effects (Gao et al., 2011). Systemic L-SPD administration attenuated the acquisition, maintenance, and reacquisition of morphine-induced CPP, with no effects on food-induced CPP or state-dependent learning (Wang et al., 2007). A recent study showed that systemic L-SPD administration blocked the reinstatement of heroin-seeking behavior, with no effects on locomotor activity and inactive responding, suggesting that L-SPD antagonizes the heroin-induced reinstatement of heroin-seeking behavior (Ma et al., 2014).

### 1.8 *Ferula sinkiangensis K.M. Shen* (*Xinjiang'awei*, 新疆阿魏)

Asafetida extract is derived from the factice resin of *Ferula sinkiangensis K.M. Shen* and *Ferula fukanensis K.M. Shen*. The active compounds of asafetida are α-pinene, 2-borneol, terpin-4-ol, D-fenchyl alcohol, pinocarveol, disulfide compounds, 2,2-dimethylthiopropane, 1,2-diethylthiopropane, and *N,N*-dimethylthioformamide. Asafetida extract significantly decreased morphine abstinence symptoms, including abnormal posture, irritation, and teeth chattering, and attenuated weight loss in morphine-dependent rats (Doosti, Dashti, Tabatabai, & Hosseinzadeh, 2013).

## 2. DISCUSSION

In view of the detrimental social and economic costs of drug abuse, numerous studies have been devoted to discovering effective treatments for substance abuse disorders. Increasingly more attention has been given to the use of traditional Chinese herbal medicines for the treatment of drug addiction. Chinese herbal treatments can supplement modern medications or other traditional medicines to treat drug addiction and prevent relapse, with lower costs and fewer side effects. Chinese herbal medicines can also improve immune function and working memory and protect against neurological disorders.

*Panax Ginseng* may be useful for preventing opioid and psychostimulant abuse and dependence and decreasing alcohol consumption. It can also reduce alcohol absorption from the gastrointestinal tract. *Rhizoma Corydalis* prevents relapse and inhibits the craving for drugs in rats by acting on the DA system. However, further clinical trials are needed to support the possibility that *Rhizoma Corydalis* may be useful for the treatment of addiction. *S. miltiorrhiza* and its main active constituent miltirone significantly ameliorated alcohol intake and withdrawal. *S. miltiorrhiza* has been shown to have excellent psychotherapeutic effects for the treatment of alcoholism and may have additional efficacy for alcoholic liver disease. *U. rhynchophylla* is an essential herbal ingredient that is used for the treatment of drug addiction. Alkaloids of *U. rhynchophylla* might have beneficial effects on methamphetamine and ketamine addiction. *Radix Pueraiae* has long been used for the treatment of alcohol-related problems and inhibits alcohol intake through actions on benzodiazepine receptors and monoamine metabolism (Shen et al., 1996). Sinomenine, a main ingredient of *C. sinomenii*, has been shown to prevent and cure opioid dependence. Pretreatment with sinomenine reduced cAMP levels in the nervous system in morphine-tolerant mice, prevented the formation of morphine-induced CPP in mice, and reversed the formation of psychological dependence. L-SPD, an alkaloid extract of the Chinese herb *S. intermedia*, attenuated the acquisition, maintenance, and reacquisition of morphine-induced CPP and antagonized the heroin-induced reinstatement of heroin-seeking behavior.

In summary, the application of Chinese herbal medicines may be promising alternatives for the treatment of drug addiction. The general therapeutic principle of traditional Chinese medicine is based on the theory of "reinforcing healthy Qi and resolving and removing effects of toxicity" (Kang et al., 2008). Chinese herbs have multiple actions, including sedative and analgesic effects. They can also alleviate fatigue, increase metabolism, and boost immune function while having fewer side effects (Shi et al., 2008). Drug addiction is interpreted as "stagnation of the circulation of healthy Qi" according to traditional Chinese medicine theory. The long-term use of addictive drugs and side effects of pharmaceutical interventions often cause deteriorating health. Chinese herbal medicines can enhance immune function and increase metabolism, which are necessary for fighting infections. Further basic research and clinical trials are essential for investigating the future use of Chinese herbal medicines for the treatment of drug addiction. A deeper understanding of the molecular mechanisms of action of Chinese herbal medicines may open new avenues for the treatment of drug addiction.

## REFERENCES

Abascal, K., & Yarnell, E. (2007). Kudzu—The miracle vine. *Alternative and Complementary Therapies*, *13*, 78–85.

Attele, A. S., Wu, J. A., & Yuan, C. S. (1999). Ginseng pharmacology: Multiple constituents and multiple actions. *Biochemical Pharmacology, 58*, 1685–1693.

Benlhabib, E., Baker, J. I., Keyler, D. E., & Singh, A. K. (2004). Effects of purified puerarin on voluntary alcohol intake and alcohol withdrawal symptoms in P rats receiving free access to water and alcohol. *Journal of Medicinal Food, 7*, 180–186.

Brunetti, G., Serra, S., Vacca, G., Lobina, C., Morazzoni, P., Bombardelli, E., et al. (2003). IDN 5082, a standardized extract of Salvia miltiorrhiza, delays acquisition of alcohol drinking behavior in rats. *Journal of Ethnopharmacology, 85*, 93–97.

Carai, M. A., Agabio, R., Bombardelli, E., Bourov, I., Gessa, G. L., Lobina, C., et al. (2000). Potential use of medicinal plants in the treatment of alcoholism. *Fitoterapia, 71*(Suppl. 1), S38–S42.

Chang, C. K., & Lin, M. T. (2001). DL-Tetrahydropalmatine may act through inhibition of amygdaloid release of dopamine to inhibit an epileptic attack in rats. *Neuroscience Letters, 307*, 163–166.

Charlson, F. J., Baxter, A. J., Cheng, H. G., Shidhaye, R., & Whiteford, H. A. (2016). The burden of mental, neurological, and substance use disorders in China and India: A systematic analysis of community representative epidemiological studies. *Lancet, 388*, 376–389.

Chen, S. W., Mi, X. J., Wang, R., Wang, W. J., Kong, W. X., Zhang, Y. J., et al. (2005). Behavioral effects of sinomenine in murine models of anxiety. *Life Sciences, 78*, 232–238.

Chi, S. M., Li, C. X., Liu, Y. L., Zhu, Y. L., & Gu, J. W. (2003). Influence of ACh on the level of protein kinase C, intracellular free Ca(2+) and cyclic AMP/cyclic GMP of cultured human pituitary adenoma cells. *Sheng Li Xue Bao, 55*, 165–170.

Christensen, L. P. (2009). Ginsenosides chemistry, biosynthesis, analysis, and potential health effects. *Advances in Food and Nutrition Research, 55*, 1–99.

Chung, T. W., Koo, B. S., Kim, K. O., Jeong, H. S., Kim, M. G., Chung, K. H., et al. (2006). Salviae Miltiorrhizae BGE Radix increases rat striatal K(+)-stimulated dopamine release and activates the dopamine release with protection against hydrogen peroxide-induced injury in rat pheochromocytoma PC12 cells. *Neurochemical Research, 31*, 109–120.

Colombo, G., Agabio, R., Lobina, C., Reali, R., Morazzoni, P., Bombardelli, E., et al. (1999). Salvia miltiorrhiza extract inhibits alcohol absorption, preference, and discrimination in sP rats. *Alcohol, 18*, 65–70.

Cui, S. Q., Wang, Q., Zheng, Y., Xiao, B., Sun, H. W., Gu, X. L., et al. (2015). Puerarin protects against damage to spatial learning and memory ability in mice with chronic alcohol poisoning. *Brazilian Journal of Medical and Biological Research, 48*, 515–522.

Ding, Z. R., Li, G. S., & Jiang, X. J. (2000). The effect of sinomenine on sodium L-type callium current in guinea pig ventricular myocytes. *The Chinese Journal of Cardiac Pacing and Electrophysiology, 14*, 39–41.

Doosti, F., Dashti, S., Tabatabai, S. M., & Hosseinzadeh, H. (2013). Traditional Chinese and Indian medicine in the treatment of opioid-dependence: A review. *Avicenna Journal of Phytomedicine, 3*, 205–215.

Ellenbroek, B. A., Zhang, X. X., & Jin, G. Z. (2006). Effects of (-)stepholidine in animal models for schizophrenia. *Acta Pharmacologica Sinica, 27*, 1111–1118.

Feltenstein, M. W., & See, R. E. (2008). The neurocircuitry of addiction: An overview. *British Journal of Pharmacology, 154*, 261–274.

Feng, H., Lu, L. M., Huang, Y., Zhu, Y. C., & Yao, T. (2005). Blockade of NMDA receptor enhances corticosterone-induced downregulation of brain-derived neurotrophic factor gene expression in the rat hippocampus through cAMP response element binding protein pathway. *Sheng Li Xue Bao, 57*, 537–544.

Gao, M., Chu, H. Y., Jin, G. Z., Zhang, Z. J., Wu, J., & Zhen, X. C. (2011). l-Stepholidine-induced excitation of dopamine neurons in rat ventral tegmental area is associated with its 5-HT(1A) receptor partial agonistic activity. *Synapse, 65*, 379–387.

Gao, J. L., He, T. C., Li, Y. B., & Wang, Y. T. (2009). A traditional Chinese medicine formulation consisting of Rhizoma Corydalis and Rhizoma Curcumae exerts synergistic anti-tumor activity. *Oncology Reports, 22*, 1077–1083.

Du, G. H., & Zhang, J. T. (2004). The general situation and progress of the modern research of red sage root (Radix Salviae miltiorrhizae). *Yiyao Daobao, 23*, 435–440.

Hassan, H. E., Kelly, D., Honick, M., Shukla, S., Ibrahim, A., Gorelick, D. A., et al. (2017). Pharmacokinetics and safety assessment of l-tetrahydropalmatine in cocaine users: A randomized, double-blind, placebo-controlled study. *Journal of Clinical Pharmacology, 57*, 151–160.

Imanshahidi, M., & Hosseinzadeh, H. (2006). The pharmacological effects of Salvia species on the central nervous system. *Phytotherapy Research, 20*, 427–437.

Jang, M. H., Shin, M. C., Chung, J. H., Shin, H. D., Kim, Y., Kim, E. H., et al. (2002). Effects of Puerariae radix on cell proliferation and nitric oxide synthase expression in dentate gyrus of alcohol-intoxicated Sprague-Dawley rats. *Japanese Journal of Pharmacology, 88*, 355–358.

Jang, M. H., Shin, M. C., Lee, T. H., Bahn, G. H., Shin, H. S., Lim, S., et al. (2003). Effect of Puerariae radix on c-Fos expression in hippocampus of alcohol-intoxicated juvenile rats. *Biological & Pharmaceutical Bulletin, 26*, 37–40.

Jin, G. Z. (1987). Progress in studies of the pharmacology of l-tetrahydropalmatine and l-stepholidine. *Yao Xue Xue Bao, 22*, 472–480.

Jin, G. Z., Zhu, Z. T., & Fu, Y. (2002). (−)-Stepholidine: A potential novel antipsychotic drug with dual D1 receptor agonist and D2 receptor antagonist actions. *Trends in Pharmacological Sciences, 23*, 4–7.

Kang, L., Li, B., Gao, L., Li, S., Wang, D., Hu, M., et al. (2008). Tai-Kang-Ning, a Chinese herbal medicine formula, alleviates acute heroin withdrawal. *The American Journal of Drug and Alcohol Abuse, 34*, 269–276.

Keung, W. M., & Vallee, B. L. (1993a). Daidzin and daidzein suppress free-choice ethanol intake by Syrian golden hamsters. *Proceedings of the National Academy of Sciences of the United States of America, 90*, 10008–10012.

Keung, W. M., & Vallee, B. L. (1993b). Daidzin: A potent, selective inhibitor of human mitochondrial aldehyde dehydrogenase. *Proceedings of the National Academy of Sciences of the United States of America, 90*, 1247–1251.

Keung, W. M., & Vallee, B. L. (1998). Daidzin and its antidipsotropic analogs inhibit serotonin and dopamine metabolism in isolated mitochondria. *Proceedings of the National Academy of Sciences of the United States of America, 95*, 2198–2203.

Kim, H. J., Kim, P., & Shin, C. Y. (2013). A comprehensive review of the therapeutic and pharmacological effects of ginseng and ginsenosides in central nervous system. *Journal of Ginseng Research, 37*, 8–29.

Kim, S. E., Shim, I., Chung, J. K., & Lee, M. C. (2006). Effect of ginseng saponins on enhanced dopaminergic transmission and locomotor hyperactivity induced by nicotine. *Neuropsychopharmacology, 31*, 1714–1721.

Koyuncuoglu, H., Genc, E., Gungor, M., Eroglu, L., & Sagduyu, H. (1977). The antagonizing effect of aspartic acid on the brain levels of monoamines and free amino acids during the development of tolerance to the physical dependence on morphine. *Psychopharmacology, 54*, 187–191.

Lam, F. F., Yeung, J. H., Cheung, J. H., & Or, P. M. (2006). Pharmacological evidence for calcium channel inhibition by danshen (Salvia miltiorrhiza) on rat isolated femoral artery. *Journal of Cardiovascular Pharmacology, 47*, 139–145.

Lee, B., Kwon, S., Yeom, M., Shim, I., Lee, H., & Hahm, D. H. (2011). Wild ginseng attenuates repeated morphine-induced behavioral sensitization in rats. *Journal of Microbiology and Biotechnology, 21*, 757–765.

Lee, J. S., Lee, K. H., & Jeong, J. H. (1999). Effects of extract of Pueraria radix on lipid metabolism in rats fed high fat diet. *Journal of the Korean Society of Food Science and Nutrition, 28*, 218–224.

Lee, C. M., Wong, H. N., Chui, K. Y., Choang, T. F., Hon, P. M., & Chang, H. M. (1991). Miltirone, a central benzodiazepine receptor partial agonist from a Chinese medicinal herb Salvia miltiorrhiza. *Neuroscience Letters, 127*, 237–241.

Lee, B., Yang, C. H., Hahm, D. H., Lee, H. J., Han, S. M., Kim, K. S., et al. (2008). Inhibitory effects of ginseng total saponins on behavioral sensitization and dopamine release induced by cocaine. *Biological & Pharmaceutical Bulletin, 31*, 436–441.

Li, J., Li, Y. H., Zhang, X. H., Zhu, X. J., Ge, Y. B., & Yuan, X. R. (2003). Changes in the phosphorylation of cAMP response element binding protein in the rat nucleus accumbens after acute and chronic ethanol administration. *Sheng Li Xue Bao, 55*, 147–152.

Li, M., Lu, Y., Hu, Y., Zhai, X., Xu, W., Jing, H., et al. (2014). Salvianolic acid B protects against acute ethanol-induced liver injury through SIRT1-mediated deacetylation of p53 in rats. *Toxicology Letters, 228*, 67–74.

Li, P., Tang, M., Li, H., Huang, X., Chen, L., & Zhai, H. (2014). Effects of ginsenosides on opioid-induced hyperalgesia in mice. *Neuroreport, 25*, 749–752.

Liang, D., Li, X., & Clark, J. D. (2004). Increased expression of $Ca^{2+}$/calmodulin-dependent protein kinase II alpha during chronic morphine exposure. *Neuroscience, 123*, 769–775.

Lin, Y. L., Wu, C. H., Luo, M. H., Huang, Y. J., Wang, C. N., Shiao, M. S., et al. (2006). In vitro protective effects of salvianolic acid B on primary hepatocytes and hepatic stellate cells. *Journal of Ethnopharmacology, 105*, 215–222.

Liu, Z. Q., Chan, K., Zhou, H., Jiang, Z. H., Wong, Y. F., Xu, H. X., et al. (2005). The pharmacokinetics and tissue distribution of sinomenine in rats and its protein binding ability in vitro. *Life Sciences, 77*, 3197–3209.

Liu, Z., Zheng, J. F., Yang, L. Q., Yi, L., & Hu, B. (2007). Effects of sinomenine on NO/nNOS system in cerebellum and spinal cord of morphine-dependent and withdrawal mice. *Sheng Li Xue Bao, 59*, 285–292.

Lu, D. Q., Li, H. G., Ye, H., Ye, S. Q., Jin, S., & Wang, D. X. (2004). Effects of histamine on endothelial nitric oxide synthase expression in pulmonary artery endothelial cells. *Sheng Li Xue Bao, 56*, 288–294.

Lu, L., Liu, Y., Zhu, W., Shi, J., Liu, Y., Ling, W., et al. (2009). Traditional medicine in the treatment of drug addiction. *The American Journal of Drug and Alcohol Abuse, 35*, 1–11.

Luo, S. Y., Guo, P., Qian, G., Yang, M. L., Lin, X., & Yang, P. R. (2012). Study on acting mechanism of anti-morphine conditioned place preference between aqueous extract of Corydalis yanhusuo and L-THP and comparison of their effects. *Zhongguo Zhong Yao Za Zhi, 37*, 3457–3461.

Ma, B., Yue, K., Chen, L., Tian, X., Ru, Q., Gan, Y., et al. (2014). L-stepholidine, a natural dopamine receptor D1 agonist and D2 antagonist, inhibits heroin-induced reinstatement. *Neuroscience Letters, 559*, 67–71.

Maccioni, P., Vargiolu, D., Falchi, M., Morazzoni, P., Riva, A., Cabri, W., et al. (2014). Reducing effect of the Chinese medicinal herb, Salvia miltiorrhiza, on alcohol self-administration in Sardinian alcohol-preferring rats. *Alcohol, 48*, 587–593.

Mah, S. J., Tang, Y., Liauw, P. E., Nagel, J. E., & Schneider, A. S. (1998). Ibogaine acts at the nicotinic acetylcholine receptor to inhibit catecholamine release. *Brain Research, 797*, 173–180.

Mantsch, J. R., Wisniewski, S., Vranjkovic, O., Peters, C., Becker, A., Valentine, A., et al. (2010). Levo-tetrahydropalmatine attenuates cocaine self-administration under a

progressive-ratio schedule and cocaine discrimination in rats. *Pharmacology, Biochemistry, and Behavior, 97*, 310–316.
Marcenac, F., Jin, G. Z., & Gonon, F. (1986). Effect of l-tetrahydropalmatine on dopamine release and metabolism in the rat striatum. *Psychopharmacology, 89*, 89–93.
Min, X., Lee, D. T., Jinhua, X., Wenjun, D., Li, C., Bin, D., et al. (2007). A database on treating drug addiction with traditional Chinese medicine. *Addiction, 102*, 282–288.
Mo, Z. X., An, S. L., & Zhou, J. Y. (2006). Effects of Caulis Sinomenii and sinomenine on morphine-induced place preference and brain histamine level in mice. *Nan Fang Yi Ke Da Xue Xue Bao, 26*, 1709–1713.
Mo, Z. X., Leung, W. N., & Yung, K. L. (2005). Changes in cAMP and cGMP levels in neonatal rat histaminergic neurons of tuberomammillary nucleus following 48-hour morphine exposure and effects of sinomenine intervention. *Di 1 Jun Yi Da Xue Xue Bao = Academic Journal of the First Medical College of PLA, 25*, 1105–1108.
Mostallino, M. C., Mascia, M. P., Pisu, M. G., Busonero, F., Talani, G., & Biggio, G. (2004). Inhibition by miltirone of up-regulation of $GABA_A$ receptor alpha4 subunit mRNA by ethanol withdrawal in hippocampal neurons. *European Journal of Pharmacology, 494*, 83–90.
Phillipson, J. D., & Hemingway, S. R. (1975). Chromatographic and spectroscopic methods for the identification of alkaloids from herbarium samples of the genus Uncaria. *Journal of Chromatography, 105*, 163–178.
Rastogi, V., Santiago-Moreno, J., & Dore, S. (2014). Ginseng: A promising neuroprotective strategy in stroke. *Frontiers in Cellular Neuroscience, 8*, 457.
Ryu, S. Y., Lee, C. O., & Choi, S. U. (1997). In vitro cytotoxicity of tanshinones from Salvia miltiorrhiza. *Planta Medica, 63*, 339–342.
Serra, S., Vacca, G., Tumatis, S., Carrucciu, A., Morazzoni, P., Bombardelli, E., et al. (2003). Anti-relapse properties of IDN 5082, a standardized extract of Salvia miltiorrhiza, in alcohol-preferring rats. *Journal of Ethnopharmacology, 88*, 249–252.
Shen, X. L., Witt, M. R., Nielsen, M., & Sterner, O. (1996). Inhibition of [3H] flunitrazepam binding to rat brain membranes in vitro by puerarin and daidzein. *Yao Xue Xue Bao, 31*, 59–62.
Shi, J., Xu, G. Z., Liu, T. T., Wang, X., Shen, L. Y., Li, J., et al. (2008). A comparative clinical study of the effects of the traditional Chinese medicine Jinniu capsules and lofexidine on acute heroin withdrawal symptoms. *The American Journal of Drug and Alcohol Abuse, 34*, 792–800.
Shi, J. S., Yu, J. X., Chen, X. P., & Xu, R. X. (2003). Pharmacological actions of Uncaria alkaloids, rhynchophylline and isorhynchophylline. *Acta Pharmacologica Sinica, 24*, 97–101.
Singh, A. K., Jiang, Y., Benlhabib, E., & Gupta, S. (2007). Herbal mixtures consisting of puerarin and either polyenylphosphatidylcholine or curcumin provide comprehensive protection against alcohol-related disorders in P rats receiving free choice water and 15% ethanol in pure water. *Journal of Medicinal Food, 10*, 526–542.
Sticher, O. (1998). Getting to the root of Ginseng. *Chemtech, 28*, 26–32.
Su, H. L., Zhu, J., Chen, Y. J., Zhao, N., Han, W., Dang, Y. H., et al. (2013). Roles of levo-tetrahydropalmatine in modulating methamphetamine reward behavior. *Physiology & Behavior, 118*, 195–200.
Sun, M., Liu, J., Lin, C., Miao, L., & Lin, L. (2014). Alkaloid profiling of the traditional Chinese medicine Rhizoma corydalis using high performance liquid chromatography-tandem quadrupole time-of-flight mass spectrometry. *Acta Pharmaceutica Sinica B, 4*, 208–216.
Sushchyk, S., Xi, Z. X., & Wang, J. B. (2016). Combination of levo-tetrahydropalmatine and low dose naltrexone: A promising treatment for prevention of cocaine relapse. *The Journal of Pharmacology and Experimental Therapeutics, 357*, 248–257.

Tokuyama, S., & Takahashi, M. (2001). Pharmacological and physiological effects of ginseng on actions induced by opioids and psychostimulants. *Nihon Yakurigaku Zasshi, 117*, 195–201.
Vacca, G., Colombo, G., Brunetti, G., Melis, S., Molinari, D., Serra, S., et al. (2003). Reducing effect of Salvia miltiorrhiza extracts on alcohol intake: Influence of vehicle. *Phytotherapy Research, 17*, 537–541.
Wang, C. Y., Mo, Z. X., & Shao, H. X. (2003). Effects of sinomenine on the psychic dependence on morphine and the brain cyclic AMP level in mice. *Chinese Pharmacological Bulletin, 19*, 575–577.
Wang, C. Y., Mo, Z. X., & Zhang, Z. H. (2002b). The effect of sinomenine on CPP of morphine dependent mice. *Chinese Journal of Drug Abuse Prevention and Treatment, 6*, 818.
Wang, C., Mo, Z., Zhu, Q., & Wen, L. (2002a). Effect of sinamine on withdrawal symptom and neurotransmitter of morphine-dependent rats. *Zhong Yao Cai, 25*, 337–339.
Wang, W., Zhou, Y., Sun, J., Pan, L., Kang, L., Dai, Z., et al. (2007). The effect of L-stepholidine, a novel extract of Chinese herb, on the acquisition, expression, maintenance, and re-acquisition of morphine conditioned place preference in rats. *Neuropharmacology, 52*, 355–361.
Wong, K. K., Ho, M. T., Lin, H. Q., Lau, K. F., Rudd, J. A., Chung, R. C., et al. (2010). Cryptotanshinone, an acetylcholinesterase inhibitor from Salvia miltiorrhiza, ameliorates scopolamine-induced amnesia in Morris water maze task. *Planta Medica, 76*, 228–234.
Wu, Z. M., Wen, T., Tan, Y. F., Liu, Y., Ren, F., & Wu, H. (2007). Effects of salvianolic acid A on oxidative stress and liver injury induced by carbon tetrachloride in rats. *Basic & Clinical Pharmacology & Toxicology, 100*, 115–120.
Xu, W., Wang, Y., Ma, Z., Chiu, Y. T., Huang, P., Rasakham, K., et al. (2013). L-isocorypalmine reduces behavioral sensitization and rewarding effects of cocaine in mice by acting on dopamine receptors. *Drug and Alcohol Dependence, 133*, 693–703.
Yamanaka, E., Kimizuka, Y., Aimi, N., Sakai, S., & Haginiwa, J. (1983). Studies of plants containing indole alkaloids. IX. Quantitative analysis of tertiary alkaloids in various parts of Uncaria rhynchophylla MIQ. *Yakugaku Zasshi, 103*, 1028–1033.
Yin, H. Q., Choi, Y. J., Kim, Y. C., Sohn, D. H., Ryu, S. Y., & Lee, B. H. (2009). Salvia miltiorrhiza Bunge and its active component cryptotanshinone protects primary cultured rat hepatocytes from acute ethanol-induced cytotoxicity and fatty infiltration. *Food and Chemical Toxicology, 47*, 98–103.
Yin, H. Q., Kim, Y. S., Choi, Y. J., Kim, Y. C., Sohn, D. H., Ryu, S. Y., et al. (2008). Effects of tanshinone IIA on the hepatotoxicity and gene expression involved in alcoholic liver disease. *Archives of Pharmacal Research, 31*, 659–665.
Yu, S. Y., Yang, P. R., Qian, G., Wu, M. S., Bai, W. F., Tu, P., et al. (2013). Study on effects of Corydalis yanhusuo and L-THP on dopamine of reward circuitry in conditioned place preference rats and comparison. *Zhongguo Zhong Yao Za Zhi, 38*, 3928–3932.
Yue, K., Ma, B., Ru, Q., Chen, L., Gan, Y., Wang, D., et al. (2012). The dopamine receptor antagonist levo-tetrahydropalmatine attenuates heroin self-administration and heroin-induced reinstatement in rats. *Pharmacology, Biochemistry, and Behavior, 102*, 1–5.
Zhang, H. N., An, C. N., Zhang, H. N., & Pu, X. P. (2010). Protocatechuic acid inhibits neurotoxicity induced by MPTP in vivo. *Neuroscience Letters, 474*, 99–103.
Zhang, Z., Li, S., Jiang, J., Yu, P., Liang, J., & Wang, Y. (2010). Preventive effects of Flos Perariae (Gehua) water extract and its active ingredient puerarin in rodent alcoholism models. *Chinese Medicine, 5*, 36.
Zhang, H. S., & Wang, S. Q. (2006). Salvianolic acid B from Salvia miltiorrhiza inhibits tumor necrosis factor-alpha (TNF-alpha)-induced MMP-2 upregulation in human aortic smooth muscle cells via suppression of NAD(P)H oxidase-derived reactive oxygen species. *Journal of Molecular and Cellular Cardiology, 41*, 138–148.

Zhao, N., Chen, Y., Zhu, J., Wang, L., Cao, G., Dang, Y., et al. (2014). Levo-tetrahydropalmatine attenuates the development and expression of methamphetamine-induced locomotor sensitization and the accompanying activation of ERK in the nucleus accumbens and caudate putamen in mice. *Neuroscience, 258*, 101–110.

Zhao, Z. Z., Liang, Z. T., Zhou, H., Jiang, Z. H., Liu, Z. Q., Wong, Y. F., et al. (2005). Quantification of sinomenine in caulis sinomenii collected from different growing regions and wholesale herbal markets by a modified HPLC method. *Biological & Pharmaceutical Bulletin, 28*, 105–109.

Zhou, L., Zuo, Z., & Chow, M. S. (2005). Danshen: An overview of its chemistry, pharmacology, pharmacokinetics, and clinical use. *Journal of Clinical Pharmacology, 45*, 1345–1359.

Zou, L. L., Liu, J., & Jin, G. Z. (1997). Involvement of receptor reserve in D1 agonistic action of (−)-stepholidine in lesioned rats. *Biochemical Pharmacology, 54*, 233–240.

CHAPTER THIRTEEN

# Effect of Chinese Herbal Medicine on Male Infertility

**Dan Jiang*[,1], Alberto Coscione[†], Lily Li[‡], Bai-Yun Zeng[§]**
*Hallam Institution of TCM in Sheffield UK, Sheffield, United Kingdom
[†]The Princess Alexandra Hospital NHS Trust, Harlow, United Kingdom
[‡]St Mary's Hospital Paddington, London, United Kingdom
[§]Institute of Pharmaceutical Science, King's College, London, United Kingdom
[1]Corresponding author: e-mail address: djiang52@hotmail.com

## Contents



## Abstract

Male infertility normally refers a male's inability to cause pregnancy in a fertile female partner after 1 year of unprotected intercourse. Male infertility in recent years has been attracting increasing interest from public due to the evidence in decline in semen quality. There are many factors contributing to the male infertility including abnormal spermatogenesis; reproductive tract anomalies or obstruction; inadequate sexual and ejaculatory functions; and impaired sperm motility, imbalance in hormone levels, and immune system dysfunction. Although conventional treatments such as medication, surgical operation, and advanced techniques have helped many male with infertility cause pregnancy in their female partners, effectiveness is not satisfactory and associated with adverse effects. Chinese herbal medicine (CHM) has been used to improve male infertility in China for a very long time and has now been increasingly popular in Western countries for treating infertility. In this chapter we summarized recent

*International Review of Neurobiology*, Volume 135
ISSN 0074-7742
http://dx.doi.org/10.1016/bs.irn.2017.02.014

development in basic research and clinical studies of CHM in treating male infertility. It has showed that CHM improved sperm motility and quality, increased sperm count and rebalanced inadequate hormone levels, and adjusted immune functions leading to the increased number of fertility. Further, CHM in combination with conventional therapies improved efficacy of conventional treatments. More studies are needed to indentify the new drugs from CHM and ensure safety, efficacy, and consistency of CHM.

## 1. INTRODUCTION

Infertility is defined by the World Health Organization as a disorder of reproductive system, characterized by failure to achieve a clinical pregnancy after ≥12 months of regular unprotected intercourse (Mascarenhas, Flaxman, Boerma, Vanderpoel, & Stevens, 2012; Zegers-Hochschild et al., 2009). Although much attention has been focused on issues related with infertility in women, male infertility, in recent years, has been attracting increasing interest due to evidence in decline in semen quality among young health men and public awareness (Alrabeeah et al., 2014). More than 90% of male infertility cases are due to low sperm counts, poor sperm quality, or both (Bensdorp, Cohlen, Heineman, & Vandekerckhove, 2007; Levine & Grifo, 2008). The remaining cases of male infertility can be linked with a number of factors including ejaculation dysfunction, immunological factor, hormonal imbalances, and genetic defects (Corona et al., 2015; Pitteloud & Dwyer, 2014; Ray et al., 2017; Zhao, Zhu, Xue, & Han, 2014). In addition, obesity and varicocele are believed to have adverse impacts on male fertility (Evers & Collins, 2003; Sallmén, Sandler, Hoppin, Blair, & Baird, 2006). Conventional treatments such as drug therapy, surgical approaches, intrauterine insemination, and intracytoplasmic sperm injection (ICSI) helped many men with fertility problems achieve clinical pregnancy (Dorjpurev et al., 2011; Ho & Tan, 2013; Johnson, Sasson, Sammel, & Dokras, 2013; Pisipati & Pearcy, 2010; Valenti et al., 2013). However, those treatments are sometimes ineffective, invasive, and expensive or associated with adverse effects and high risks (Cissen et al., 2016; Palomba, Homburg, Santagni, La Sala, & Orvieto, 2016).

Chinese herbal medicine (CHM), as an important part of traditional Chinese medicine (TCM), has been used to treat infertility for both women and men in China for a long time (Xu, Yin, Tang, Zhang, & Gosden, 2003). CHM helped improving sperm quality (Liao et al., 2013; Zeng et al., 2003) and rebalance immune system (Yu, Zhang, & Zhang, 2004). When used in combination with conventional medicine, CHM enhanced the efficacy of conventional medicine and reduced its side effect (Ma & Jia, 2011). However, the mechanism of action underlying the therapeutic effect of CHM is not fully understood. In this chapter, we reviewed the published studies that concerned effect of CHM on male infertility from online databases including PubMed, Medline, EMBASE, CNKI, CIAHL, and Wanfan for the past 20 years and summarized the development of basic research and clinical studies of CHM on the male infertility.

## 2. PRECLINICAL RESEARCHES BY ANIMAL MODELS

A number of studies looked at established animal models of sexual dysfunction, or sperm disorders, or in particular TCM differentiation patterns relevant to infertility; and assessed the effect and the underlying mechanism of action of CHM on these models.

### 2.1 Improving Sperm Abnormalities and Quality

Recently, adenine (6-amino-purine), a nitrogen-containing heterocyclic purine compound, was used to develop male infertility model in rat (Wang, Chen, Zhou, Cao, & Jia, 2008; Wang, Huang, & Liang, 2008). Adenine may reduce spermatogenesis and testosterone synthesis through the large amount of free radicals generated by the xanthine oxidase reaction; or may be related to transforming growth factor-beta (TGF-β)'s inhibition of spermatogenesis and promotion of spermatocyte apoptosis (Huang, Xia, & Ren, 2003). Adenine-induced male infertile rat model was used to investigate the effect and mechanisms of some CHM formulas, because the symptoms adenine-induced is similar to the human Shen-yang (kidney-Yang) deficiency pattern according to TCM theory (Yue, Chen, & Dai, 1997). Bushen Shengjing Decoction (BSSJD 补肾生精汤) was one of the CHM formulas earlier used in the male infertile model to explore its mechanism of therapeutic effect in treating infertility (Yue et al., 1997). Administration of BSSJD to adenine-treated rats for 30 days significantly increased the number of sperm, improved quality of sperm, and normalized serum levels of luteinizing hormone (LH), testosterone compared with non-treatment group (Yue et al., 1997). This indicated that BSSJD formula could improve testicular function.

Ma, Wang, and Lu (2006) investigated the effect of Wenyang Shengjing Decoction (WSD 温阳生精汤) on Smads proteins and the relevant genes expression. Smads are intracellular signal transduction molecules, belonging to the TGF-β family, in the testis of adenine-induced male infertile rats. It was observed that in the normal rats, Smad 1 and Smad 2 were expressed in cytoplasm of spermatogenic cells in rats' testis. Positive Smad 2 expression could be found in cytoplasm of Sertoli's cell. Both Smad 1 and Smad 2 were not expressed in Leydig's cell. On the other hand, Smad 4 was positively expressed in cytoplasm of Leydig's cell but not expressed in spermatogenic cell and Sertoli's cell (Ma et al., 2006). Compared with the normal control, in adenine-induced male infertile rats, Smad 1 expression was markedly lower, but expression of Smad 2 and Smad 4 was markedly higher; WSD treatment significantly reversed these abnormal changes. Further, WSD treatment significantly neutralized adenine-induced changes in body weight, sperm number and serum testosterone level, and levels of follicle-stimulating hormone and LH (Ma et al., 2006). In another study, Liu, Wang, and Nan (2012) assessed the effects of WSD containing serum on the estradiol (E2) secretion, the synthesized cytochrome P450 aromatase (P450arom) activities, and the expression of its encoding gene CYP19 in Leydig cells of adenine-induced male infertility rats. The results showed that the E2 secretion of Leydig cells obviously decreased accompanied with the inhibition of P450arom activities and significantly reduced protein and mRNA expressions of CYP19 in the model groups. Following administration of WSD containing serum to adenine-induced infertile model, the E2 secretion in the Leydig cells significantly increased, the P450arom activities upregulated, and the CYP19 expressions partially upregulated at the protein and mRNA in a dose-dependent manner (Liu et al., 2012). Together, those studies showed that WDS formula could improve male infertility via multiple action mechanisms. It could increase the sperm number through elevating serum testosterone level, and adjust the levels of sex hormones via regulating Smads genes expression to promote the production of sperm; or could effectively elevate the E2 secretion in Leydig cells, through upregulating P450arom expression.

Using the same model, Wang, Li, Zhang, and Dang (2014) studied the influence of Yougui Capsule (右归胶囊) on Kidney Yang deficiency-related reproductive dysfunction in infertile rats. They found that administration of Yougui Capsule to adenine-treated rats partially reversed decrease in the weight of reproductive organs, with increasing spermatogenic cells and mesenchymal cells compared with control group. The morphology of seminiferous tubules of testis improved in Yougui Capsule group compared with control group. This implied that reproductive function might be improved although no data were presented.

## 2.2 Modulating Immune Function

Antisperm antibody is believed to prevent fertility by reducing sperm concentration, sperm motility, and sperm liquefaction (Cui et al., 2015). Lai, Song, and Liu (1997) investigated whether CHM formula Tai-bao (胎宝) could inhibit antisperm antibody in an infertile mouse model. Treatment with Tai-bao dose dependently and significantly increased pregnant rates and implantation rates compared with control group. Detection of cytotoxic antibody with enzyme-linked immunosorbent assay showed that Tai-bao formula significantly reduced the level of cytotoxic antibody to sperm compared with control group (Lai et al., 1997). In a rabbit model of immune infertility, CHM formula Yi Kang Ling (YKL 抑抗灵) was used to test its effect on serum and seminal plasma antisperm antibody. Results showed that YKL dose dependently and significantly reduced the levels of antisperm antibody in both serum and seminal plasma compared with prednisone-treated and nontreatment groups (Cui et al., 2003). The studies from both of the CHM formulas suggested that they may exert a regulatory effect on immune system function via inhibition of expression of antisperm antibody in serum and seminal plasma, leading to improved sperm quality and mobility.

## 2.3 Improving Blood Supply

Arteriogenic erectile impotence (AED) also known as erectile impotence is caused by insufficient arterial blood supply to the cavernous bodies regardless of the arterial disease or abnormality responsible for the insufficiency (Ghanem & Shamloul, 2008). Wang et al. (2012a) and Wang et al. (2012b) conducted a series of studies to investigate the underlying mechanism of CHM formula Shugan Yiyang capsule (SGYY 疏肝益阳胶囊) in treating AED in a rat model. First, they investigated the effect of SGYY on the expression of nitric oxide synthase (NOS)/cyclic guanosine monophosphate (cGMP) molecules in AED rat model and found that the gene and protein expression of three subtypes of NOS-neuropathic (nNOS), inducible (iNOS), and endothelial (eNOS) and cGMP concentrations in cavernous tissue were significantly decreased in the model. Following SGYY treatment expression of NOS/cGMP and nNOS, iNOS and eNOS levels were significantly elevated compared with nontreatment control group (Wang et al., 2012a). Then, authors studied the effect of SGYY on the expression of cytokines, such as vascular endothelial growth factor (VEGF), insulin-like growth factors (IGFs), and Akt1 (a serine-threonine protein kinase) and found that SGYY dose dependently elevated the decrease in the mRNA level and plasma concentrations of VEGF and IGF in treated group compared with nontreatment group (Wang et al., 2012b). These results suggested that the therapeutic effect of SGYY seen in clinical studies (Wang, Jixiang, & Li, 2004) may be due to its synergetic effect by increasing the expression of NOSs and cGMP, modulating the NOS–cGMP pathway, and upregulating the expressions of VEGF, IGF, and Akt1 in the corpus cavernosum penis of AED rats, leading to improve the function of blood vascular endothelium.

### 2.4 Antiinfection

*Escherichia coli* is the most common cause of urogenital infection and is implicated in the genesis of male infertility (Pellati et al., 2008). Recently an in vitro study was conducted to assess effect of CHM formula Qinglishengjing pills (清淋生精片) on spermatozoa infected with *E. coli* (Yu et al., 2011). It was observed that *E. coli* infection caused significant detrimental changes in sperm motility parameters, phosphatidylserine externalization, and morphology of spermatozoa. Treatment of Qinglishengjing pills to infected cells almost normalized all changes caused by *E. coli* compared with control groups (Yu et al., 2011). This study supported the clinical effect of Qinglishengjing pills as an anti-*E. coli* agent.

## 3. CLINICAL OBSERVATION ON MALE INFERTILITY TREATED WITH CHM

### 3.1 Improving Blood Supply

Cui et al. (2007) conducted a clinical study to assess the therapeutic effect of CHMs Danshen (丹参) and Chaihu (柴胡) acupoint injection therapy on functional and mild arterial low-level blood supply erectile dysfunction (ED). A total of 150 patients with functional and arterial low-level blood supply ED were recruited for randomized controlled trial and divided into three groups, a Chinese herbs acupoint injection group, a saline acupoint injection group, and a HuichunRuyi capsules group as a control group, 50 patients in each group. The acupoints used for injection were Guanyuan (CV 4), Zuwuli (LR 10), and Huiyin (CV 1). The changes of the International Index of Erectile Function (IIEF) questionnaire, symptoms and signs, serum sexual hormones, and corpus spongiosum peak systolic velocity (PSV) before and after the treatment were monitored. Results revealed that acupoint injection of Danshen injection and Chaihu injection had a definite beneficial effect on functional and arterial low-level blood supply ED. This therapy raised IIEF Questionnaire II score and improved corpus spongiosum PSV. The overall effect of Chinese herb acupoint injection group was better than that of saline acupoint injection or oral HuichunRuyi capsule group (Cui et al., 2007).

## 3.2 Treating Sperm Disorders—Severe Oligospermatism and Azoospermi

Zhang, Zhao, and Zhang (2007) investigated the pregnancy-promoting effect of Bushen Shengjing Decoction (BSSJD 补肾生精汤) combined with ICSI in treating male infertile patients with severe oligospermatism and azoospermia (SOA). ICSI was applied on 164 patients; among them, 82 patients were assigned for additional BSSJD treatment for 2–3 months before ICSI, and the other 82 were assigned as the control group and received ICSI alone. The density, motility, viability, and deformity of sperm; semen level of reactive oxygen species (ROS); number of eggs retrieved, Metaphase II eggs, and mean transplanted fetus; rates of fertilization, cleavage, available embryo, and clinical pregnancy in the two groups were monitored before and after treatment. It was found that compared with those in the control group, the density, motility, and viability of sperm were higher, and the deformity rate and ROS level were lower in ICSI plus BSSJD group. Moreover, rates of fertilization and clinical pregnancy were higher in ICSI plus BSSJD group than those of ICSI alone (Zhang et al., 2007). This suggests that BSSJD helped decrease semen ROS levels and improve the quality of sperm. This combined therapy could be helpful for enhancing the natural fertilization ability of patients with SOA and improving the viability of their sperm to increase the ovarian fertilization and clinical pregnancy rate in ICSI cycles.

Wang, Chen, et al. (2008) and Wang, Huang, and Liang (2008) compared the therapeutic effects of Chinese herb formula WuziYanzong Pill (五子衍宗片) alone with that of combined treatment with acupuncture in patients with oligospermia and asthenospermia. Two hundred and thirty-one patients with infertile conditions were divided into three groups: WuziYanzong Pill group ($n=82$), acupuncture alone group ($n=71$), and WuziYanzong Pill plus acupuncture group ($n=78$). Acupuncture was applied at acupoints Qihai (CV 6), Guanyuan (CV 4), Zhongji (CV 3), and other optional acupoints. Semen routine test and the acrosome enzyme activity were observed before and after treatment. The data showed that the effective rate was 68.3% in the WuziYanzong Pill, 67.6% in the acupuncture group, and 84.6% in the WuziYanzong Pill plus acupuncture group. The semen density, vitality, and the acrosome enzyme activity were increased in all three groups, with more obvious increase in the WuziYanzong Pill plus acupuncture group (Wang, Chen, et al., 2008; Wang, Huang, & Liang, 2008). This showed that combined treatment of Chinese herbal formula WuziYanzong Pill with acupuncture was more effective than acupuncture or herb therapy alone in improving the semen quality and pregnancy rate of male interfile patients with oligospermia and asthenospermia.

## 3.3 Improving Sperm Quality

Liao et al. (2013) evaluated the effect of short-course kidney-invigorating therapy on near-term semen quality in patients with asthenozoospermia (reduced sperm motility). A total of 121 patients with asthenozoospermia were divided into groups A (kidney-yin deficiency), B (kidney-yang deficiency), and C (spleen and kidney deficiency), and treated with Yougui Decoction (右归饮) plus WuziyanzongPills (五子衍宗丸) for group A, Jinkuishenqi Pills (金匮肾气丸) plus Wuziyanzong Pills (五子衍宗丸) for group B, and ShiziPills (十子丸) for group C once daily for 4 weeks. Sperm parameters of the patients were analyzed with the computer-assisted sperm analysis system before and after treatment and compared among the three groups. All the three groups showed significant increase in sperm motility after treatment as compared with the baseline levels. The cure rate and total effectiveness rate were significantly higher in group B than in A, but had no significant difference between either A and C or B and C. No adverse effect was reported (Liao et al., 2013). This suggested that CHM formulas with short-term kidney-invigorating property could significantly improve near-term semen quality in patients with asthenozoospermia and kidney asthenia, especially in those with kidney-yang deficiency.

In an in vitro study Zeng et al. (2003) investigated the effect of Shouwu-Huanjing Decoction (SWHJR 首乌还精煎)-medicated serum on human sperm motility and fertility. Human sperm was cocultured with SWHJR-medicated serum in vitro. Data from computer-assisted semen analysis revealed that the cocultured sperm with SWHJR-medicated serum significantly increased the sperm motion velocity, the amplitude of lateral head movement and the beat frequency of flagellum, the density of progressive motility sperms, and the acrosome reaction rate. Further, the fertilization rate and the fertilization index in sperm penetration assay test were improved. The stimulation of SWHJR-medicated serum occurred in a dose-dependent manner (Zeng et al., 2003). This showed that SWHJR could improve human sperm motility and fertility. In another study Han, Liu, Wang, and Guo (2015) investigated the effect of the YishenZhongzi Pill (益肾种子丸) on the sperm motility in patients with Shen-essence (kidney essence) deficiency syndrome (SEDS) according to TCM theory, in comparison with CHM formula WuziYanzong Pill (五子衍宗丸). A total of 134 patients were randomly assigned to YishenZhongzi Pill group ($n=67$) and WuziYanzong Pill group ($n=67$). The therapeutic course for both groups was 1 month. Changes in sperm motility, clinical efficacy, and symptoms scores were observed between the two groups before and after treatment. Results showed that percentages of sperm progressive motility were enhanced in the two groups after treatment in comparison with baseline. However, rate of rapid sperm progressive motility was higher in YishenZhongzi Pill group than in WuziYanzong Pill group. The total effective rate in YishenZhongzi Pill group was significantly higher than those of WuziYanzong Pill group. Scores for symptoms were improved in YishenZhongzi Pill group compared with WuziYanzong Pill group (Han et al., 2015). This suggested that YZP was more effect in treating patients with asthenospermia and classified as SEDS, and could effectively elevate the motility of sperm.

## 3.4 Therapeutic Effect on Male Immune Infertility

Autoimmune reaction against sperm cells is defined by WHO (2000) as one of the causes of infertility in men. Antisperm antibody (ASA) is believed to be a possible causative factor in infertility, with significant levels of ASA detected in the semen of 5%–15% of infertile men (Adeghe, 1993) but in only 1%–2% of fertile men (Bates, 1997). Studies of ASA showed that ASA has adverse effect on male fertility by influencing the sperm concentration, sperm motility, and sperm liquefaction (Cui et al., 2015; Garcia, Rubio, & Pereira, 2007). Many CHM formulas have been used to treat immune infertility in men. Sun and Bao (2006) conducted a clinical study to assess a CHM formula, Yikang decoction (益抗汤) for male immune infertility, in which 100 patients with immune infertility were treated with Yikang decoction, while another 100 patients were treated with prednisone as control. At the end of the treatment, it was found that although both Yikang decoction and prednisone significantly decreased the levels of ASA in serum and semen, they markedly improved sperm density and semen liquidation compared with the baseline. The effect of Yikang decoction was more significant in all parameters tested compared with prednisone treatment (Sun & Bao, 2006). In another study, CHM formula Huzhangdanshenyin (虎杖丹参饮) was used to treat patients with immune infertility (Lu et al., 2006). Sixty patients were treated with Huzhangdanshenyin and another 30 patients were treated with prednisone for 3 months. Improvement of clinical symptoms, immunologic indexes (anti-sperm antibodies in serum and seminal plasma), and sperm indexes (semen liquefied duration, motility, viability, density, and abnormal morphology rate) was monitored before and after the treatment. It was found that at the end of the treatment, ASA levels in serum and seminal plasma were significantly decreased, and sperm motility and viability and clinical symptoms were markedly improved in Huzhangdanshenyin-treated group compared with prednisone-treated group. However, there was no significant difference in sperm density and semen liquefied duration between the two groups (Lu et al., 2006).

CHM in combination with conventional therapies improved the efficacy of conventional treatment. For example, Yu et al. (2004) investigated the combined effect of CHM formula Zhuanyindan (ZYD 转阴丹) with hormone in treating infertile men with positive ASA and its influence on nitric oxide (NO) level in a randomized study. Eighty-two patients were divided into ZYD-alone group, hormone-alone group, and combined group (ZYD + hormone). It was found that all treatments significantly improved serum NO level, sperm motion parameters, such as linear motion speed, linearity, propulsion, sperm vitality, and mean moving angle, and quality of semen compared with pretreatment baseline. However, the total effective rate was significantly higher in combined group compared with ZYD-alone and hormone-alone treatment (Yu et al., 2004). In another study, Lian, Zhang, and Zhang (2002) compared the effect of combined therapy of CHM formula Zhenqi Zhuanyin decoction (ZQZYD 贞芪转阴汤) and timely intrauterine insemination (IUI) with that of using ZQZYD or IUI alone in treating patients with immune infertility. A total of 103 patients were randomly divided into three groups, group A (ZQZYD + IUI, $n=34$), group B (ZQZYD alone, $n=34$), and group C (IUI alone, $n=35$). At the end of the treatment, it was shown that the negative conversion rate of ASA was 82.35% in group A, 76.47% in group B, and 8.57% in group C. Pregnancy rate was 41.18% in group A, 20.59% in group B, and 11.43% in group C. T-lymphocyte subsets CD4 and CD8 were significantly decreased in groups A and B compared with group C (Lian et al., 2002). Those studies showed that CHM formulas alone or in combination with conventional therapies reduced the expression of ASA, improved immunologic indexes and sperm indexes, and enhanced pregnancy rate.

## 3.5 Treating Disorders of Accessory Organs

### 3.5.1 Infertility Caused by Varicocele

Varicocele is an abnormally enlarged and twisted (varicose) vein in the spermatic cord that connects to the testicle. Varicoceles are found in about 15% of all men and in about 40% of infertile men (Jarow, 2001). Although it is not clear how much they affect fertility or by what mechanisms, they can raise testicular temperature, which may have adverse effects on sperm production, movement, and sperm shape. Qi, Lu, and Kan (2001) conducted a study to compare the effect of the CHM formula Tongjing granule (TJG 通精冲剂) in treating varicocele-induced infertility with surgical operation. A total of 75 infertile men caused by moderate or severe varicocele were divided into two groups: TJG group ($n=44$) and surgery group (high ligation of spermatic vein and/or inferior epigastric venous bypass, $n=31$). Data showed that there was no significant difference in the measurement of various parameters between the two groups (Qi et al., 2001). This indicated that TJG treatment achieved the same effect as that of surgery in treating the varicocele-caused infertility and might be an alternative therapy to surgery. In another study, Yan, Jiang, and Shao (2004) investigated the effect of Jingling oral liquid (JLOL 精灵口服液) in treating infertile patients with recurrent varicocele after varicocelectomy. Sixty patients were randomly divided into two groups: JLOL treatment group ($n=30$) and injection group treated with intramuscular injection of human chorionic gonadotropine (hCG) ($n=30$). Results revealed that in JLOL-treated group, levels of superoxide dismutase (SOD) and zinc in semen were significantly increased at the end of the treatment compared with baseline, and this was not found in injection group (Yan et al., 2004). Further, the quality of semen and the pregnant rate of the patient's partners were markedly higher in JLOL group than in injection group (Yan et al., 2004). This suggested that JLOL improved the quality of semen and sperm in infertile patients with recurrent varicocele after varicocelectomy, and increased the pregnancy rate of their spouses.

### 3.5.2 Male Infertility Caused by Accessory Gland Infection

Sun, Zhou, and Ding (2006) studied the effect of CHM formula QingreYulin Decoction (QYD 清热毓麟汤) on male infertility caused by accessory gland infection (AGI). Sixty infertile patients with AGI were divided into two groups: QYD-treated group and medication group treated with antibiotic plus vitamin E for 3 months. At the end of the treatment, results showed that in QYD-treated group cure rate was 26.7%, the markedly effective rate was 43.3%, the effective rate was 16.7%, and the total effective rate was 86.7%, whereas in medication group it was 6.7%, 30.0%, 40.0%, and 76.7%, respectively. Data showed a higher cure rate and total effective rate in the QYD-treated group than in the medication group. Additionally, sperm quality, average liquefaction time, and survival rate and vitality of sperm were markedly improved in both groups, but the efficacy was better in the QYD-treated group than in the medication group (Sun et al., 2006). The study showed that CHM formula QYD not only has antiinfectious effect but also improved the quality of sperm, leading to enhance the pregnant rate of patient's spouses.

## 4. CONCLUSION

There have been many researches to investigate the scientific basis to understand the traditional uses of CHM formulas for male infertility. Our studies from literature review showed that CHM formulas could significantly improve sperm abnormalities and sperm quality, modify immune system, and treat many aspects of andrological conditions, via different action mechanisms depending on the formula compositions. CHM not only could treat many male infertile conditions and increase pregnancy rate but also improve the efficacy of conventional treatments. However, in addition to the need for more extensive pharmacological studies and randomized controlled trials to investigate the efficacy of promising CHM formulas for male infertility, evaluation of their safety is also needed. Other challenges for the investigation of CHM formulas for their relevance in male infertility include their standardization to known active constituents, understanding polyvalent and synergistic effects, which is particularly difficult to elucidate for CHM formulas composed of a mixture of different herb species. It is clear, under these circumstances, that the appropriate quantitative and qualitative chemical profiles of specific CHM formulas should be investigated to determine their safety and efficacy.

## ACKNOWLEDGMENT

This work was partly supported by ATCM (The Association of Traditional Chinese Medicine and Acupuncture UK). Authors thank for the IT support from the King's College London.

## REFERENCES


Adeghe, J. H. (1993). Male subfertility due to sperm antibodies: A clinical overview. *Obstetrical and Gynecological Survey, 48*, 1–8.

Alrabeeah, K., Yafi, F., Flageole, C., Phillips, S., Wachter, A., Bissonnette, F., et al. (2014). Testicular sperm aspiration for nonazoospermic men: Sperm retrieval and intracytoplasmic sperm injection outcomes. *Urology, 84*, 1342–1346.

Bates, C. A. (1997). Antisperm antibodies and male subfertility. *British Journal of Urology, 80*, 691–697.

Bensdorp, A. J., Cohlen, B. J., Heineman, M. J., & Vandekerckhove, P. (2007). Intra-uterine insemination for male subfertility. *Cochrane Database Systematic Review, 17*. CD000360.

Cissen, M., Bensdorp, A., Cohlen, B. J., Repping, S., de Bruin, J. P., & van Wely, M. (2016). Assisted reproductive technologies for male subfertility. *Cochrane Database Systematic Review, 26*. CD000360.

Corona, G., Rastrelli, G., Limoncin, E., Sforza, A., Jannini, E. A., & Maggi, M. (2015). Interplay between premature ejaculation and erectile dysfunction: A systematic review and meta-analysis. *Journal of Sexual Medicine, 12*, 2291–2300.

Cui, Y., Feng, Y., Chen, L., Zhou, Y., Yang, X. Q., He, J., et al. (2007). Randomized and controlled research of Chinese drug acupoint injection therapy for erectile dysfunction. *Zhongguo Zhen Jiu, 27*, 881–885.

Cui, D., Han, G., Shang, Y., Liu, C., Xia, L., Li, L., et al. (2015). Antisperm antibodies in infertile men and their effect on semen parameters: A systematic review and meta-analysis. *Clinica Chimica Acta, 444*, 29–36.

Cui, Y. X., Huang, Y. F., Wang, Y. M., Hu, Y. A., Xu, J. P., Yao, B., et al. (2003). Chinese traditional medicine yi kang ling to treat immunity infertility caused by anti-sperm antibody: The experimental research. *Zhonghua Nan Ke Xue, 9*, 628–631.

Dorjpurev, U., Kuwahara, A., Yano, Y., Taniguchi, T., Yamamoto, Y., Suto, A., et al. (2011). Effect of semen characteristics on pregnancy rate following intrauterine insemination. *Journal of Medical Investigation, 58*, 127–133.

Evers, J. L., & Collins, J. A. (2003). Assessment of efficacy of varicocele repair for male subfertility: a systematic review. *Lancet, 361*, 1849–1852.

Garcia, P. C., Rubio, E. M., & Pereira, O. C. M. (2007). Antisperm antibodies in infertile men and their correlation with seminal parameters. *Reproductive Medicine and Biology, 6*, 33–38.

Ghanem, H., & Shamloul, R. (2008). An evidence-based perspective to commonly performed erectile dysfunction investigations. *Journal of Sexual Medicine, 5*, 1582–1589.

Han, Q., Liu, D. C., Wang, F., & Guo, J. (2015). Treatment of asthenozoospermia men with Shen-essence deficiency syndrome by YishenZhongzi pill: A clinical study. *ZhongguoZhong Xi Yi Jie He Za Zhi, 35*, 1050–1053.

Ho, C. C., & Tan, H. M. (2013). Treatment of the hypogonadal infertile male—A review. *Sexual Medicine Reviews, 1*, 42–49.

Huang, T., Xia, M., & Ren, K. (2003). The correlation research of adenine-induced male infertility and chronic renal failure in the rats. *Chinese Journal of Andrology, 17*, 171–173.

Jarow, J. P. (2001). Effects of varicocele on male fertility. *Human Reproduction Update, 7*, 59–64.

Johnson, L. N., Sasson, I. E., Sammel, M. D., & Dokras, A. (2013). Does intracytoplasmic sperm injection improve the fertilization rate and decrease the total fertilization failure rate in couples with well-defined unexplained infertility? A systematic review and meta-analysis. *Fertility and Sterility, 100*, 704–711.

Lai, A. N., Song, J. F., & Liu, X. J. (1997). An experimental study on inhibitory effect of Chinese medicine tai-bao on antisperm antibody. *Zhongguo Zhong Xi Yi Jie He Za Zhi, 17*, 360–362.

Levine, B. A., & Grifo, J. A. (2008). Intrauterine insemination and male subfertility. *Urologic Clinics of North America, 35*, 271–276.

Lian, F., Zhang, N., & Zhang, J. W. (2002). Clinical observation on effect of zhenqizhuanyin decoction combined with intrauterine insemination in treating spleen-kidney deficiency type patients of sterility with positive anti-sperm antibody. *ZhongguoZhong Xi Yi Jie He Za Zhi, 22*, 95–97.

Liao, Z. J., Zhou, J. F., Yin, Z. C., Zhang, J. J., Xiang, S. T., Wang, S. S., et al. (2013). Short-course kidney-invigorating therapy improves near-term semen quality in asthenozoospermic men with kidney deficiency. *Zhonghua Nan Ke Xue, 19*, 159–162.

Liu, M. L., Wang, Z. R., & Nan, Y. Y. (2012). Effects of WenyangShengjing decoction containing serum on the estradiol secretion of Leydig cells of sterile rats of shen-yang deficiency. *ZhongguoZhong Xi Yi Jie He Za Zhi, 32*, 248–252.

Lu, T. K., Ouyang, H. G., Jin, G. Y., Hong, Y. S., Zou, Q., Lin, Z. Y., et al. (2006). Clinical study on the treatment of male immune infertility by Huzhangdanshenyin. *Zhonghua Nan KeXue, 12*, 750–755.

Ma, W. G., & Jia, J. M. (2011). The effects and prospects of the integration of traditional Chinese medicine and Western medicine on andrology in China. *Asian Journal of Andrology, 13*, 592–595.

Ma, J., Wang, Z. R., & Lu, Z. F. (2006). Regulatory effect of WenyangShengjing decoction on Smads expressions in testis of sterile rats with Shen-yang deficiency. *ZhongguoZhong Xi Yi Jie He Za Zhi, 26*, 1107–1113.

Mascarenhas, M. N., Flaxman, S. R., Boerma, T., Vanderpoel, S., & Stevens, G. A. (2012). National, regional, and global trends in infertility prevalence since 1990: A systematic analysis of 277 health surveys. *PLoS Medicine, 9*, e1001356.

Palomba, S., Homburg, R., Santagni, S., La Sala, G. B., & Orvieto, R. (2016). Risk of adverse pregnancy and perinatal outcomes after high technology infertility treatment: A comprehensive systematic review. *Reproductive Biology and Endocrinology, 14*, 76.

Pellati, D., Mylonakis, I., Bertoloni, G., Fiore, C., Andrisani, A., Ambrosini, G., et al. (2008). Genital tract infections and infertility. *European Journal of Obstetrics & Gynecology and Reproductive Biology, 140*, 3–11.

Pisipati, S., & Pearcy, R. (2010). The role of urological surgery in male infertility. *Human Fertility, 13*, 233–241.

Pitteloud, N., & Dwyer, A. (2014). Hormonal control of spermatogenesis in men: Therapeutic aspects in hypogonadotropic hypogonadism. *Annales d Endocrinologie, 75*, 98–100.

Qi, G. C., Lu, J. K., & Kan, Q. L. (2001). Comparative clinical study on treatment of varicocele caused infertility by tongling granule and surgical operation. *ZhongguoZhong Xi Yi Jie He Za Zhi, 21*, 412–415.

Ray, P. F., Toure, A., Metzler-Guillemain, C., Mitchell, M. J., Arnoult, C., & Coutton, C. (2017). Genetic abnormalities leading to qualitative defects of sperm morphology or function. *Clinical Genetics, 91*, 217–232. http://dx.doi.org/10.1111/cge.12905. PMID: 27779748.

Sallmén, M., Sandler, D. P., Hoppin, J. A., Blair, A., & Baird, D. D. (2006). Reduced fertility among overweight and obese men. *Epidemiology, 17*, 520–523.

Sun, Z., & Bao, Y. (2006). TCM treatment of male immune infertility—A report of 100 cases. *Journal of Traditional Chinese Medicine, 26*, 36–38.

Sun, J., Zhou, A. F., & Ding, C. F. (2006). Clinical observation of qingreyulin decoction in treatment of male infertility caused by accessory gland infection. *ZhongguoZhong Xi Yi Jie He Za Zhi, 26*, 877–880.

Valenti, D., La Vignera, S., Condorelli, R. A., Rago, R., Barone, N., Vicari, E., et al. (2013). Follicle-stimulating hormone treatment in normogonadotropic infertile men. *Nature Review Urology, 10*, 55–62.

Wang, J. H., Chen, D., Zhou, J. G., Cao, Y. G., & Jia, J. M. (2008). Construction of the male rat model of adenine-induced infertility with kidney-yang deficiency. *Zhonghua Nan Ke Xue, 14*, 565–570.

Wang, Z. Q., Huang, Y. Q., & Liang, B. (2008). Clinical observation on electroacupuncture and Chinese drug for treatment of oligospermia and asthenospermia of the male infertility patient. *Zhongguo Zhen Jiu, 28*, 805–807.

Wang, Q., Jixiang, Y., & Li, G. (2004). Multi-center and randomly controlled trial of shuganyiyang capsules for the treatment of erectile dysfunction (Chinese). *Journal of Beijing University of Traditional Chinese Medicine, 27*, 72–75.

Wang, X. J., Li, H. S., Zhang, H., & Dang, J. (2014). Influence of Yougui capsule on kidney Yang deficiency infertility rats' reproductive function. *Global Traditional Chinese Medicine, 7*, 669–672.

Wang, J., Wang, Q., Liu, B., Li, D., Yuan, Z., & Zhang, H. (2012a). A Chinese herbal formula, Shuganyiyang capsule, improves erectile function in male rats by modulating Nos-CGMP mediators. *Urology, 79*, 241.e1–241.e6.

Wang, J., Wang, Q., Liu, B. X., Li, D. H., Zhang, H. M., & Yuan, Z. J. (2012b). ShuganYiyang capsule increases the expressions of VEGF, IGF and Akt1 in the rat model of arteriogenic erectile dysfunction. *Zhonghua Nan KeXue, 18*, 184–188.

WHO. (2000). Manual for the standardized investigation and diagnosis of the infertile couple. In WHO (Ed.), (3rd ed., p. 83). Cambridge, United Kingdom: Cambridge University Press.

Xu, X., Yin, H., Tang, D., Zhang, L., & Gosden, R. G. (2003). Application of traditional Chinese medicine in the treatment of infertility. *Human Fertility, 6*, 161–168.

Yan, L. F., Jiang, M. F., & Shao, R. Y. (2004). Clinical observation on effect of jingling oral liquid in treating infertile patients with varicocele after varicocelectomy. *ZhongguoZhong Xi Yi Jie He Za Zhi, 24*, 220–222.

Yu, A., Li, X., Deng, W., Zhu, X., Zhuang, D., Liu, L., et al. (2011). Influence of traditional Chinese medicine on spermatozoa infected in vitro with Escherichia coli. *Andrologia, 43*, 321–326.

Yu, A. L., Zhang, F. Z., & Zhang, F. X. (2004). Clinical observation on effect of combination of zhuanyindan and hormone in treating male infertility with positive antisperm antibody. *ZhongguoZhong Xi Yi Jie He Za Zhi, 24*, 223–226.

Yue, G. P., Chen, Q., & Dai, N. (1997). Experimental study on effect of bushen shengjing decoction on kidney yang and testicular dysfunction in rats. *Zhongguo Zhong Xi Yi Jie He Za Zhi, 17*, 289–291.

Zegers-Hochschild, F., Adamson, G. D., de Mouzon, J., Ishihara, O., Mansour, R., Nygren, K., et al. (2009). International Committee for Monitoring Assisted Reproductive Technology (ICMART) and the World Health Organization (WHO) revised glossary of ART terminology, 2009. *Fertility and Sterility, 92*, 1520–1524.

Zeng, J. X., Dai, X. H., Liu, J. H., Yang, J. H., Chen, J. F., & Lin, M. N. (2003). Effects of shouwu-huanjing recipe on human sperm motility and fertility in vitro. *Zhonghua Nan Ke Xue, 9*, 476–479.

Zhang, H. Q., Zhao, H. X., & Zhang, A. J. (2007). Male infertility with severe oligospermatism and azoospermia treated by Bushen Shengjing Decoction combined with intracytoplasmic sperm injection. *ZhongguoZhong Xi Yi Jie He Za Zhi, 27*, 972–975.

Zhao, S., Zhu, W., Xue, S., & Han, D. (2014). Testicular defense systems: Immune privilege and innate immunity. *Cellular & Molecular Immunology, 11*, 428–437.

Kolibri 360 GmbH
Ferdinand-Jühlke-Straße 7
99095 Erfurt, Deutschland
produktsicherheit@kolibri360.de